W0091469

Helping Students Become Powerful Mathematics Thinkers

Case Studies of Teaching for Robust Understanding

Alan Schoenfeld, Heather Fink, and Sandra Zuñiga-Ruiz, with Siqi Huang, Xinyu Wei, and Brantina Chirinda

Routledge
Taylor & Francis Group

NEW YORK AND LONDON

Designed cover image: © Getty Images

First published 2023
by Routledge
605 Third Avenue, New York, NY 10158

and by Routledge
4 Park Square, Milton Park, Abingdon, Oxon, OX14 4RN

Routledge is an imprint of the Taylor & Francis Group, an informa business

© 2023 Alan Schoenfeld, Heather Fink, Sandra Zuñiga-Ruiz, Siqi Huang, Xinyu Wei, and Brantina Chirinda

The right of Alan Schoenfeld, Heather Fink, Sandra Zuñiga-Ruiz, Siqi Huang, Xinyu Wei, and Brantina Chirinda to be identified as authors of this work has been asserted in accordance with sections 77 and 78 of the Copyright, Designs and Patents Act 1988.

All rights reserved. No part of this book may be reprinted or reproduced or utilised in any form or by any electronic, mechanical, or other means, now known or hereafter invented, including photocopying and recording, or in any information storage or retrieval system, without permission in writing from the publishers.

Trademark notice: Product or corporate names may be trademarks or registered trademarks, and are used only for identification and explanation without intent to infringe.

Library of Congress Cataloging-in-Publication Data
Names: Schoenfeld, Alan H., author. | Fink, Heather, author. |
Zuñiga-Ruiz, Sandra, author. | Huang, Siqi (Researcher in mathematics education), author. | Wei, Xinyu, author. | Chirinda, Brantina, 1978– author.
Title: Helping students become powerful mathematics thinkers:
case studies of teaching for robust understanding / Alan Schoenfeld,
Heather Fink, Sandra Zuñiga-Ruiz, with Siqi Huang, Xinyu Wei, and Brantina Chirinda.
Description: New York, NY: Routledge, 2023. |
Series: Studies in mathematical thinking and learning series |
Includes bibliographical references.
Identifiers: LCCN 2022057511 (print) | LCCN 2022057512 (ebook) |
ISBN 9781032450629 (hardback) | ISBN 9781032441689 (paperback) |
ISBN 9781003375197 (ebook)
Subjects: LCSH: Mathematics—Study and teaching (Primary)—Case studies. |
Mathematics—Study and teaching (Secondary)—Case studies.
Classification: LCC QA11.2 .S357 2023 (print) |
LCC QA11.2 (ebook) | DDC 510.71/2—dc23/eng20230429
LC record available at https://lccn.loc.gov/2022057511
LC ebook record available at https://lccn.loc.gov/2022057512

ISBN: 978-1-032-45062-9 (hbk)
ISBN: 978-1-032-44168-9 (pbk)
ISBN: 978-1-003-37519-7 (ebk)

DOI: 10.4324/9781003375197

Typeset in Sabon
by codeMantra

Helping Students Become Powerful Mathematics Thinkers

This book supports teacher educators, teachers, coaches, administrators, math-ed faculty, and researchers in understanding and using the Teaching for Robust Understanding (TRU) Framework to improve instruction. Detailed case studies take readers on deep dives into five essential dimensions of classroom practice: The Mathematics; Cognitive Demand; Equitable Access; Agency, Ownership, and Identity; and Formative Assessment.

Three case studies form the core of the book. Each case uses the TRU Framework to pose conversational questions to the reader on different aspects of the lessons, focusing on the ways that students are led to engage with mathematics and how they make sense of it. These include "What's important in this classroom episode?," "What might students be experiencing?," or "What might the impact of alternative teaching decisions have been in this situation?". The book concludes with guides for planning, observation, and reflection that readers can use in their own work, continuing the journey toward the ambitious and equitable instruction that each case study describes.

This book will support all mathematics educators in developing deeper understandings of mathematics classrooms and in problematizing their own mathematics instruction. By exploring the challenges students face, the decisions teachers make, and the ways that students learn, readers will experience TRU as a powerful way of thinking about instruction – one that can shape lesson planning and reflection and make teaching more impactful and equitable.

Alan Schoenfeld is a Distinguished Professor of Education and Mathematics at the University of California, Berkeley, USA.

Heather Fink is a postdoctoral researcher at Portland State University, USA.

Sandra Zuñiga-Ruiz is an Assistant Professor in the Teacher Education Department at San José State University, USA.

Siqi Huang is a PhD student in the Graduate Group of Science and Mathematics Education (SESAME program) at the University of California, Berkeley, USA.

Xinyu Wei is a Master's student studying Learning Sciences at the University of California, Berkeley, USA.

Brantina Chirinda is a Mathematics and Mathematics Education Lecturer in the Faculty of Education at Cape Peninsula University of Technology, South Africa.

Studies in Mathematical Thinking and Learning

Alan H. Schoenfeld, Series Editor

Recent Publications

Hulbert/Petit/Ebby/Cunningham/Laird
A Focus on Multiplication and Division: Bringing Research to the Classroom

Ebby/Hulbert/Broadhead
A Focus on Addition and Subtraction: Bringing Mathematics Education Research to the Classroom

Clements/Sarama
Learning and Teaching Early Math: The Learning Trajectories, Third Edition

Takahashi
Teaching Mathematics Through Problem-Solving: A Pedagogical Approach from Japan

Horn & Garner
Teacher Learning of Ambitious and Equitable Mathematics Instruction: A Sociocultural Approach

Petit/Laird/Ebby/Marsden
A Focus on Fractions: Bringing Mathematics Education Research to the Classroom, Third Edition

Schoenfeld/Fink/Sayavedra/Weltman/Zuñiga-Ruiz
Mathematics Teaching On Target: A Guide to Teaching for Robust Understanding at All Grade Levels

Schoenfeld/Fink/Zuñiga-Ruiz/Huang/Wei/Chirinda
Helping Students Become Powerful Mathematics Thinkers: Case Studies of Teaching for Robust Understanding

Contents

Acknowledgments *vii*
About the Authors *viii*
Welcome! *x*

Part I   The Big Ideas in Teaching and Learning   1

1   An Introduction to the TRU Framework   3
1.1   Where Did These Ideas Come from and How Can They Be Useful?   3
1.2   The Five Dimensions of TRU   6
1.3   How the Dimensions Fit Together – in Practice, in
Professional Development, and in this Volume   15

Part II   Reflecting on Images of Practice (The Case Studies)   19

Introduction to Part II   19
Norms for Visiting Classrooms   19
The Primary Focus of Each of the Three Cases   20
The Structure of Our Sessions – and of Chapters 2–4   22

2   The Car Value Problem: A Focus on Cognitive Demand   25

2.0   Introduction   25
2.1   Thinking about the Math in the Car Value Problem   27
2.2   Some Background about the Classroom We're Visiting   38
2.3   What Happened, in Detail (With Some Reflection Questions and
Comments Along the Way)   40
2.4   Reflecting on "The Car Value Problem"   73

3   Where is the Ten?: A Focus on Equitable Access and Agency, Ownership, and Identity   82

3.0   Introduction   82
3.1   The Mathematical Task   83
3.2   Some Background about the Classroom We're Visiting   97

Contents

3.3 What Happened, in Detail (with Some Reflection Questions and Comments Along the Way) 98

3.4 Reflecting on "Where is the ten?" 127

4 Graphing Quadratic Functions: A Focus on Formative Assessment 137

4.0 Introduction 137

4.1 Thinking about the Mathematics in the Quadratic Functions Formative Assessment Lesson 139

4.2 Some Background about the Classroom We're Visiting 153

4.3 What Happened, in Detail (With Some Reflection Questions and Comments along the Way) 154

4.4 Reflecting on "Graphing Quadratic Functions" 185

Part III Conclusions and Next Steps 195

Introduction to Part III 195

5 Looking Back and Looking Forward: TRU in the Case Studies, TRU in Our Classrooms 197

5.1 Reflecting on the Three Cases 197

5.2 Reviewing the Big Ideas about TRU, in Preparation for Applying it to Our Teaching 201

5.3 Tools that Can Help and Possible Next Steps 203

5.4 Where Might You Go from Here? 207

6 Resources and Tools for Moving Forward 218

6.0 Overview 218

6.1 Resources 218

6.2 The TRU Conversation Guide 224

6.3 The TRU Observation Guide 234

6.4 A Brief Coda 242

References 243

Index 251

Acknowledgments

This book rests on a foundation of ideas born in partnerships and collaborations with a wonderful collection of friends, teachers, and scholars. The ideas in TRU emerged from our work in the Algebra Teaching Study, a partnership with Bob Floden at Michigan State University, and our work in the Mathematics Assessment Project with Hugh Burkhardt, Malcolm Swan, and colleagues at the University of Nottingham. David Foster at the Silicon Valley Mathematics Initiative was a thought partner throughout the development of TRU, and Michael Driskill at Math for America has supported TRU in similar ways. From the early onset of the project through the final version of this book, Abraham Arcavi has been a critical friend and contributor, problematizing and deepening our ideas.

Our partnership with the Strategic Education Research Partnership (SERP) Institute involved classroom collaborations with the San Francisco and Oakland Unified School Districts. SERP partners Suzanne Donovan, Matt Ellinger, and Karen Tran, along with Phil Daro, contributed in many ways. San Francisco's district team, led by Jim Ryan, developed the first version of the TRU Observation Guide; Evie Baldinger and Nicole Louie were the lead authors of the TRU Conversation Guide. Many of the ideas behind the case studies included in this book and our conceptualization of TRU-Lesson Study emerged in the partnership with Oakland Unified district immediately preceding our work on this book. Alyssa Sayavedra contributed in multiple roles, as teacher leader, developer, and analyst. Anna Weltman shaped TRU-Lesson-Study efforts at a partner school and contributed to our discussions of the essential character of all five dimensions. Diana Casanova contributed to design, PD, and analysis.

We are truly grateful.

About the Authors

Alan Schoenfeld is a Distinguished Professor of Education and Mathematics at the University of California, Berkeley. His research focuses on mathematical thinking and problem-solving, teaching, and learning. Throughout his career, he has worked in partnership with school districts, teachers, instructional designers, and professional learning communities, with the goal of enriching mathematics instruction and supporting equitable and ambitious instruction. This book builds on more than four decades of such partnerships. (For more detail, see https://bse.berkeley.edu/alan-h-schoenfeld and https://truframework.org/.)

Heather Fink is a Postdoctoral Researcher at Portland State University working on an NSF-funded project supporting equitable and ambitious teaching practices. Before earning her doctorate in Mathematics Education at the University of California, Berkeley, she worked for 11 years as a middle school math teacher and instructional coach in the San Francisco Bay Area. Heather is a national board-certified teacher (early adolescence/mathematics) who is committed to promoting educational equity through her teaching and research by challenging persistent injustices and unfair power distributions in mathematics classrooms. Her research focuses on understanding how inequities are constructed through classroom interactions and how, at a micro level, interactions shape the opportunities students have to engage in rich content and to build positive identities as thinkers, learners, and members of communities.

Sandra Zuñiga-Ruiz is an Assistant Professor in the Teacher Education Department at San José State University. Prior to earning her doctorate in Mathematics Education at UC Berkeley, she taught college mathematics at universities and community colleges in the Central Coast of California with an emphasis on issues of equity and social justice. Her research and teacher education work takes an emancipatory approach that seeks to cultivate dignity affirming mathematical learning communities for teacher candidates and children in the K-12 educational system. Her dissertation work investigated how four prospective maestras Mexicanas committed to justice developed understandings related to race, justice, and mathematics through the co-construction of a counterspace via critical conversations, pláticas.

Siqi Huang is a PhD student in the Graduate Group of Science and Mathematics Education (SESAME) at the University of California, Berkeley. She is an active member of Alan Schoenfeld's Teaching for Robust Understanding (TRU) and Functions research groups. Siqi is interested in the teaching and learning of mathematics, particularly as it relates to (a) understanding students' thought processes and problem-solving practices and (b) engaging research that seeks to create safer, more inclusive spaces for students to agentically explore, evaluate, reflect, and take risks (as opposed to merely memorize) in expanding their understanding of mathematical and social issues. Siqi is a Graduate Student Instructor for various Calculus courses at UC Berkeley. She is pursuing a Master's degree in mathematics in order to advance her mathematical expertise in support of her research.

Xinyu Wei is a Master's student studying Learning Sciences at the University of California, Berkeley. Growing up in Luoyang, China, she earned bachelor's degrees in Education and in English from Southwest University. Before moving to the U.S. Bay Area, Xinyu secured three patents as an award-winning science learning product designer, and she was funded by the Ministry of Education of China to lead a two-year educational innovation project. As an emergent researcher in mathematics education, Xinyu desires to foster students' conceptual and procedural understanding through the reinvention of culturally grounded mathematical instruments, where she leverages students' everyday experiences and sociocritical thinking to help them interpret the real-world meaning of knowledge as much as mathematicians do.

Brantina Chirinda is a Mathematics and Mathematics Education Lecturer in the Faculty of Education at Cape Peninsula University of Technology, South Africa. Dr Chirinda holds a PhD in Mathematics Education from the University of Witwatersrand, South Africa. She is interested in the teaching and learning of mathematics in contexts of disadvantage, specifically focusing on mathematical problem-solving and equitable access to content in the mathematics classroom. She has taught Mathematics and Mathematics Education courses at various Southern African institutions for over 20 years. She has published several conference proceedings, book chapters, and articles in accredited local and international journals. She has presented her work at national, regional, and international conferences.

Welcome!

What this book is about, in Q&A form

This book takes readers on three "deep dives" into mathematical classrooms, exploring aspects of teaching and learning that contribute to ambitious and equitable instruction. We introduce the Teaching for Robust Understanding Framework (TRU), which describes the properties of classrooms in which students become mathematically powerful and empowered thinkers. In the three case studies readers use TRU to examine the mathematics at the heart of the lesson and then the lesson itself, focusing on the ways that students are led to engage with the mathematics and how they make sense of it. The cases include questions and commentaries, inviting dialogue. We've done our best to have conversations with readers, as if we were discussing the lessons together in person.

By exploring the challenges students face, the decisions teachers make, and the ways that students learn, readers experience TRU as a powerful way of thinking about instruction – one that can shape lesson planning and reflection and make teaching more powerful and equitable. The book concludes with a set of validated tools (guides for planning, observation, and reflection) that teachers and teacher educators can use in their own work, continuing the journey toward the kind of powerful instruction described in the book.

Who is this book for?

This book is for anyone who cares about improving mathematics teaching. Learning about the Teaching for Robust Understanding Framework and working through the case studies in the book will deepen your understanding of the kinds of classroom experiences students need in order to become mathematically powerful and empowered learners. Whether you teach or you support teachers as a teacher educator, coach, or administrator, this book will help you think in richer ways about how students can be supported to engage more deeply and powerfully with mathematics. All readers, including researchers, can see how student thinking is revealed by a very close look at classroom interactions.

This book starts by introducing the Teaching for Robust Understanding (TRU) Framework. The ideas in TRU aren't simply abstract. They come to life when we use them to think about what happens in classrooms – and then use what we've learned in our own classrooms. The core of the book consists of three case studies that serve as "deep dives" into mathematics classrooms, where we use TRU to think deeply about what took place. Following the three cases we discuss ways in which the ideas we'd worked through can be applied to our own teaching, and provide some tools for planning and reflecting on our own teaching.

As discussed in Chapter 1, the TRU Framework provides a comprehensive research-based response to the question,

> What are the attributes of equitable and robust learning environments – environments in which all students are supported in becoming knowledgeable, flexible, and resourceful disciplinary thinkers?

We've chosen the three case studies in this volume for a number of reasons. First, they're examples of real, every-day practice, with teachers being faced by challenging pedagogical dilemmas. They're not meant to be exemplars or "model instruction." We picked them because they present interesting and challenging situations that are worth thinking through. Second, they reflect a diversity of classroom contexts – small group, whole class, a mix of both. Third, they reflect different windows on content. One case shows students learning something brand new. One shows students building on what they know. And one shows a review/synthesis lesson. In each case study (in fact, in all teaching!) all five dimensions are important. We've chosen them so that each case highlights particular dimensions of TRU. Working through all three cases provides opportunities to reflect deeply on a wide range of secondary mathematics instruction. You are invited to think with us about the mathematics involved, how the students interact with the mathematics and with each other, and what that might mean for student learning. Our goal is to help you get good at planning, teaching, and reflecting on instruction with the five TRU dimensions in mind. That's a way to achieve continuous improvement.

There's a lot to be learned by reading this book by yourself. But, the experiences in this book are much more powerful if you read and discuss them with colleagues. The transcripts of the lessons we present are given verbatim; they reflect our attempt to portray what took place as accurately as possible. (Our videotape sources did not have the permissions required for us to make the tapes widely available.) Then, in the case discussions, we've done our best to re-create the kinds of conversations that we (researchers, teacher educators, and teachers) have with groups of teachers when we've just watched a lesson together. Sometimes these conversations take place right after a teacher has given a sample lesson – maybe in a lesson study group, maybe as a practice lesson in a teaching seminar. Sometimes we watch a lesson on videotape. We ask questions to get the process going, but things really take off when the group generates and discusses ideas, using the TRU Framework to focus on essential aspects of teaching and learning. So...

If you can read and discuss the classroom stories in this book with colleagues, your discussions will be enriched. That makes this book ideal for professional learning communities and for pre- or in-service teaching seminars. If you're a district leader, this book shows you what matters in classrooms and how to support it. And if you're a researcher, the discussions in the cases should help to deepen your understanding of the TRU dimensions[1].

The big picture: what matters in mathematics classrooms?

Here's a one-sentence summary. What matters in teaching is knowing your students, knowing mathematics, and connecting the two – building classroom environments in which all students have opportunities to engage meaningfully with mathematics and come to see themselves as powerful mathematical thinkers.

This book is based on the Teaching for Robust Understanding (TRU) Framework. We'll have a lot to say about it in Chapter 1, but Figure I.1 gives you the essence of TRU. The bottom line is simple: the more things go well along the five TRU dimensions, the more the students are likely to emerge as powerful thinkers and learners.

The Five Dimensions of Powerful Mathematics Classrooms

The Mathematics	Cognitive Demand	Equitable Access to Mathematics	Agency, Ownership, and Identity	Formative Assessment
The extent to which classroom activity structures provide opportunities for students to become knowledgeable, flexible, and resourceful mathematical thinkers. Discussions are focused and coherent, providing opportunities to learn mathematical ideas, techniques, and perspectives, make connections, and develop productive mathematical habits of mind.	The extent to which students have opportunities to grapple with and make sense of important mathematical ideas and their use. Students learn best when they are challenged in ways that provide room and support for growth, with task difficulty ranging from moderate to demanding. The level of challenge should be conducive to what has been called "productive struggle."	The extent to which classroom activity structures invite and support the active engagement of all of the students in the classroom with the core mathematical content being addressed by the class. Classrooms in which a small number of students get most of the "air time" are not equitable, no matter how rich the content: all students need to be involved in meaningful ways.	The extent to which students are provided opportunities to "walk the walk and talk the talk" – to contribute to conversations about mathematical ideas, to build on others' ideas and have others build on theirs – in ways that contribute to their development of agency (the willingness to engage), their ownership over the content, and the development of positive identities as thinkers and learners.	The extent to which classroom activities elicit student thinking and subsequent interactions respond to those ideas, building on productive beginnings and addressing emerging misunderstandings. Powerful instruction "meets students where they are" and gives opportunities to deepen their understandings.

Figure I.1 The TRU Framework, in a nutshell

This book is devoted to fleshing out what Figure I.1 really means, with classroom examples. It addresses the following questions related to the five TRU dimensions:

1. What does powerful mathematical understanding look like?
2. How can we increase the likelihood that our students are engaged in meaningful sensemaking, building on what they know and being stretched in reasonable ways?
3. How can we provide opportunities for each and every student to engage with centrally important mathematical content and practices?
4. How can we do so in ways that students feel safe venturing ideas and contributing to collective understandings, contributing to their sense of self as productive mathematical thinkers and learners?
5. How can we organize instruction so that student thinking is made visible, so that we can make adjustments that provide students opportunities for deep experiences along all of the dimensions in Figure I.1?

In this brief introduction, we want to stress one key feature of TRU. Our primary focus when we plan for and reflect on instruction is on the students' classroom experiences. That's because the key question about any learning environment is: *what opportunities do the students have to learn and to grow?* If you look back at Figure I.1, you'll see that TRU dimensions 2 through 5 (Cognitive Demand; Equitable Access to Content; Agency, Ownership, and Identity; Formative Assessment) are concerned with the opportunities students have to engage in mathematical sense making. We've tried to capture that in Figure I.2, where we represent the five dimensions from the student's point of view.

Observe the Lesson Through a Student's Eyes

The Content	• What's the big idea in this lesson? • How does it connect to what I already know?
Cognitive Demand	• How long am I given to think, and to make sense of things? • What happens when I get stuck? • Am I invited to explain things, or just give answers?
Equitable Access to Content	• Do I get to participate in meaningful math learning? • Can I hide or be ignored? In what ways am I kept engaged?
Agency, Ownership, and Identity	• What opportunities do I have to explain my ideas? In what ways are they built on? • How am I recognized as being capable and able to contribute?
Formative Assessment	• How is my thinking included in classroom discussions? • Does instruction respond to my ideas and help me think more deeply?

Figure I.2 What's important, from the student perspective

Talking with students can help to answer these questions, but that is not always possible in a professional development setting. We can learn a lot by trying to imagine ourselves in the students' places, seeing what we can understand from what they say and do, and thinking about how students might have answered these questions.

What are our goals?

Short answer: improving teaching, improving learning. Of course, that's a lifelong project. We see this book as a catalyst to ongoing professional growth. There's compelling evidence that, the more what takes place in a classroom is consistent with the principles underlying TRU, the more mathematically powerful and empowered the students in that classroom become (see, e.g., Schoenfeld 2013, 2104. Additional evidence is given in Chapter 1 and the publications section of the TRU website, at https://truframework.org/.) Our intention in this volume is to introduce you to TRU, to provide enough experience through reflecting on the cases so that you can begin to see instruction through a TRU lens, and to provide you with tools that you can use to enhance planning, in-the-moment instruction and reflection. Our long-term goal is to build an expanding collection of research-based tools that supports the continuous improvement of teaching through planning, teaching, and reflection.

Is this a "how to" book?

In a word, no. Although this book is focused on effective teaching, we refuse to tell you how to teach. Why? It's insulting – and it doesn't work. Teaching contexts differ, and there is no "one size fits all" style of teaching. Some of the best teachers we know have very different styles – why would we want them to fit one mold[2]?

What we will do is offer you some examples of teaching and teaching dilemmas, and some instances of instruction to reflect on, individually and collectively. We use the TRU Framework to help you think through the examples we present, and we offer a number of TRU tools (See Chapter 6), which you can use for planning and reflecting on your own teaching. Our hope is that by the end of the book, thinking with and using TRU to plan, teach, and reflect will feel helpful and natural to you.

What's in the book?

The book is divided into three parts. Part I introduces you to the TRU Framework. Part II presents and discusses three cases that illustrate interesting teaching situations. These case studies, taken from real classrooms and transcribed verbatim, illustrate teaching situations in which the teachers face choices when planning and implementing day-to-day lessons: what tasks do they select, what questions do they ask, what might they do next to help their students?

Our goal is for you to experience these cases as workshops, engaging with the content and thinking about the students as though you were teaching the lessons. We'll use the TRU Framework to raise questions about student thinking and learning, and we'll invite you to think about them. Then we'll discuss some of the issues we see in the cases and the implications of some possible choices. Part III indicates how you can take the ideas from Part II and incorporate them into your own work, whether that work involves teaching, coaching, mentoring, or researching. It summarizes the "lessons" we hope you take from the cases and how you can build on them, either by yourself or with colleagues. It continues by providing you with two major tools, the TRU Conversation Guide and the TRU Observation Guide, and rounds things off with a list of resources you may find useful.

Where can I find more resources?

We'll point to resources throughout the book, and discuss a range of resources in Part III. The place to start is https://truframework.org/.

Welcome to the TRU community!

How to use this book

Part I of this book introduces the TRU Framework, which identifies the key dimensions of powerful learning environments. Part I provides essential background. We've tried to keep it very terse, so that we can begin to dig into the case studies together. (There's lots more information about TRU on the "publications" page of the TRU website, at https://truframework. org/publications/. The collection of downloadable papers on the website provides the theoretical foundations for TRU and lots of evidence substantiating its validity.)

Part II is the "hands-on, minds on" core of this book, in which you'll dig deeply into three case studies of classroom instruction. These cases are not meant to be read passively! We provide multiple opportunities for reflecting on how the classroom activities (transcribed from videotapes of real mathematics lessons[3]) affect what students learn and how they see themselves as mathematics learners.

In discussing each activity, we've tried to reproduce – as much as we can in print – the kinds of activities that would take place if we were working together in person. Many of the authors' professional development activities have involved watching and talking about episodes of classroom practice. Much of that has involved looking at videotapes of teaching (often selected by the teacher, for purposes of reflection). Here's what we'd do if we were together in the same space.

We'd start our work together by doing the featured math task and discussing the mathematics at the heart of the video clip, so that all of us develop a sense of the challenges the students might encounter. Then we'd tell you about the class in the video we'd be watching. After talking about our norms for watching tapes[4], we'd watch the clip once through completely, to get a "big picture" sense of what happens – what the important events seem to be, and how things wind up at the end. That top-level view would set us up for a closer look together.

Then we'd watch the clip again. We'd watch a few minutes at a time, stopping to discuss each segment so that we could agree on what happened – especially in the sections that seem important. Some of the questions we'd ask are: How do we see or hear students engaging with important mathematical ideas and practices? Which students get to explain their ideas? How are those ideas evaluated? What happens to incorrect answers or "ideas in progress?" How are students positioned, by the teacher and by each other? Here too, we'd focus on how students seem to experience whole class and small group interactions. The five TRU dimensions are the lenses through which we view the interactions.

Finally, once we had a solid sense of what we think happened, we'd explore some of the issues raised by the clip. How can we understand what happened? What issues do classroom interactions, whether they take place between students or between students and the teacher, raise for consideration? What might be the impact of particular interactions on students? What reasons might a teacher have for doing X (as in the clip) and not Y (an alternative, including possibly doing nothing at all)? What are the risks and possible payoffs for students of doing X, or Y, or something else? We'd use the TRU Framework to open up questions along each of the five dimensions – typically with an emphasis on one or two particular dimensions at a time. (In classroom episodes, some dimensions often stand out more than others.)

That's what we'd do if we were all watching and reflecting on a video together. Here's what we'll try in the text.

We'll start by introducing the mathematics task or tasks that are the focus of each case, giving you a chance to work through the mathematics and reflect on it. What matters in this body of mathematics? What interesting ideas might be explored? What are the essential ideas that we want students to learn? We'll then provide an overview of the events that occurred during the classroom episode that's the focus of the case, so you have a rough idea of what's coming. With that as background, we'll dig more closely into what happened based on what we deciphered from the video. We'll break the case into segments, give you a fair amount of detail about what happened in each segment, and ask you to focus on key questions like the ones mentioned above. Then we'll explore more deeply, digging into and reflecting on the content of the case. Each case is divided into 3–5 episodes. For each episode we'll raise some "focus questions" that address key aspects of the episode – what happened? What might the implications be, what options might the teacher have for proceeding? Our hope is that you'll take the time to think through and discuss these questions – that's the "workshop" aspect of reading this text. Then we'll provide our thoughts about the issues.

Our hope is that each of the "deep dives" in Part II will help you explore the case closely – and that the questions we ask will become part of your repertoire for thinking about teaching and learning.

Cumulatively, Part II provides you with opportunities to dig deeply into mathematics thinking, teaching, and learning, in ways consistent with the TRU Framework. Once thinking about instruction in those ways becomes second nature, planning, teaching, and reflecting on instruction can be a lifelong habit – and a source of ongoing professional growth.

Part III begins with a discussion of different ways in which the kinds of planning, teaching, and reflecting on your teaching discussed in the cases can become a part of your ongoing professional work. Whether you've read this book by yourself, are working together with a few colleagues, or are part of a professional learning community, there exist models by which you can make "thinking with TRU" an ongoing part of your teaching and learning. Part III concludes with some essential resources, the *TRU Conversation Guide* and *the TRU Observation Guide*. These too are "hands on" documents for planning and reflecting. And, we offer pointers to some additional resources and tools. We hope you'll take advantage of these, and make them part of your ongoing professional growth.

Notes

1 Research indicates that the TRU Framework applies to all learning environments – to classrooms at all levels, in all content areas. *Any* classroom video can be the focus of the kinds of conversations we model in this book; we hope you will be inspired to use your own. In order to focus our conversations, we have chosen three videos whose mathematics content is at the middle and secondary level in mathematics. We hope to see other volumes, with different content and grade level emphases, in the future.

2 Different styles are a matter of their underlying understandings and what they feel comfortable with personally. What great teachers *do* have in common is that their pedagogical beliefs and practices are consistent with the five dimensions of TRU. In a sense, those dimensions serve as principles for powerful instruction. Teachers can live up to those principles in different ways, consistent with their own personal styles.

3 We are extremely grateful to the teachers for their permission to make use of the videos in this way.

4 Our goal is to be respectful and supportive, and to build a climate of collective inquiry. The simplest way to think about things: Imagine that the teacher in the video is a colleague who's in the room, watching the video with you. We'll elaborate on norms for visiting classrooms in the introduction to Part II.

Part I

The Big Ideas in Teaching and Learning

Chapter 1

An Introduction to the TRU Framework[1]

This chapter provides a brief overview of the Teaching for Robust Understanding (**TRU**) Framework. The framework provides a comprehensive research-based response to the question,

> What are the attributes of equitable and robust learning environments – environments in which all students are supported in becoming knowledgeable, flexible, and resourceful disciplinary thinkers?

The essence of TRU is given in Figure 1.1. Research (see, e.g., the publications listed at https://truframework.org/publications/) shows that the quality of a learning environment depends on the extent to which it provides opportunities for students along these five dimensions:

1. The richness of discipline-related concepts and practices available for learning
2. Student sense-making and "productive struggle"
3. Meaningful and equitable access for each and every student to central concepts and practices
4. Opportunities to construct positive disciplinary identities through presenting, discussing, and refining ideas
5. The responsiveness of the environment to student thinking.

1.1 Where Did These Ideas Come from and How Can They Be Useful?

There is a huge literature on teaching and learning and on the "good things" that should happen in classrooms. In fact, there's too much to look at and use; it's impossible to keep it all straight. One of our main goals was to create a framework that covered everything essential but was also easy to remember and use[2]. The five dimensions of TRU have the following properties:

1. They are comprehensive. If a learning environment supports student learning along these dimensions, then the students who emerge from that environment will be knowledgeable, flexible, and resourceful thinkers and learners.

    In short: if you learn to think with TRU, you'll be focusing on everything that counts.

DOI: 10.4324/9781003375197-2

The Five Dimensions of Powerful Mathematics Classrooms				
The Mathematics	**Cognitive Demand**	**Equitable Access to Mathematics**	**Agency, Ownership, and Identity**	**Formative Assessment**
The extent to which classroom activity structures provide opportunities for students to become knowledgeable, flexible, and resourceful mathematical thinkers. Discussions are focused and coherent, providing opportunities to learn mathematical ideas, techniques, and perspectives, make connections, and develop productive mathematical habits of mind.	*The extent to which students have opportunities to grapple with and make sense of important mathematical ideas and their use. Students learn best when they are challenged in ways that provide room and support for growth, with task difficulty ranging from moderate to demanding. The level of challenge should be conducive to what has been called "productive struggle."*	*The extent to which classroom activity structures invite and support the active engagement of all of the students in the classroom with the core mathematical content being addressed by the class. Classrooms in which a small number of students get most of the "air time" are not equitable, no matter how rich the content: all students need to be involved in meaningful ways.*	*The extent to which students are provided opportunities to "walk the walk and talk the talk" – to contribute to conversations about mathematical ideas, to build on others' ideas and have others build on theirs – in ways that contribute to their development of agency (the willingness to engage), their ownership over the content, and the development of positive identities as thinkers and learners.*	*The extent to which classroom activities elicit student thinking and subsequent interactions respond to those ideas, building on productive beginnings and addressing emerging misunderstandings. Powerful instruction "meets students where they are" and gives them opportunities to deepen their understandings.*

Figure 1.1 The Teaching for Robust Understanding (TRU) Framework

2. TRU focuses on the classroom experience, as seen from the student perspective.

    The fundamental way in which TRU differs from most other frameworks is that TRU focuses on the student's classroom experience – it's truly student-centered. If you look at Figure 1.1, you'll see that TRU is primarily concerned with making sure that students experience instruction in ways that are personally meaningful and powerful. In fact, the key idea behind TRU is to learn to see instruction from the perspective of the student. That's why Figure 1.2 is so central.

3. Together, the five TRU dimensions provide a language and a framework for inquiring into instruction and improving it.

    The five dimensions provide a way of thinking and talking about teaching that focuses on what matters. As you'll see, implementing TRU involves asking questions – Dimension 1, how can the math be made richer and deeper? Dimension 2, how can instruction be arranged so that more students are engaged in sense-making, building on what they know? Dimension 3, how can more students be invited to engage meaningfully and equitably with core material? Dimension 4, how can we provide students with more opportunities to see themselves as mathematical thinkers? Dimension 5, how can student thinking be made public and built on so that instruction meets the students where they are?

    Once you start thinking and talking in these ways, it's easy to talk about teaching in ways that connect to other teachers and administrators, in ways that focus on what counts. A district leader told us that her kindergarten, fourth-grade, eighth-grade, and tenth-grade teachers along with district observers and administrators could now talk to each other about what was going on in their classrooms. The central ideas in TRU guided coaches' and administrators' classroom visits and professional development – teachers and administrators

Observe the Lesson Through a Student's Eyes

The Content	• What's the big idea in this lesson? • How does it connect to what I already know?
Cognitive Demand	• How long am I given to think, and to make sense of things? • What happens when I get stuck? • Am I invited to explain things, or just give answers?
Equitable Access to Content	• Do I get to participate in meaningful math learning? • Can I hide or be ignored? In what ways am I kept engaged?
Agency, Ownership, and Identity	• What opportunities do I have to explain my ideas? In what ways are they built on? • How am I recognized as being capable and able to contribute?
Formative Assessment	• How is my thinking included in classroom discussions? • Does instruction respond to my ideas and help me think more deeply?

Figure 1.2 What's important, from the student perspective

were on the same page regarding what to look for and focus on. (Of course, we offer some tools that help to do this; see Part III of this volume.)

4. TRU is *not* a set of "recipes" that tells you what you should do in the classroom.

TRU contains no "thou shalts." That's because there is no one "right way" to teach. In fact, you should distrust anyone who says there is. Think of the best teachers you know. They do many things differently, in their own ways – but, we can guarantee you that despite surface dissimilarities, what they do is consistent with the five TRU dimensions. That's because the five dimensions act as *principles of high-quality instruction*. "Getting" what TRU is all about means making it your own, and living the five dimensions in ways that are comfortable for you.

5. Each dimension of TRU can serve as the focus of coherent professional development. Departments, schools, and districts can organize themselves in ways to make systematic improvements.

There are various ways to do this. There's a TRU version of Lesson Study, for example (Schoenfeld et al. 2019a). Different school districts have implemented TRU-based professional development (PD) in very different ways (Schoenfeld et al. 2019b). What matters is that the PD fits into the local context and helps focus on what counts.

In this book, we work through one particular model – using the five TRU dimensions as lenses through which we examine and reflect on extended episodes of classroom practice. Doing so gives us opportunities to see, close up, how students experience the math classroom, and to reflect on how to enhance those experiences. Our hope is that having done so "with us" by working carefully through the three case studies in this book, you'll be in a position to continue doing so – preferably with your colleagues.

6. TRU doesn't compete with other initiatives; it enriches them.

All teachers and administrators have too much on their individual and collective plates. Depending on your school or grade level, you may be involved with complex instruction, various equity initiatives, other disciplinary frameworks, Lesson study, programs for math fluency, problem-based learning, readers/writers workshops, SEAL, GLAD, and more. How in the world could you make room for yet one more thing, without something falling off the plate?

The good news is that you don't have to. Once you come to see TRU as a method of questioning what you're doing so you can improve it, then TRU *enriches* every initiative you're engaged in, rather than competing with it. That happens because you ask questions similar to those we discussed in point 3 above. Suppose you're implementing "Program X." Once TRU is part of the way you think, it's natural to ask questions like this:

Dimension 1, how can the math in Program X be made richer and deeper? Dimension 2, how can instruction be arranged so that student sense-making in Program X is more within reach? Dimension 3, how can more students be invited to engage meaningfully with core material in Program X, in ways that (Dimension 4) provide students with more opportunities to see themselves as mathematical thinkers? Dimension 5, how can student thinking in Program X be made public and built on so that instruction meets the students where they are?

That is, life gets simpler when you view all classroom instruction through the same (TRU) lens.

Our bottom line: No matter how good classroom instruction may be, it can always be enriched. It will be enriched if – as a matter of routine and habit – teachers, coaches, and administrators take the five dimensions of TRU into account when planning, implementing, and reflecting on instruction[3,4].

There are no quick fixes, no magic methods. Improving our teaching is not easy. But, knowing what to focus on can be a big help. When teachers engage in systematic and collaborative reflection on: (Dimension 1) disciplinary ideas; (Dimension 2) ways to open up those ideas to students; (Dimension 3) ways to provide *all* students opportunities for sense-making; (Dimension 4) providing opportunities for students to express their understandings and build on their own ideas and those of others; and (Dimension 5) adjusting instruction in the light of the understandings that students reveal, the result is the ongoing enhancement of instruction. Professional learning communities that focus on what counts produce sustained improvement in teaching and in student understanding.

The pages that follow provide brief introductions to the five TRU dimensions.

1.2 The Five Dimensions of TRU

Dimension 1: The Discipline (With a focus on Mathematics)

The extent to which classroom activity structures provide opportunities for students to become knowledgeable, flexible, and resourceful disciplinary thinkers. Discussions are focused and coherent, providing opportunities to learn disciplinary ideas, techniques, and perspectives, make connections, and develop productive disciplinary habits of mind.

Students' understanding of every discipline is shaped in fundamental ways by their classroom experiences with it. If, for example, a reading class focuses on decoding text, a history class focuses on memorizing the dates of major events, or a mathematics class focuses on memorizing procedures, there is little chance that the students in that class will emerge from it with either an appreciation of the discipline or the conceptual understandings they need.

Learning to "think like a historian," or like a scientist, or a practitioner of any discipline, means coming to grips with the concepts and practices of that discipline – approaching phenomena through a disciplinary lens, with a broad spectrum of knowledge and tools at one's disposal. Historians "place themselves in context" to understand the motivations and actions of historical figures. Writers have a sense of purpose and a sense of audience when writing, as well as relevant factual and grammatical knowledge. Scientists and mathematicians inquire into "what makes things tick," using reason, equations, representations, and models in the service of sense-making. This combination of disciplinary orientations, knowledge (including concepts and tools), practices, and habits of mind is what we refer to in shorthand as the "content" of the discipline. Students need to experience that content in its full richness if they are to become disciplinary thinkers.

Mathematical content and practices matter. It's not just the lists of content in various mathematical standards: we'd like students to have a deep sense of mathematical concepts, to seek and see mathematical connections, and to have a wide range of representations at their disposal. We'd like for them to have a sense of mathematical initiative and agency and fluency in mathematical practices – conjecturing, problem posing and problem-solving, reasoning and proving, representing, etc. Classroom mathematics should not simply be about "answer-getting." It should provide students opportunities for thinking mathematically.

Notions of "thinking mathematically" go back a long way. Virtually all contemporary work on problem-solving is grounded in Pólya's (1945, 1954, 1962, 1965/81) foundational ideas. Those were expanded on by Schoenfeld (1985, 1992), who identified various components of mathematical thinking: the knowledge base, problem-solving strategies, metacognition (monitoring and self-regulation), and belief systems. Mason, Burton, and Stacey (1982) provided a lovely collection of problems to work on, as well as a "play by play" commentary on the process of solving the problems. More recent studies of mathematical thinking have focused first on *processes* (e.g., the two volumes of NCTM Standards (1989, 2000) focused on processes such as problem-solving, reasoning, making mathematical connections, communicating mathematically, and using representations) and then on *practices* (e.g., the 2010 Common Core State Standards identified eight key mathematical practices, beginning with problem-solving). For an international perspective on "Essential mathematics for the next generation: What and how students should learn" see McDougal (2017). For varied conceptions of the role of mathematicians' practice in mathematics education, see Weber and Dawkins (2020). For a recent view of mathematical exploration, see Schoenfeld (2023).

Similarly, content goals for student learning have evolved through the years – and they will surely continue to evolve. Statistics and data analysis play a much more significant role in today's curriculum than they did a few decades ago, for example, and mathematical modeling is becoming more prominent (see, e.g., Lamb 2021). As we will see in Chapter 2 (the car value problem), modeling or "contextual" problems provide expanded opportunities for sense-making and making connections outside-the-classroom aspects of students' lives (see, e.g., Berry et al. 2020). Mathematizing experiences that reflect their lived experiences also provides expanded opportunities for developing positive mathematical identities (NAEd/NCTM Sub-Committee on Civic Reasoning and Discourse 2023).

Whatever the current curriculum is or will be, we want the mathematical content and practices that students encounter to be as rich as possible. That's why, in each of the three case studies in Part II, we explore the mathematical potential of the lessons we study in great depth. The more we see the mathematical possibilities potentially available in a lesson, the more we can make them available to students! It's also why Part IV of this volume offers the TRU Conversation and Observation Guides, which offer ways to inquire into and reflect on the richness of the disciplinary content that we offer students.

But … rich content and practices are just a beginning. The primary idea behind TRU is that what counts in instruction is how students engage with the mathematical content and practices – how they are or are not positioned to take advantage of the riches the discipline has to offer. We have all been in classes where, for example, the content was "over our heads" or we failed to connect to it for some reason; no matter how beautiful it may have been, we were lost. That is why dimensions 2 through 5 of the TRU Framework – how students themselves experience the discipline – are so important.

Dimension 2: Cognitive Demand

The extent to which students have opportunities to grapple with and make sense of important disciplinary ideas and their use. Students learn best when they are challenged in ways that provide room and support for growth, with task difficulty ranging from moderate to demanding. The level of challenge should be conducive to what has been called "productive struggle."

If students are given work that is too easy, there is little for them to learn. Moreover, they're likely to be bored or frustrated. If students are given work that is too distant from their current understanding and they can see no pathways to progress, then there is no pathway to learning. They're likely to be bored or frustrated as well. As Stein and Smith (1998) put it,

> Tasks that ask students to perform a memorized procedure in a routine manner lead to one type of opportunity for student thinking; tasks that require students to think conceptually and that stimulate students to make connections lead to a different set of opportunities for student thinking.

Our challenge is to find tasks and classroom activities that provide students with meaningful opportunities for learning and that support their growth through active engagement with the content.

Researchers use the term "cognitive demand" to describe the level of difficulty, relative to what they know, of the work that students are asked to engage in. The goal is to find a middle ground, where students have opportunities to build on what they know and to stretch their current understandings. In order to make sense of rich content, students need to engage in what has been called "productive struggle" (Stein & Smith 1998, Hess 2006). Webb (1997, 2002) discusses four levels of knowledge: Recall & Reproduction, Skills & Concepts, Strategic Thinking & Reasoning, Extended Thinking (see also Hess 2013). At various times, students need to engage at all of these levels.

When students get stuck – when they experience difficulty dealing with complex issues – it's uncomfortable, and there's a strong tendency to help them out by telling them what to do. The challenge is that doing so deprives the students of opportunities for productive struggle and sense-making (Henningsen & Stein 1997). The challenge we face is to provide clarifications and other support (e.g., offering heuristic advice, raising issues, or suggesting alternative approaches) without telling students precisely what to do. This is by no means easy (but see Dimension 5, *formative assessment,* for a discussion of providing the right levels of scaffolding and support). But, it's through making progress on problems that are challenging yet within reach that students build their mathematical muscles – and the sense that they can make progress on difficult problems. (See Dimension 4, Agency/Ownership/Identity.)

There are many ways that we can offer challenging mathematical activities and work to maintain appropriate levels of cognitive demand. For example,

- In designing and selecting tasks, we can avoid providing detailed step-by-step instructions for solving problems, repetitive exercises, or detailed "recipes" for completing tasks that allow little room for students to build on their current understandings.
- We can actively support students in individual work, group work, and whole-class discussions by asking clarifying questions and providing scaffolds, rather than moving directly to suggesting specific ways to go about assigned tasks.
- We can employ a range of techniques to support students in "getting their ideas on the table" and working through them. See, for example, SERP (2016) on academically productive talk.

- We can encourage students' productive struggle in a general way by discussing ideas of malleable intelligence and a growth mindset (Dweck 2007), making it clear that learning is not a matter of memorization and that one gets better at any discipline by working hard at it.

Classroom practices that support rich mathematics discourse place students in their zones of proximal development (Vygotsky 1978), allowing for learning through productive struggle. The *5 Practices for Orchestrating Productive Mathematics Discussions* highlighted in Smith and Stein (2018) –

- *Anticipating* students' solutions to a mathematics task,
- *Monitoring* students' in-class, "real-time" work on the task,
- *Selecting* approaches and students to share them,
- *Sequencing* students' presentations purposefully, and
- *Connecting* student responses during the discussions,

provide a robust structure for building dialogic classrooms that focus on mathematical substance in a way that adjusts conversations to fit student understandings, a primary goal for cognitive demand. Smith and colleagues (Smith, Bill, & Sherin 2019, Smith & Sherin 2019, Smith, Steele, & Sherin 2020) exemplify and analyze lessons using the five practices at the elementary, middle, and secondary levels, respectively. We also note that the sequence, in general, is very close to the general structure of classic Lesson Study lessons (see, e.g., the geometry video from the TIMSS (2022) video study, at http://www.timssvideo.com/japan-mathematics-lessons, or the discussions of Lesson Study in Fernandez & Yoshida 2004, Lewis & Hurd 2011, Takahashi 2015, Takahashi et al. 2022).

As we've said, all of the dimensions of TRU are interrelated. Before moving on, we want to point to some connections between cognitive demand and the other dimensions of TRU. Fundamentally, "productive struggle" is a mechanism for developing deep and meaningful understandings of mathematical content; struggles to understand, connect, and explain are themselves productive mathematical practices (Dimension 1). Productive struggle and the sense-making involved in it are essential for *all* students (Dimension 3), not only for meaningful participation but so that students engage with the content in ways that they come to "own" it and develop positive disciplinary identities (Dimension 4). And, the best way to arrange for students to be working at the right levels of challenge is to make their thinking publicly accessible, so instruction can "meet the students where they are" in order to support their moving forward (Dimension 5; see Burkhardt & Schoenfeld 2019, 2022). It's essential that we work on making sure students are engaged in sense-making – and that as we do, all the other dimensions of TRU come into play.

Dimension 3: Equitable Access to Content and Practices

The extent to which classroom activity structures invite and support the active engagement of all the students in the classroom with the core disciplinary content and practices being addressed by the class. Classrooms in which a small number of students get most of the "air time" are not equitable, no matter how rich the content: all students need to be involved in meaningful ways.

Equitable classrooms provide every student with meaningful opportunities to engage in and with disciplinary concepts and practices, supporting the students in developing their own understandings and building productive disciplinary identities.

This dimension, *Equitable Access to Content and Practices,* focuses on the question of whether, within the classroom, there is differential access to the mathematics being addressed.

There may be rich discussions or other productive activities taking place – but, who participates in those discussions or activities?

There is a long history of differential achievement ("performance gaps") by students who come from varied racial, ethnic, and economic backgrounds. And, there are strong arguments that such differences can be tied to differential access to opportunities to learn (Ansalone 2003, Oakes 2005, Oakes, Joseph, & Muir 2001). While one clear source of this differential access is tracking, which is outside of the scope of classroom improvement efforts, another is the pattern of discourse within classrooms. Do all students have frequent opportunities to discuss important ideas? In *How Schools Shortchange Girls* (American Association of University Women 1992), for example, research revealed that in math classrooms boys were being called upon far more often than girls. Moreover, when girls were called upon, they were often asked questions that were less conceptually oriented than the questions that were asked of boys. That kind of pattern denied girls meaningful opportunities to engage in key mathematical practices. Such patterns persist today, and they result in differential performance: as Reinholz et al. (2022) indicate, "women's participation predicts gender inequities in mathematical performance." Similar inequalities take place when students are separated from others to do remedial work while the rest of the class advances – the students who are supposedly "catching up" are denied opportunities to progress with the rest of the class.

Such inequities occur with regard to gender, ethnicity, and race – see, for example, Reinholz & Shah (2018). Patterns of racial inequities can be significant – and, because the discussion of race-related issues can be challenging, such inequities may go unmentioned and unaddressed by participants, even when objective data show that race-related opportunity gaps are substantial (Fink 2022). As Moses (2001) makes clear, the consequences for children's lives are significant. Persistent "performance gaps" are in part a consequence of differential treatment and in part a consequence of the structural issues that both provide limited opportunities for certain racial, ethnic, and socioeconomic groups (Schoenfeld 2022). There is, as above, a long history of attempts to conceptualize and address such inequities (see, e.g., Aguirre, Mayfield-Ingram & Martin 2013, Boaler & Staples 2008, Cohen & Lotan 2014, Cohen, Lotan, Scarloss, & Arellano 1999, Darling Hammond 2010, DIME Center 2007, Gutstein 2006, Gutstein & Peterson 2005, Horn 2012, Kozol 1992, Ladson-Billings 1997, Martin 2009a, 2009b, 2013, Nasir & Cobb 2007, Nasir et al. 2014, Reinholz & Shah 2018, Schoenfeld 2003, Shah 2017).

We should always ask, what opportunities do students who are labeled and placed as English language learners have for full participation, or students from differing demographic or racial groups? Do multiple opportunities exist for students to engage with the content, to develop and display competence (Cohen 1994), and to build understanding based on the knowledge they bring with them into the classroom (see, e.g., Moll et al. 1992, Shah 2017)? What tools can we offer that provide English language learners (and everyone else) access to the mathematics under discussion, both informally and in terms of formal mathematical language (Daro 2021, Moschkovich 2012, 2013)?

Research indicates that effective teachers encourage participation by all students in the intellectual community of the classroom (Boaler 2008, Cohen & Lotan 1997; Schoenfeld 2003). They select and make use of tasks that enable all students to engage in challenging content, and they establish and reinforce expectations for various ways to participate in and contribute to classroom activities.

There are numerous ways in which students can be supported in access to disciplinary content and practices.

- In choosing and designing activities, and in launching activities, teachers can provide multiple access points to the relevant material, supporting the expectation that *all students* are able and expected to participate.

- Tasks that can be approached in multiple ways or from multiple perspectives, and in which approaches can be compared and contrasted, provide access to students who choose different pathways into the activity. In addition, they provide opportunities for making connections between student approaches.
- Teachers can encourage the generation and refinement of ideas rather than mainly critiquing or ignoring comments that are only partially correct.
- Teachers can support the use of multiple language registers by, for example, asking one student to restate another's contribution in more precise academic language, or, perhaps, in more informal language.
- During discussions, teachers can use a variety of strategies to encourage broad participation, for example: choosing to call only on students who have not yet spoken; allowing time to talk to a partner before responding publicly; and randomly selecting students to contribute.
- Teachers can use tasks with language and contexts that connect to students' lived experiences and provide windows into unfamiliar experiences, being mindful of power and privilege.

The point we want to stress in summary is that what matters is not simply that students get to participate – it's *how* they participate. One form of inequity is when the teacher calls on a small number of students repeatedly, and the rest of the class does not get to participate. But, another more subtle form of inequity is when students are divided into "ability groups" or otherwise clustered, and the students in the "high ability" groups work on challenging problems while the students in the "low ability" groups do remedial work or practice skills. The goal for equitable instruction is for each and every student to engage meaningfully with the core concepts and practices being studied.

Needless to say, equitable access is connected in fundamental ways to the other TRU dimensions. If students are not provided opportunities to engage with powerful mathematical content and practices (Dimension 1), they're not going to become mathematically powerful! Similarly, meaningful access means students being put in a position where they can learn – where they can grow by means of building on what they know, a matter of cognitive demand (Dimension 2). Perhaps most important, access, while essential, is not enough: the question is, how are students engaged with the mathematics? Do they have opportunities to participate meaningfully in mathematical discourse, to get their ideas on the table, to explain their thinking? These kinds of activities are what shape the development of mathematical identities (Dimension 4). And, of course, the way we work to adjust classroom activities so that each student has meaningful opportunities to participate is through formative assessment.

Dimension 4: Agency, Ownership, and Identity

The extent to which students are provided opportunities to "walk the walk and talk the talk" – to contribute to conversations about disciplinary ideas, to build on others' ideas and have others build on theirs – in ways that contribute to their development of agency (the willingness to engage), their ownership over the content, and the development of positive identities as thinkers and learners.

Your mathematical identity includes how you see yourself with regard to mathematics. Do you think you're good at math? Do you like it? Are you willing to jump in and try things? Or, does math make you uncomfortable, and do you shy away from it if you can?

People's disciplinary identities in general – for example, "I am a reader," or "I'm a history buff" – are shaped by their experiences both outside and inside school. In math, stereotypes about gender, race, and ethnicity play a significant role. Stereotypes such as "girls are bad at math," "Asians are good at math," and "Blacks and Latinos are bad at math" shape both how people see others and (because they see themselves reflected in others' eyes) how they see themselves. (Martin 2009a, Nasir & Shah 2011). Such biases shape people's mathematical senses of self and their actions, even before – and as – they enter the classroom. (Shah (2017) relates the story of an African-American student who was asked "what are you doing here?" as he entered a high school calculus class that was otherwise populated by whites and Asian-Americans.) The issue of opportunity is exacerbated by tracking – both within schools and, de facto, by economic segregation and differential opportunities for learning. (The "Savage Inequalities" chronicled by Kozol (1992) are still very much with us.) But no matter what the ethnic/gender/racial distribution of students in a school, societal myths about different groups' mathematical abilities permeate school walls. If students enter our classrooms being stereotyped as non-performers, or if they are positioned as such by other students, that has a profound impact on who they are and how they act mathematically (Schoenfeld 2022).

At the same time, the vast majority of a student's experience with mathematics takes place within our classrooms. The way you think about yourself mathematically is shaped by your personal history with mathematics, and by current circumstances – for example, how you are positioned by your fellow students and the teacher, and whether you feel it's safe to venture ideas and what happens when you do. Over time, our classroom experiences are the major factors shaping our mathematics identities.

Dimension 4 (Agency, Ownership, and Identity, which we refer to as AOI for short) focuses on the extent to which students have opportunities to generate and share ideas, both in whole-class and small-group settings; the extent to which student contributions are encouraged, recognized and supported as part of regular classroom activity; and the extent to which student ideas are valued and built upon as the classroom constructs its collective understandings. These properties shape how students see themselves, how they see others, and how they see the very nature of mathematics and their opportunities to be involved with it.

One fundamental aspect of mathematics identity is an individual's sense of *agency* or initiative – their willingness to engage in and with mathematics. This comes from the perception that they can progress on challenging issues by working away at them and trust in the conclusions that they draw. Engle (2011) writes,

> Learners have intellectual agency when they … share what they actually think about the problem in focus rather than feeling the need to come up with a response that they may or may not believe in, but that matches what some other authority like a teacher or textbook would say is correct.

Ownership refers to the feeling that you've made sense of mathematical ideas, rather than parroting or memorizing those of others. It is the difference between saying "I've reasoned this through and I'm confident it makes sense" and relying on external authority.

A key issue is the extent to which the classroom provides students with opportunities to develop these aspects of their disciplinary and personal identities. Supportive teachers recognize and capitalize on the strengths of each student, finding ways to help individual students enter into the learning community when they do not easily enter it on their own (Boaler 2008, Cohen & Lotan 1997). There are multiple ways to do this. Teachers can create opportunities for public recognition of students' contributions to disciplinary discussions, help students work together in small groups, and attend to students who are struggling by building on the strengths in their thinking. For example, Resnick, O'Connor, and Michaels (2007) identify powerful talk moves by teachers such as revoicing (repeating, paraphrasing, or summarizing a student contribution for the whole class to react), asking students to restate others' reasoning, to build on what other students have said, and prompting for explanations.

Above and beyond teacher moves, however, is the very nature of the classroom environment. Do students feel safe making contributions to classroom conversations? Have norms been established for making contributions? For building on contributions from others? For critiquing contributions from others? Fink (2022) demonstrates the subtle ways in which inequities manifest themselves in classroom discourse.

There is a large literature on "accountable talk," the kind of classroom discourse that supports students in responsibly and respectfully co-constructing ideas. For a large portfolio of resources, see Institute for Learning (2016).

To give just one example, a technique for shaping classroom discourse productively is the use of "sentence stems" or "sentence starters" aimed at promoting accountable talk. Students learn to use phrases such as the following:

- I disagree (or agree) with that because _____
- I still have questions about _____
- This is the same because _____
- I observed _____
- I'm confused by _____
- To expand on what _____ said _____

Such neutral language (as opposed to "that's wrong!") supports reasoned discussions and collaboration. It's when students feel comfortable contributing to the collective development of mathematical ideas they have opportunities to develop a sense of academic and mathematical agency, ownership of the ideas discussed, and positive mathematical identities.

There are a wide range of ways to open things up for students. See *Mathematics Teaching On Target* (Schoenfeld et al. 2023) for a systematic approach to expanding tasks and activities to make them both more rich mathematically and more open to student participation. Generally speaking, the mathematics curriculum is seen by many students as being divorced from their personal lives; this can be addressed both in curricular terms (NAEd/NCTM Sub-Committee on Civic Reasoning and Discourse 2023) and by drawing upon students' funds of knowledge in various ways (Moll et al. 1992, Zuñiga-Ruiz 2022).

Foundational work on identity was done by Wenger (1988). As Turner et al. (2013) note,

> Researchers have referred to [the] connection between participation and identity in various ways, including *participatory identities* (Greeno 2006), *discourse identities* (Koole 2003), *positional identities* (Holland, Lachiotte, Skinner, & Cain, 1998), and *identities in practice* (Holland et al., 1998, Wenger 1998). Common to these perspectives is the proposition that an individual's participation in settings that are socially, culturally, discursively, historically, and politically constituted is directly related to that individual's identity development.
> (Turner et al. 2013, p. 201)

Aguirre, Mayfield-Ingram, and Martin (2013) explore the impact of identity in K-8 mathematics learning and teaching and provide ways of rethinking equity-based practices. Gutierrez (2002, 2009) explores the complexities of language in instruction and the tensions involved in equity-focused instruction. Hodge and Cobb (2019) explore the ways in which conceptions of culture shape mathematics teaching and learning. Martin (2009b) provides a systematic exploration of the role of race in mathematics education and ways to explore it in research.

As always, the mathematics that students encounter (Dimension 1) needs to be rich and connected, with opportunities for students to develop mathematical ideas and practices – we're talking about students' mathematics identity, after all! Positive disciplinary identities come from meaningful engagement and growth within the discipline; thus a reasonable degree of cognitive demand and productive struggle (Dimension 2) is essential. It goes without saying that this applies to all students (Dimension 3). But, as we noted when discussing Dimension 3, mere access to content and practices isn't enough; what matters is *how* the students engage with the mathematics. Access and AOI (Agency, Ownership, Identity) are closely tied; we think of Dimensions 3 and 4 as two sides of the same coin. Given that there are many tools for providing access, it's worth having and focusing on as a separate dimension. But, we can never think of access without thinking "access to what?" And we can never think of issues related to the development of positive mathematical identities without asking, "who has such opportunities?" Finally, it's only by providing opportunities for student thinking and exchanges to be made public and making suitable adjustments (Dimension 5) that we can adjust classroom activities in ways that provide students with the opportunities to develop powerful mathematical identities.

Dimension 5: Formative Assessment

The extent to which classroom activities elicit student thinking and subsequent interactions respond to those ideas, building on productive beginnings and addressing emerging misunderstandings. Powerful instruction "meets students where they are" and gives them opportunities to deepen their understandings.

Formative assessment involves the ongoing monitoring of student thinking and adjusting instruction in response to it, *during the learning process*. The idea is to arrange for the kinds of feedback that provide midcourse corrections as the students are building their understandings, rather than focusing on testing after the unit is over – when it's too late to do anything about student learning.

There are numerous ways to bring student thinking out into the open. There are pretests, which provide a sense of what students know before instruction begins. Homework and quizzes provide useful information during a unit. But, much of the real action related to formative assessment takes place during instruction.

Teachers can pose questions designed to bring out into the open incorrect assumptions or ideas that need to be challenged, or that help students realize that they need to dig more deeply into the content. Challenging problems can be assigned for group work, and teachers can circulate through the classroom listening to students as they work on the problems. This "real-time" monitoring provides opportunities to see what is making sense to the students, what ideas they're generating, and what challenges they face.

What teachers pick up about student thinking should play a major role in shaping the classroom activities that follow (Black et al. 2003, Black & Wiliam 1998a, 1998b, 2009, Shepard 2000). This may seem daunting at first – who knows what students will say, once students feel free to say what's on their minds? – but it's essential to see and hear what students are thinking and understanding in order to meet them where they are and to build on their understanding. It's a skill, like riding a bike: Once you start providing students opportunities to think and talk about mathematics, you get better at doing so, and it becomes an enjoyable habit. And, you don't have to figure out everything on your own. There are large literatures on student misconceptions, or "alternative conceptions," that document the kinds of partial understandings students typically develop in specific content areas. Knowing about these typical patterns of student reasoning helps teachers to be prepared to deal with them[5].

Implementing formative assessment consistently can make a big difference. (For a general overview, see Black & Wiliam 1998a, 1998b.) The Mathematics Assessment Project produced 100 Formative Assessment Lessons, known as FALs, for grades 6–10. They're available at no cost from https://www.map.mathshell.org/lessons.php. The FALs provide extended lesson plans that structure small-group and whole-class work, indicating where students are likely to face challenges and suggesting what the teacher might do when students get stuck or display misconceptions. Independent research analyses indicated that students who studied from FALs for an average of 10–12 days over the course of a year gained 4.6 *months* in comparison to a control group of students who studied from the regular curriculum only (Hermann et al. 2014, Research for Action 2015). How could that be possible? It turns out that in teaching from the FALs, teachers got better at asking questions and listening to what students say, *in their regular lessons*. These changes in their teaching – starting to use formative assessment in their regular instruction – produced their students' knowledge gains.

In general, through deliberately attending to student reasoning and understanding, and then shaping instruction in response, teaching "becomes clearer, more focused, and more effective" (National Research Council 2001, p. 350). Furthermore, hearing student reasoning provides the information that allows teachers to adjust the level of cognitive demand so that students are better positioned to engage in meaningful sense-making.

Burkhardt and Schoenfeld (2019), building on many of the ideas in Swan (2006, 2017) and Swan & Burkhardt (2014), provide a list of design tactics for formative assessment. Those include:

- having students make posters and presenting their work to the class
- working sorting and matching tasks, where their choices often reveal their thinking. (see, e.g., the formative assessment lesson "interpreting distance-time graphs" at https://www.map.mathshell.org/lessons.php?unit=8225&collection=8)
- critiquing sample student work (see, e.g., the formative assessment lesson "evaluating statements about length and area," at https://www.map.mathshell.org/lessons.php?unit=9310&collection=8)
- role shifting (having teachers and students play various roles, such as counselor, fellow student, or resource).

As we'll see throughout this volume, formative assessment can enhance all of the other TRU dimensions. When you have powerful mathematical goals (Dimension 1), formative assessment helps you see how well you're making progress toward them and make midcourse corrections that move your students further toward deep understandings. Specifically, it focuses on student sense-making and whether students are engaged in productive struggle, the key to cognitive demand (Dimension 2). A close look at student thinking and engagement indicates who's engaged and in what ways. This positions the teacher to focus more effectively on equitable access (Dimension 3). And, because formative assessment supports meaningful engagement, it works to the benefit of students' mathematical agency and initiative, ownership of the mathematics, and their mathematical identities (Dimension 4).

One final comment. Formative assessment *is* complex; it's possible to look at the challenge and be overwhelmed, thinking "I can't possibly keep track of all my students' different understandings and address them." It's true that you can't do that unless you have a superhuman memory. But, you don't have to. When you look at the case studies in this book, you'll see that a lot of the learning takes place in small groups, when students work collaboratively to sort out each other's partial understandings. The Mathematics Assessment Project's Formative Assessment Lessons provide numerous examples of how this can be done, providing tasks that invite student collaboration and critique and descriptions of "common issues" students face when trying to get their heads around the material.

1.3 How the Dimensions Fit Together – in Practice, in Professional Development, and in this Volume

In this opening chapter, we've talked about each of the five TRU dimensions as if they were somewhat separate things. Well, they are and they're not. By way of analogy, think of a musical quintet. The music that emerges from the quintet is a beautiful combination of the contributions of the five instruments in it. Each instrument makes its own contribution, maybe even in a solo – but the whole is more than the sum of the parts. (And, of course, if any of the parts was missing or not very well played, the overall performance would suffer.) At the same time, each of the players has devoted great effort to learning the nuances of their own instrument, and to being part of the ensemble.

It's the same in teaching. In the classroom, everything is happening at once! A teacher sees or hears something, and decides to make a particular move. That move may serve to highlight an important aspect of the mathematics the students are dealing with, position students so that they're better able to engage with it productively, and open up access for some students who hadn't yet found a way to engage – if things go well in ways that allow students to make progress and demonstrate some proficiency. Such things don't happen magically. The odds are that this teacher, over time, developed their noticing skills, their always deepening grasp of what matters in the mathematics, and their knowledge of their students and how to nurture the students' engagement and understanding.

What TRU offers is a description of the building blocks for that proficiency. There's plenty of evidence (see, e.g., Schoenfeld 2013, 2014, 2015, Schoenfeld, Floden, & the Algebra Teaching Study and Mathematics Assessment Project 2018) that when things go well along the five dimensions of TRU, students do very well – personally and mathematically. So, it makes good sense to become increasingly proficient at supporting students' growth in those five dimensions. The five TRU dimensions have been developed in ways that allow each to be the focus of professional growth; we can get better at each one. But they're never completely isolated. The dimensions are the five instruments in the musical quintet: you can focus on each one to get better at it *and* attend to the harmonies when all the instruments come into play at the same time.

There's no one "right way" to do this. Various TRU-based professional learning communities have taken different approaches, designed to fit local contexts and needs (see Schoenfeld et al. 2019a). In Part III of this volume, we'll describe some ways in which that's been done.

Despite the diversity of approaches, all of the groups that have built their own TRU-based PD share two main things in common. One is the desire to develop deeper understandings of the TRU dimension so that they can use TRU more effectively in practice. Two essential tools for doing so are the *TRU Conversation Guide* and the *TRU Observation Guide*, which are given in full in Part IV. We make use of the ideas in them throughout this volume.

A second, which is essential, is the ongoing examination of real instances of instruction, in cycles of planning, teaching, and reflection. This can be done by individuals or by collectives – in fact, the TRU conversation and observation guides are set up for those kinds of cycles. Some of the collective work is done in the tradition of Lesson Study, and some of it consists of group members bringing videotapes of their lessons to the group for discussion. (This is one way to structure a video club.) The question for discussion is *not* "what did I do well, what did I do badly?" Rather, the question is, "Here's a segment of instruction. How can I develop a deeper understanding of (a) what happened, (b) what some of the pedagogical options at various points might be, and (c) what the risks and opportunities of those options might be?"

That's the focus of Part II of this book. By the time you've worked your way through the three case studies we present there, you'll be in a better position to do so on your own.

Notes

1   This chapter draws from and builds on: Schoenfeld, A. H., & the Teaching for Robust Understanding Project. (2016). *An Introduction to the Teaching for Robust Understanding (TRU) Framework*. Berkeley, CA: Graduate School of Education. Retrieved from http://tru.berkeley.edu.

2   For the history and some of the details, see Schoenfeld 2013, 2014, 2015, Schoenfeld, Floden, and the Algebra Teaching Study and Mathematics Assessment Project 2018.

3   Making a practice of reviewing what counts can result in significant improvements. For example, Gawande (2007, 2009) has shown that checklists that remind doctors and nurses of things they know they should be doing result in significant improvements in hospital recovery and mortality rates. If reminders to wash one's hands before interacting with patients can improve medical results, then it stands to reason that instruction can be enhanced by routinely asking (for example) where in a lesson students have opportunities to engage in sense-making at an appropriate level of cognitive demand.

4   A quick note: The five dimensions of TRU were chosen so that they could be thought about and worked on independently. In practice, they're interrelated. All five dimensions are "alive" at the same time, and they interact – for example, the choice of a very challenging problem that can only be solved in one way raises challenges with regard to cognitive demand, equity (who is ready to engage with it?), students' identities (it's hard on you when you're clueless!), and formative assessment (what can I do to adjust so that more students can engage meaningfully?)

5   Of course, teachers develop some of this kind of knowledge (called "pedagogical content knowledge") from experience over time. The process can be accelerated if you engage in it deliberately and self-consciously.

Part II

Reflecting on Images of Practice (The Case Studies)

Introduction to Part II

In Part II of this book, we invite you to join us in three detailed case studies of mathematics classrooms[1]. The cases in Chapters 2–4 focus on middle and secondary school mathematics lessons, featuring content that is typically found in eighth through tenth grade in the US. To show the range of TRU, the cases feature different class configurations, different mathematical content, and different pedagogical issues[2]. Our goal in presenting each case is to come as close as we can to creating the kinds of conversations that we have in professional development meetings.

Imagine for a moment that we've gotten together to discuss a lesson you've taught.

It's scary, isn't it? Inviting someone into your classroom, whether in person or by way of video, puts you in a very vulnerable position. Thoughtful or helpful comments can help us improve our teaching – and that's the goal! – but thoughtless or disrespectful comments can be harmful. There needs to be a climate of trust in order to have productive, supportive conversations.

That's why we always begin by discussing the ground rules for our conversations.

Norms for Visiting Classrooms

In our view, visiting someone's classroom is a privilege. When teachers and students open their classrooms to us, they are inviting us to learn with them. We can see things that they cannot – for instance, we can often see and hear things when the teacher is in another part of the room. It's also the case that teachers and students know things that we never will, things that are important for making full sense of what happened. For instance, we don't know what happened the day before a video was taken. We also don't know the students' and teacher's histories – for example, a teacher might ask a question in a particular way because of something the student had said the previous day or because this framing helps to connect to insights the group had come up with before we joined the class. We try to honor these understandings as we learn about teaching with the students and teacher in this vignette.

DOI: 10.4324/9781003375197-3

Here are our norms for respectful observation and reflection:

- The goal is not to judge what happened or say how we might do things. It's to understand and reflect on what took place.
- To the degree you can, approach what happened from the perspectives of the students.
- Talk about the teacher and students as if they are in the conversation with us.
- Assume that the teacher and students acted with good intentions.
- Recognize that there are many things that we don't know about the classroom, the teacher, and the students, that matter in our reflections.
- Problematize what you see, but productively. Do not accuse; instead, question, reason, and provide evidence for any assertions you make.

The Primary Focus of Each of the Three Cases

Even in the same course, different lessons offer different opportunities for reflection. One day we may see students working really hard to make sense of an important idea. In that case, while all five dimensions of TRU are relevant, issues of Cognitive Demand are particularly noticeable. On another day, conversations between students may have the potential to include or exclude particular students and to position them as being competent (or not). In that case, issues related to Equitable Access and Agency, Ownership, and Identity in that lesson may deserve very close attention. Yet another day, the information revealed by student comments may raise issues of how to adjust the lesson plan productively. In that case, Formative Assessment is a useful focus for discussion.

We've chosen our cases accordingly. In our first case, in Chapter 2, issues of cognitive demand (Dimension 2) are particularly important. In our second case, in Chapter 3, we'll focus significantly on Dimensions 3 and 4. And, Chapter 5 has a focus on formative assessment. This will not be an exclusive focus, however. All five dimensions of TRU are active all the time: when we review each lesson as a whole, we will seek answers to the questions in Figure 1.2 ("What's important, from the student perspective") for all five dimensions.

In Chapter 2 we visit an Algebra II classroom[3]. The students have been assigned a "contextual" or modeling problem concerning a used car's declining value. The class has studied exponential growth, but in this lesson the students need to make sense of a problem in which there is exponential decay rather than growth. Their task is to develop a model representing the value of a car that depreciates at a steady rate and then to use that model to address some applied questions.

The teacher in this case had developed and assigned the task because it builds on the students' previous work in new ways and because information from the problem context supports sensemaking by providing a way for the students to check the models they develop – the students can judge whether or not the values produced by their formulas correspond to reasonable expectations about car values. The mathematics in this vignette is complex and interesting. There are multiple ways to look at the task and to make connections.

The students find the task very challenging. At times they get stymied, at times they seem lost. The teacher, who circulates through the room while the students work in groups and who convenes the class together at times, is aware of the challenges. A main question is, what kinds of support will keep the students engaged, not overly frustrated, and support productive struggle? For this reason, our main attention is on issues of cognitive demand – and the teacher's decision-making in the light of her observations of student struggle (formative assessment). In addition, questions of participation (equitable access) are also apparent in this vignette. There are always issues regarding who volunteers, who gets called on, and how students and student ideas are positioned both by the teacher and by other students. In consequence issues of agency, ownership, and identity (AOI) are also worth noting.

Chapter 3 highlights the closely related dimensions of Equitable Access and AOI. This deep dive focuses on small group work in an introductory algebra "sheltered mathematics" classroom. The students are Spanish speakers with varied degrees of proficiency in English; a goal of instruction is for the students to refine their use of English while they are learning mathematics. In the lesson, the students grapple with a non-standard introductory algebra problem, which asks them to find the perimeter of a complex figure whose sides are given in x's and 1's. There are some interesting issues of mathematical representations involved, and getting to the bottom of the questions that the students are asked to address is challenging. We'll spend a long time working through the mathematical potential of the task they're working on.

In this case, as always, there are issues of cognitive demand and, because of this, issues of formative assessment. How do you arrange things so that instruction meets students where they are and supports them in moving forward? But, in fundamental ways, this extended vignette concerns issues of equitable access and the students' AOI. One of the three students is responsible for explaining the group's work to the teacher, and she has some difficulty getting her head around the material. Given these challenges, there are questions both of how she can be supported and what the implications of the group interactions are, not only for her but for all three of the students in her working group.

Chapter 4 visits an Algebra II classroom in which the students are working on a "Formative Assessment Lesson" devoted to graphing quadratic functions. This review/synthesis lesson covers a substantial amount of mathematical territory. The mathematics is once again of significant interest, although the kinds of questions that arise here are somewhat different than in the more narrowly focused vignettes in Chapters 2 and 3 – this is a major content review. A fundamental question for any review lesson is, what are the really important ideas in this whole topic?

As always, looking at instruction with an eye toward each of the five TRU dimensions helps to unpack what took place in the lesson. The math is rich and connected; we'll discuss it as suggested in the previous paragraph. Whenever a lesson involves both whole class and small group work there are a host of questions regarding who participates and in what ways, how they're positioned, and so on. That is, there's plenty of opportunity to look at issues of equitable access and AOI. There are, as always, questions of how much sensemaking and productive struggle the students are engaged in. For us, the design of this Formative Assessment Lesson provides some very nice opportunities to dig into TRU Dimension 5. In fundamental ways, this lesson shows how formative assessment enhances all of the other dimensions. It points to mathematical riches and, when it reveals places where students are challenged, provides opportunities for all students to engage in mathematically rich and personally affirming ways. Moreover, some of the things that students say, and the ways that the teacher picks up on them and brings them to the class, give us a close look at formative assessment in action.

Each case, then, has somewhat different emphases. The mathematics at the heart of each case is important, in different ways – our mathematical explorations in the three chapters provide three different ways of thinking about mathematical richness. The student populations in each class are different – one point being that TRU can be informative when you look at any lesson, with any group of students. And the foci in each lesson are somewhat different, giving us a chance to delve deeply into each of the five TRU dimensions.

We've chosen these three classrooms to look at because we find what happens in them to be genuinely interesting – there's a lot to think about in each lesson. We have *not* chosen them because we think what happens in them is exemplary or because we want to present "model" instruction. What we like about the cases is that they raise issues to think about. Our hope is that in working through these deep dives together – in identifying moments of instruction where interesting things happen, in looking closely at what happened, in thinking about what other instructional options might be, and what the consequences of those actions might be – that you'll exercise and sharpen your observational, planning and reflection skills. If all goes well, working through these cases will sharpen your planning, observational and reflection skills, and lead to increasingly powerful instruction for your students. If you're part of a learning community, we hope that the book will serve as a catalyst for looking collaboratively at each other's videos. These cases will prepare you to do that.

The Structure of Our Sessions – and of Chapters 2–4

It helps to understand the context for any lesson before observing it. So, in all of our professional development work, we work through the mathematics in a lesson and learn about the students before we observe the lesson itself. Whether our observations are in person or on video, things go by so fast! With video, we have the advantage of instant replays. When we watch the lesson, we often stop the tape to make sure we understand what's taking place. Once we have a solid understanding of what actually happened in the case, then we reflect on the case as a whole.

We've tried to replicate that structure in each of our deep dives. Each chapter is structured as follows.

Section 1, The Mathematics

We always begin our PD sessions with extended discussions of the mathematical task at the heart of the lesson. We start by working on the problem individually or in pairs. Having done so, we look for the mathematical riches that discussions of the task can reveal. What's important about the task – how does it tie to big ideas? Did participants find different ways to solve it? Which solutions seem more accessible? Which reveal interesting connections? How do the solutions tie to underlying conceptual ideas? What can we learn from comparing and contrasting the solutions? The more deeply we understand the mathematical potential of the task, the more we're in a position to (a) anticipate what our students might say and help them build on it and (b) capitalize on the potential of the task to help the students develop deeper understandings and engage in productive mathematical practices. These discussions go on for many pages!

Section 2, Some Background About the Classroom

Here, we provide key contextual information about the classroom we're about to visit. As we've noted, the three classrooms we visit are very different. That matters.

Section 3, What Happened, in Detail

Here's where we lay out the substance of what happened. The case is segmented into "episodes" that correspond to places where, if we were watching the video of this classroom together, we'd stop the tape to talk about what happened. That might be because there's a natural break (the teacher called the class together or had the students work in small groups) or because something complex happened, and we want to understand it before going on. At the end of each episode we ask a few questions that focus on how the students experienced what was happening. After you think about those questions, either alone or with colleagues, we offer our commentary.

Section 4, Reflecting and Problematizing

After you've finished reading the entire case presented in Section 3, we invite you to think and/or talk about some big issues related to how TRU helps us understand what took place. We review the role of each of the five dimensions in helping us see and understand students' experiences more clearly, taking a holistic view of the vignette. We'll share our thoughts and some questions we still have.

You can read each chapter and think about the questions we ask on your own. Or you can read and answer the questions with one or more colleagues. As you move through the chapter, we encourage you to think about the questions before you read our commentary. Do the task, reflect on what happened, and collect your own conclusions and questions. Our commentary is part of a conversation we're having with you and that we hope you are having with others.

That said, let's turn to the three cases.

Notes

1   We'll also refer to these case studies as cases, deep dives, and vignettes, just to vary the language.
2   As we noted in Chapter 1, TRU can be used to think about what happens in any classroom, at any grade level. We've chosen three middle/secondary math classrooms so that the discussions come together as a whole. It's hard to see the commonalities in lessons in different disciplines or widely different grade levels.
3   For a broad overview of US curriculum coverage, which applies to all three of our cases, see the Common Core State Standards for mathematics (2010).

Chapter 2

The Car Value Problem

A Focus on Cognitive Demand

2.0 Introduction

We are about to join an Algebra II class as it delves into the complexity of representing exponential decay in symbolic terms. The teacher, Ms. Sierra[1], has asked the students to model the decreases in a car's value over time. She has designed this task, which she calls a "contextual" problem, so that the students can draw upon their general knowledge of the context (what they know about car costs, depreciation, etc.) as they model the changes in the car's value. The idea is that their real-world knowledge can provide "reality checks" on the models they create.

We'll be looking at what happens through the lens of the TRU Framework. In this and every classroom, the five dimensions of TRU are active all the time. During any particular classroom episode, issues related to one or more of the other TRU dimensions may come into sharp focus. Such occasions provide us with opportunities to reflect deeply on those issues. Specifically, the "car value problem" in this classroom episode provides us with an opportunity to shine a spotlight on issues of Dimension 2, Cognitive Demand. Here's a reminder:

> The key phrase with regard to Cognitive Demand (Dimension 2) is "productive struggle." There's not much to be learned if students aren't challenged. At the same time, if the challenge is too great, then learning may be out of reach! Here's a weightlifting analogy. If you give someone a 1-pound weight to lift (even with multiple repetitions), they're unlikely to get appreciably stronger. But the same is the case if you give them 500-pound weights! They'd struggle without results. What matters is using weights that are within people's current capacity – and adjusting when they get stronger.

This chapter highlights questions regarding the ways we frame classroom problems so that students have opportunities for sense-making, in ways that they are stretched but still able to make good progress. Then, assuming that the students are challenged in meaningful ways, there are tons of decisions to be made as the lesson evolves. For example: Is it time to provide some scaffolding? Of what type? What are the possible advantages and disadvantages related to any action the teacher might take?[2]

It's important to stress that there are no "right answers" to these questions. The purpose of our case studies is not to present "exemplary" instruction or to suggest "this is how to do

DOI: 10.4324/9781003375197-4

things." Nor is it to judge what the teacher does. We've chosen the classroom scenarios in Chapters 2–4 because they provide opportunities to think deeply about the issues that come up during classroom discussions.

The teacher who developed and taught this lesson is very explicit about her reasons for focusing on contexts, in a way that goes beyond the oft-times artificial "modeling" problems found in typical textbooks. In describing her approach to us, she wrote that her intention is to use

> one carefully chosen context for several days of instruction to study one important mathematical idea. The curriculum would refer back to variations of the same context in different stages of learning, such as problem posing, cuing intuitions, exploration, concept development, and practice. Ideally, each context should be
>
> 1. Compelling and powerful to many stakeholders. Students should be able to go home and say, "we are learning about fair loan practices" rather than "we are learning about exponential growth by a percent."
> 2. Familiar to all students and/or worth spending time familiarizing students with.
> 3. Mathematically productive for cuing student intuitions and/or as an instructional analogy.
>
> The goal is that each problem context is something to be explored in some depth, providing expansive framing and a sense of "why?" to learning a powerful mathematical idea. This can support student engagement both by sparking curiosity and by signaling to students that they can draw on outside of school knowledge when beginning to explore the mathematical idea. They do not need to wait to be told a method for solving the problem. Once students are familiar with the context, it can continue to be leveraged as an instructional analogy throughout the learning process. Students can be encouraged to contextualize and decontextualize their mathematical work in terms of this same context. This can foster student agency by giving students a wider mathematical arena in which to brainstorm strategies and test their conclusions, rather than waiting to ask a teacher if they are correct.

Often, when introducing topics that extend beyond what our students know, we begin by explaining what is new. Then we demonstrate how to solve one or more sample problems and assign similar problems for practice. That sets the students on track, and it's efficient. However, proceeding in that way means that students have less opportunity for putting things together for themselves than they might otherwise have. "Demonstrate and practice" lessons thus limit possibilities for student sense-making – and in limiting sense-making opportunities, they also limit opportunities for students to develop agency, ownership, and identity (AOI). At the same time, opening things up and giving students the opportunity to develop some of the mathematics themselves can be a risky proposition. Doing so raises issues of formative assessment and cognitive demand. Is the task within reach, and what might the teacher do if it doesn't seem to be? Moreover, if students fail or get frustrated, that's not good for AOI either! We're always balancing things and making judgment calls.

Ms. Sierra designed the lesson to give the students a lot of room for exploration and sense-making. The task asks students to develop a formula that models the decreasing value of a car over time. Having worked on problems that deal with exponential growth, the students are in a position to tackle exponential decay. But as you will see, doing so is a challenge.

To introduce the topic Ms. Sierra poses the problem and tells the students that it's related to, but different from, the problems on exponential growth that they've worked on. As she frames the problem for the students during class, she makes it clear that their explicit task is to engage in sense-making:

> You are gonna write an equation to model the value of the car over time. This is not one where I'm gonna check and tell you whether the equation is right or wrong ... Today, I really want you to take a second to [reading from the problem statement on the board]

"find a way to check or justify that your equation is realistic." If you think your equation is right, try to prove it to yourself and prove it to your group mates, [that] your equation makes sense before you go on. Is it gonna be an equation that tells you after one year, after two years? Is it gonna be reasonable? Ok? Then...

As we think about this problem setup, there are multiple issues to consider. First, there's the act of modeling itself. What does it take for a modeling task to be meaningful to students? Will they take the task to be an "academic exercise," or will it feel real in some way? Second, when students are given the responsibility of sorting out the mathematics themselves, what kinds of scaffolding or support will help them make meaningful progress, and when should it be offered? The goal is to support sense-making, but what do you do if the students seem stuck? If you offer help right away, you haven't given them the opportunity for productive struggle. If you wait too long, they may get frustrated. There are no easy answers to such questions – and no "right answers!" But if you know your students and you're paying attention to both their mathematical progress and how they seem to be dealing with the challenge, then you have grounds for either letting them continue to grapple with a problem or providing some scaffolding to help them along the way. That's the kind of issue we explore in this case study.

While this case study highlights Cognitive Demand (TRU Dimension 2), it's important to remember that all five dimensions of TRU are always in action; we will discuss all five.

We begin with a detailed examination of the Mathematics (Dimension 1) involved in the lesson. Why so much detail? Our metaphor is that we think of our mathematical explorations as creating a topographic map of the mathematical terrain for our students. The more we know the territory, the more we can help the students explore it. As you'll see, there's a lot to explore here.

The five dimensions interact. For example, the task sets up the potential for engagement with the other four dimensions. If the math that students experience is rote or mechanical, for example, students will not have opportunities to be invested in it or to be excited about having done it. If there's no challenge, students won't experience a sense of growth or ownership if they succeed; but if the task is simply beyond reach, students will be frustrated and disheartened. Thus, the level of cognitive demand (Dimension 2) sets up the potential for Agency, Ownership, and Identity (Dimension 4). Again, we stress that Equitable Access (Dimension 3) means that *every* student has opportunities to engage with the central mathematical ideas not only in ways that contribute to their understanding but also in ways that provide opportunities for the development of AOI. The key question related to access is, who participates, in what ways? Is the task or activity open to participation from a wide range of students? Formative Assessment (Dimension 5) is the mechanism that helps to make things go well along the other dimensions. When teachers attend to what students are thinking and make adjustments when needed, they can provide every student with opportunities for productive struggle – and to take agency, own mathematical ideas, and build a more positive mathematical identity.

As you read on, please engage actively with the text! In the math section, we invite you to explore. In the narrative of what happened, we ask questions about the case. Please think about the issues (and if you're part of a working group, discuss them) before reading our interpretations. Then feel free to discuss the issues some more! The same goes for our summary discussion. The purpose of our conversations is to deepen our individual and collective understandings of teaching and learning, not to pass judgment or describe "how to's."

2.1 Thinking about the Math in the Car Value Problem

Many complex skills (for example, driving a car or riding a bike) look easy once you've become skilled at them – but that smooth performance obscures the work it took to learn! When it comes to bike riding, for example, there are underlying issues of balance and perception. Feeling that you need to shift your weight or turn the wheel in one direction is one thing, but knowing how much to do so takes a lot of practice. In the case of the car value problem that follows, there are lots of little things that come together to deliver a clean solution. But when you take a close look, there are lots of opportunities to topple sidewise. Some pieces of mathematics look very straightforward once you've seen how everything fits together. They can be very challenging when you're learning.

Here's the problem at the core of this lesson:

Anay buys a car for $5000. The car loses 15% of its value every year.

a. How much is the car worth after 1 year?
b. Write an equation to model the value of the car over time.

Before you go on, find a way to check/justify that your equation is realistic and show your work.

c. How long before the car is worth half of its original value?
d. After owning the car for 10 years, it breaks down. Anay finds out that she will need to replace the clutch to be able to drive the car again. Is it worth it?

The 20-minute segment of instruction we examine was devoted to parts (a), (b), and (c) of the problem. (We've included part (d) above so that you can see where the mathematics was headed, and how it fits with the teacher's view of using contextual problems.)

Later in this section, we'll work through the problem with you. We'll think about the mathematics involved and the connections to other important mathematics, and we'll look at the mathematics as a student might. But, since this is a contextual or modeling problem, it's worth thinking about such problems in general before we dig into the problem itself.

The purpose of mathematical modeling is to capture selected aspects of "real world" contexts in mathematical terms, using mathematics to help make sense of complex situations that can be seen through a mathematical lens. But there are many ways this process can go wrong when modeling situations are turned into exercises for classroom use. All too often the situations involved serve as "cover stories" designed to give students practice using mathematical concepts, with the result that the mathematical solutions to the classroom tasks don't make sense in the real-life situations that are supposedly being modeled. You've probably seen problems like this. There are, for example, textbook problems using the Pythagorean theorem in which two runners who took off at right angles to each other, with one moving one mile per hour faster than the other, turn out to be running at paces of 3 and 4 miles per hour! (In reality, a jogger averaging the somewhat slow pace of 10-minute miles would cover 6 miles in an hour.) There are "age problems" in which a grandmother turns out to be 26 years old and purchasing problems where a notebook costs $5 but an ordinary pencil costs $4. After years of experience dealing with such problems, many students come to believe that school math has nothing to do with the real world – and that the answers that they obtain when working on textbook exercises don't have to make sense.

There are many examples of this type, but here's one that makes the point. In a research study, upper elementary school students were asked to solve this problem:

John wants to make wooden bookcases that are two feet wide. He has two five-foot long boards. How many two-foot long boards can he cut from them?

A full 70% of the students responded "five."

What seems to have happened here is that once students set up the problem as a division, they no longer took the context into account. If they'd kept the board lengths in mind, they might have realized that a 5-foot board yields just two 2-foot pieces, with a 1-foot piece remaining.

That means you can only get four 2-foot lengths from the two 5-foot boards. But even by upper elementary school students have learned to ignore the context once the relevant computations have been set up. More generally, they've picked up a widespread belief – that the mathematics studied in school isn't about the real world and that the problems worked in class use real-world contexts as "window dressing."

So, there are two fundamental questions as we consider any modeling or contextual problem. First, are the assumptions and information in the problem realistic? Second, can students engage in the problem in ways that enable them to do real sense-making, so that they see the modeling enterprise as something meaningful rather than artificial? We'd like to try to do things so that the answers to these questions are "yes" whenever possible. The more students engage in mathematical sense-making in the classroom, the more likely it is that they'll see mathematics as meaningful and think to use what they've learned outside the classroom.

In the lesson we're discussing, the teacher has chosen the numbers so that the problem is realistic. When we googled "how much does a typical car lose in value each year?" the first answer that came up (from nerdwallet.com) was,

> Your car's value decreases around 20% to 30% by the end of the first year. From years two to six, depreciation ranges from 15% to 18% per year, according to recent data from Black Book, which tracks used-car pricing. As a rule of thumb, in five years, cars lose 60% or more of their initial value.

Most people know that new cars cost a lot more than $5000 (though we can check with students to see if they know!), so a car that costs $5000 has been around for a while. As a result, the 15% rate of depreciation is plausible.

In sum, the problem as written is potentially meaningful as a modeling problem, with a "real world" context – although, of course, whether the students will perceive it as meaningful will depend on the way that it's discussed in the classroom and what they know.

Next, we'll look at the mathematics involved in the solution. We'd like you to work on part (a) of the task before reading our discussion.

Part (a) of the Task

 Anay buys a car for $5000. The car loses 15% of its value every year.

(a) How much is the car worth after 1 year?

After you have completed part (a) of the task, answer these questions:

- How many ways can you find to determine the car's value? What are they?
- Is there a reason to prefer any of the methods you found, if you are just doing part (a)? Please explain.
- Is there a reason to teach more than one method? Please explain.

Here's how we think about these questions.

How many ways can you find to determine the car's value? What are they?

We'd like to share two ways to solve part (a).

Method 1. The most direct way to approach this problem is to follow the information in the problem statement, step by step. The car loses 15% of its value the first year. It's worth $5000 to start, and 15% of $5000 is 750. So, the car loses $750 in value the first year; at the end of year 1, the value of the car is $5000–$750 = $4250.

Method 2. When you use method 1, you wind up using two steps – calculating the loss in value and subtracting loss from the starting cost. Those calculations are straightforward, especially if you use a calculator. But it's worth thinking about mathematical connections. Is there another way to think about something losing 15% of its value? Well, yes: the car *keeps* 85% of its value. So the problem can be re-framed as,

> Anay buys a car for $5000. Each year after he buys it the car's value is 85% of its value the previous year.
> How much is the car worth after 1 year?

Once you've framed the problem this way, part (a) no longer requires two steps. The car starts out at a value of $5000. After 1 year the car is worth 85% of $5000, or $4250.

Is there a reason to prefer any of the methods you found, if you are just doing part (a)? Please explain.

We don't see much reason for preferring either of the methods we discussed above if you're only looking to find the value of the car after 1 year. Using method 1, calculating 15% of $5000 is undemanding, as is subtracting $750 from $5000. Using method 2, finding 85% of $5000 – that is, multiplying $5000 by .85 – is also straightforward, especially with a calculator.

Is there a reason to teach more than one method? Please explain.

We think so; here are some reasons. (You may have more.) First, we think it's good in general to make mathematical connections between solution methods. Second, we'd like for students to be flexible thinkers. There will be times when method 2 makes more sense to use than method 1. For example, consider this problem:

> A textbook company is going out of business. They're holding a "Monster Sale – 90% off everything." A textbook was listed at a cost of $129. How much will it be on sale?

Computing 10% of $129 is straightforward – if the students are comfortable with percentages they can just move the decimal one place to get 10% of $129 = $12.90. In contrast, computing 90% of $129 and then subtracting that value from $129 is a bit of a pain. We want our students to be flexible, adaptive, and strategic thinkers. When it's simpler for them to use an alternative method, it's good for them to know of it and reach for it.

There's a classic example in the math education literature, where the vast majority of elementary school students given a problem like

$$\frac{273 + 273 + 273 + 273}{4}$$

found the answer by adding up the four terms in the numerator and then dividing by 4. That does produce the right numerical answer, but doing the computations that way misses the point. We think of this as a sense-making task, where students should see that the computation calls for first multiplying 273 by 4 and then dividing the result by 4. Developing flexibility, and looking beyond the first computation that comes to mind, can be important.

In sum, there's reason in general to think about as many ways as you can to calculate a 15% decrease in value and to connect them.

Now let's move on to part (b) of the problem and some of the issues involved.

Part (b) of the Task

 Anay buys a car for $5000. The car loses 15% of its value every year.

(b) Write an equation to model the value of the car over time.

Before you go on, find a way to check/justify that your equation is realistic and show your work. After you have completed part (b) of the task, please answer these questions:

- How did you interpret "over time?" Can you imagine different ways students might interpret "over time"?
- Did you arrive at a formula or equation directly, or do so year by year? What challenges did you face?
- Let's revisit the question we asked at the end of part (a): Is there a reason to teach more than one method? Please explain.

Here's how we think about these questions.

How did you interpret "over time?" Can you imagine different ways students might interpret "over time"?

We posed this question because the phrase "over time" may hide some complexity. We assume that the teacher intended part (b) of the task to mean

Write an equation to model the value of the car after *n* years have passed.
(That is, the answer will be a function of the variable *n*)

But that's a leap. For students who are new to this kind of problem, it may not be obvious. What's the time frame, for example? Is it months or years? Whichever unit is selected, Is it clear that the process of finding values year after year will be iterative? There's a lot going on here mathematically.

Did you arrive at a formula or equation directly, or do so year by year? What challenges did you face?

Knowing how to find a general formula can be a challenge for students. There is a general problem-solving strategy. If you want to develop a formula for something the *n*th time it happens, it can be useful to think about patterns:

What is the value of the car after 1 year? After 2 years? After 3 years? Can you see a pattern that can be written as a general formula?

Here too, luck or previous experience may help. We knew that method 2 from part (a) had potential, so we focused on that.

Using Method 2 from Part (a)

To calculate the value of the car after 1 year, we computed 85% of $5000 = (.85)($5000), which is $4250. The question is how to proceed. One way is to compute 85% of the year 1 value = (.85)($4250) = $3612.50. We could have continued that way, taking 85% of that value to get the car's worth after 3 years ($3070.63), etc. The problem is that that string of numbers –

$5000, $4250, $3612.50, $3070.63,...

doesn't seem very informative! But we're multiplying by .85 each time. Maybe we shouldn't carry out the multiplications, but just keep track of how many times we multiply.

- At the start, the car is worth $5000.
- After 1 year, the car is worth $(.85)(5000).
- After 2 years, the car is worth $(.85)[(.85)(5000)] = $(.85)^2(5000).
- After 3 years, the car is worth $(.85)[(.85)^2(5000)] = $(.85)^3(5000)
- After 4 years, the car is worth $(.85)[(.85)^3(5000)] = $(.85)^4(5000)
- and so on ... so ...
- After n years, the car is worth $(.85)^n(5000)

Writing the year-by-year results this way enables us to see the pattern, leading us to a formula for the value of the car after n years as a function of n. We say this kind of formula, written in terms of n, expresses the answer to part (b) of the question "in closed form." It's what mathematicians usually look for. Having a formula of that type is much more efficient than calculating the change in value one year at a time.

"Unpacking" the solution in this way shows how complex the thinking behind the solution really is. That's where the bicycle analogy comes in – seeing the answer as an exponential function of n seems easy, once you know what the answer is. But in order to get there, you have to know:

1. how to interpret the phrase "over time"
2. to look for the pattern and
3. *not* to multiply out

in order to be in a position to see the pattern.

Using Method 1 from Part (a)

To be completely honest, the method we've just described was the only method we thought of when we first worked on part (b) of the problem. Multiplying again and again by .85 provided a straightforward way to get the value of the car after n years. We thought it was useful to know method 1 (subtracting 15% of the value each year), but we weren't sure it led anywhere. Then we spoke to the teacher and looked at the videotape of her lesson.

What we were reminded of is that if you have a table of values for an exponential function, the ratio of subsequent values, year after year, is the multiplicative factor that gives the exponential. Here's an example with "nice" numbers.

Consider the function $f(n)$ with the following values (Table 2.1):

In the case of this function, we can see what's happening: as n increases by 1, the value of $f(n)$ doubles. Since we started with a value of 3, the value after n years will be $f(n) = 3 \cdot 2^n$.

To make this more obvious (it's not necessary here, but it's useful in the next example), suppose we add an extra column to the table, where we calculate the ratio $f(n)/f(n-1)$. Then the table looks like this (Table 2.2):

The third column tells us that there is a constant ratio. And, it will reveal the constant ratio even if the numbers aren't as nice looking as our example.

Let's return to method 1 from part (a).

We started with $5000.

To compute the value after 1 year, we'd calculate 15% of $5000 = $750 and subtract that from $5000 to get $4250.

To compute the value after 2 years, we'd calculate 15% of $4250 = $637.50, and subtract that from $4250 to get $3612.50.

And so on! Here's a table (Table 2.3).

Table 2.1 The function f(n)

n	$f(n)$
0	3
1	6
2	12
3	24
4	48
5	96

Table 2.2 Computing $f(n)/f(n-1)$

n	$f(n)$	$f(n)/f(n-1)$
0	3	–
1	6	2
2	12	2
3	24	2
4	48	2
5	96	2

Table 2.3 Values for the first seven years

Years Elapsed (n)	Starting value that year ($f(n)$)
0	$5000.00
1	$4250.00
2	$3,612.50
3	$3,070.63
4	$2,610.03
5	$2,218.53
6	$1,885.75
7	$ 1,602.89
...	...

Table 2.4 The year-to-year ratios of values

Years Elapsed (n)	Starting value that year (f(n))	f(n)/f(n−1)
0	$ 5000.00	–
1	$ 4250.00	0.85
2	$ 3,612.50	0.85
3	$ 3,070.63	0.85
4	$ 2,610.03	0.85
5	$ 2,218.53	0.85
6	$ 1,885.75	0.85
7	$ 1,602.89	0.85
...	...	...

The numbers don't look that pretty. But if we add a third column, as we did above, the pattern becomes clear (Table 2.4).

The starting value was $5000, and in the subsequent year the value of the car was .85 of the value the previous year. So, the value of the car after n years – this time inferred from the table – is $5000(.85)^n$.

This brings everything full circle. If you wonder "where did the .85 come from?" after having found the function by using the table, you might make the connection that 85% of the value is what's left over after you've taken away 15%. In that way, both methods fit together.

Let's revisit the question we asked at the end of part (a): Is there a reason to teach more than one method? Please explain.

If we were simply interested in "answer-getting," we might respond "no" to this question. As described above, using the formula $(.85)^n(5000)$ provides a neat and clean solution to the problem. We could present it and have students solve a bunch of similar exercises. Why do more?

We think there are at least three reasons to have the students pursue both methods. The first is that, in general, making mathematical connections is a good thing! Seeing different ways to approach problems is valuable – you never know which method will help you the next time you face a problem that comes "out of the blue." Being able to move flexibly from one approach to another, or from one representation to another, is an important part of mathematical understanding. The more connected one's understanding is, the more flexible – and effective! – a problem solver one can be. That has implications for the development of students' mathematical identities.

The second reason has to do with the nature of memory and the importance of connected understandings. In general, we know that information we memorize as isolated facts or procedures don't remain in memory for very long. That's what the "summer slump" is all about. What makes something memorable, in general, is the web of connections it has to other things. Unless you use it a lot, any particular formula is likely to fade from memory. Will you remember the "two point" formula for the equation of straight lines, along with the point-slope formula, the two-intercept formula, and perhaps others? Probably not, we tend to forget them. But if you remember that any two key pieces of information determine the equation of a line and some basic information about transforming equations, you may be able to figure out the equation you need no matter what information you're given. That is, individually memorized formulas tend to vanish with time. But, connected understandings stay around. Engaging in sense-making pays off in the long run.

Third, although calculating year-by-year values is a lot of work, it feels more "real" in a certain sense. As we'll see in the case study below, it's easy for students to generate formulas that produce numbers that don't fit with reality. If classroom mathematics isn't anchored in numbers that feel real, then we can get the absurdities like those discussed earlier – runners who run at 3 miles per hour, 26-year-old grandmothers, board pieces of impossible lengths. The point of the contextual problem was that it was intended to provide a way of anchoring the students in their attempts at sense-making. (And, in the real world, the models come from data. The 15% annual depreciation wasn't simply made up; people collected resale data and saw that the numbers shrank exponentially.)

That brings us to our next set of mathematical questions.

Here are some further issues for reflection, looking at potential opportunities and challenges as the students might work through the problem.

 *This case study features student work on the problem, so we'd like to anticipate some opportunities and challenges that might arise.*

- This is a contextual modeling problem. What issues or challenges might come up as the students try to make sense of the problem?
- What mathematical content does this problem build on, and what connections might be made?
- The way a task is posed, or the possible connections it offers, can provide varying opportunities for students along each of the TRU dimensions. What opportunities do you see for Dimensions 2 through 5?
- Finally, imagine you're a ninth-grade student engaging with this kind of problem for the first time. You've done some examples of exponential growth (e.g., compound interest) but you've never seen an exponential context where things got smaller. The teacher is giving you the opportunity to make sense of the context, and the mathematics, by yourself. What possible content confusions might arise?

Take some time to discuss these questions before reading our thoughts.

Here's how we think about these questions.

This is a contextual modeling problem. What issues or challenges might come up as the students try to make sense of the problem?

We began our description of this problem by saying that the modeling assumptions seemed to make sense – car values do decrease and a steady rate of about 15% depreciation seems reasonable, for example. That's from our perspective. Students may or may not know these things.

A question is, how does student knowledge about car values and depreciation position them to deal with the modeling aspects of the problem? If the students have relevant resources and are positioned so that they can access them, then the class is ready to grapple with the problem in meaningful ways. But if there's a mismatch between what they know and the way the problem is set up (for example, that car values decrease, or how big a 15% decrease might be), then the problem might actually reinforce a belief that school math is not about the real world!

What mathematical content does this problem build on, and what connections might be made?

This task builds on a body of arithmetic and algebraic knowledge. At the simplest arithmetic level (say part (a) of the problem) the problem calls for the arithmetic modeling of the situation (a 15% loss of the initial value, equivalent to maintaining 85% of that value), which calls for some degree of number sense. There's the idea of iterated operations, which the students have seen in exponential growth. Those connections involve the arithmetic of exponents, and then the algebraic symbolization of repeated operations. Ultimately, one would like for the students to see that this kind of consistent decrease in value is closely related to the kinds of consistent increases in value they studied in cases of exponential growth – that there is a constant multiplier, but that the magnitude of the multiplier is less than 1. If the students can make these connections, then they'll have a coherent way to look at this kind of problem and issues of exponential growth and decay in general. That includes the symbolic representation of exponential growth/decay in algebraic terms.

The way a task is posed, or the possible connections it offers, can provide varying opportunities for students along each of the TRU dimensions. What opportunities do you see for Dimensions 2 through 5? (The next paragraph elaborates a bit on this question. If this question makes more sense after you read the next paragraph, then please stop and reconsider the question at that point.)

To give one example, a set of rote exercises provides little opportunity for cognitive demand. More generally, issues of cognitive demand and formative assessment (TRU Dimensions 2 and 5) hinge on many things, starting with what the students know! We'll find out what they do know as we watch the lesson play out. Will students find the task too easy, in which case, are there natural extensions to make the problem more challenging? (Note also that part (d) of the problem is a real-world extension of the task, so the teacher has provided some opportunities for further connections.) What options are available if the students find the task too challenging? There is a scaffolding structure for the problem, at least potentially. If the students find the task challenging they can be asked to work on things year by year, looking for a pattern; and they can be asked to make connections to exponential growth, which they've studied. Of course, how effectively this takes place depends on how much the classroom dialogue reveals what students are actually understanding, and whether and how the teacher picks up on what students are understanding (or not!) and frames things so that the students can make progress.

As we've seen, there are two fully correct ways to solve part (b) of the problem. Using method 2, noting that a loss of 15% per year is equivalent to a revelation of 85% each year, struck us as highly efficient; as we've said, when we first worked the problem we didn't note the potential to infer the formula from a table of values. That helped us see the mathematics in a more complete fashion! In addition, the fact that there are different ways to dig into a problem (and in this case, reality checks from the context) allow for comparing, contrasting, and building on others' ideas. These are potential supports for the development of agency and mathematical identity.

Likewise, the car value task offers students potential opportunities for discovery and sense-making. In this case study the teacher frames it as such, saying that they haven't done problems of this type (where the value of an object is shrinking rather than growing) and that she really wants them to figure out what's happening by themselves. The fact that it's a modeling problem means that there's a possibility of "natural" feedback as they work the problem – unreasonable answers may cause them to challenge their assumptions and provide an opportunity for dialogue. Beyond that, opportunities for making the math their own will depend on the norms in the classroom itself. Are the students accustomed to collaborating, testing, and refining each other's ideas? If they function as a collaborative team, collective work can result in rich interactions. That too remains to be seen as the lesson plays out.

Finally, imagine you're a ninth-grade student engaging with this kind of problem for the first time. You've done some examples of exponential growth (e.g., compound interest) but you've never seen an exponential context where things got smaller. The teacher is giving you the opportunity to make sense of the context, and the mathematics, by yourself. What possible content confusions might arise?

Let's think about the value of the car after one year. You know there's a 15% loss. How do you figure out what the value should be?

There are at least three ways to think about it! Two correct ways were discussed in our solution to part (a), above: you can figure out the loss over 1 year and subtract it from the original value of the car, or multiply the original value by .85. But let's start with a blank slate.

Imagine you're a student in this class. You've recently worked on problems where, with a gain such as 15%, you multiplied the original value by 1.15. In that case, the value of something that cost $5000 and that grew 15% in 1 year would be 5000(1.15). In 2 years it would be 5000(1.15)^2, and so on. That's when the original value was getting larger. In this case, the original value is getting smaller over time. So, maybe the connection between things getting larger by 15% and getting smaller by 15% is that you should divide by 1.15 instead of multiplying by 1.15. After all, this fits with the (sometimes true) idea that multiplication makes things larger, and division makes things smaller.

In fact, you'll hear one student muse out loud about that: "Is it subtraction? Or is it division? It's division, right?" If you do divide $5000 by 1.15, the result is $4347. That seems pretty reasonable for 1 year's depreciation.

We know that's not the right answer – but, in context, $4347 might not trip any alarms for you as a student. You've lost a lot but you haven't lost $1,000, which you might recognize as 20%. The result is plausible, and it fits with their recent mathematical experience. "Multiply to make things larger, divide to make them smaller" is not an unreasonable strategy.

That's one place things might go wrong. And there are others, even if you start off in the right way. Suppose you used method 1 and arrived at $4250 at the end of year 1. Now it's time to calculate the value after 2 years. Do you take 15% of $4250 and subtract that? Or, since you know the car is losing something in value every year, might you assume that the $750 you've calculated is the annual loss, so you figure the car will be worth $(4250–750) at the end of year 2? Or, if you're not solidly anchored in the problem, you might lose track of which number is which – you have numerical values for the starting value, the first year's loss, the next year's value, the next year's loss, and so on. There's lots of opportunity for confusion.

Finally, we presented method 2 – computing 85% of the original value instead of subtracting 15% of the original value from that value – as being straightforward. It is, for us. But if you're a student, what will it take to move from "15% off" to "85% of"?

Here again, the bicycle and driving metaphors are useful. Driving a car or riding a bike seems straightforward once you know how to do it, but it can be quite a challenge when you're figuring out how!

Because of the way things played out in the classroom, the focus of this case study is on dimensions 1, 2, and 5: the mathematics, cognitive demand (productive struggle), and formative assessment. We've just gone through some content-related issues – two correct ways to conceptualize the mathematics involved, and a number of places where things might go awry when students engage with the mathematics. That's part of Dimension 1, the mathematics. Let's remember that the space of mathematics includes content, processes, and habits of mind – so, we want to think about the different connections they might make and the kinds of ways students approach the problem. Is their approach formulaic, or are they trying to put things together in ways that reflect sense-making?

Dimensions 2 and 5 concern the ways in which the students engage with the mathematics in this lesson. We know that the students' learning will be most powerful if they put things together for themselves, and if they're stretched in doing so. But how do we arrange for that? First, it involves establishing a climate in which students understand that they are supposed to be engaged in sense-making and feel safe trying out ideas. As you'll see, the teacher sets the task up with that expectation. As instruction progresses, formative assessment involves listening carefully to the students and coming to understand their thinking. Then, it involves providing some guidance. At times you might walk away and let the students continue to grapple, unaided, with ideas. At times you might rephrase what they said or point out productive aspects of what they've done. At times you may feel a need to provide some guidance – perhaps hints, perhaps some instruction. Any of these decisions have consequences. They may leave the students confused, nudge them along, or make things too easy, depriving them of opportunities. These are all judgment calls. In what follows, we'll see how one teacher deals with them.

2.2 Some Background about the Classroom We're Visiting

This instructional episode takes place in an untracked Algebra II classroom in an urban public high school. The school district has faced financial challenges through the years. Minoritized students comprise about 90% of the district's enrollment, with approximately 30% of the students being English Language Learners and about 70% of the students being eligible for free and reduced lunch.

The students in this classroom are 10th, 11th, and 12th graders. Historically, Algebra II plays a fundamental gatekeeping role: failure to earn a passing grade functionally ends a student's academic career, and delay in success slows down academic progress in significant ways. By state regulations, Algebra II is the most advanced class that students must pass in order to graduate from high school. In this sense, the course is a potential obstacle to students' graduation and future plans. At the same time, the course serves as an entry point into advanced mathematics. It is a prerequisite for all of the mathematics electives available to students at this school (Statistics, Precalculus, and Calculus), as well as supporting students' learning in Physics and Engineering electives. Success in Algebra II is thus essential.

A major mathematical focus of the course is on exponential functions, logarithms, and applications. Prior to the focal lesson, the students have completed a unit on exponential functions of the form $y = a\, b^x$. Their work included becoming familiar with terms such as initial value and growth rate, coming to understand negative and fractional values of x such as "going back in time" and "in between values," and studying the properties of negative and fractional exponents. The students had considered exponential growth by a percentage in some detail. Their work included creating equations for simple and compound interest and making as many connections as possible between a multiplier such as 1.06 and percentage growth (e.g., adding 6% to the initial value or seeing the change as growth *to* 106% of the original value).

The teacher avoided having students memorize a formula for these situations, instead giving students time to create tables and equations and then interpret the multiplier in the context of the original problem; in this way, students became familiar with these types of problems and began to solve them more efficiently. A few problems involving exponential decay by a percentage were also introduced at that time, but not explored in detail. As we join the students in this classroom, they are studying a unit on creating and solving various types of equations – in particular polynomial and exponential equations – and interpreting the solutions that they generate in the original problem contexts. Creating equations to model contextual problems continues to be a goal that the teacher emphasizes. As we've discussed, in this focal episode students are given a contextual problem about a used car whose value decreases by a constant percentage each year that can be solved by creating and solving an exponential equation. The episode focuses on creating an equation for exponential decay by a percentage and testing whether this equation is correct.

The lesson at the heart of this case study includes both whole-class discussion and small-group work. The teacher framed the problem for the whole class (both at the board and with a handout given to all students). Students worked in small groups much of the time; however, the teacher convened the class as a whole when she thought all the students would profit from discussing a particular point. The camera caught all of the whole-class discussion and some of the small-group discussions, focusing mostly on one group. Our discussion features that one group, with students we'll call Ayra, Blanca, Caleb, and Devin, along with the teacher's interactions with that group and the conversations the teacher directed to the whole class.

Ayra, Blanca, Caleb, and Devin are 10th graders. They are all making good progress through the mathematics curriculum at the school, with the intention of enrolling in calculus in their senior year. Of course, all four have their own personalities and histories. One student tends to be quiet, but has his strengths, as do the others; some members of the group take the lead in mathematical conversations, while others take more subsidiary (but still active) roles. Three of the four students went on to take calculus, and all four students graduated. Their position as 10th graders in a class composed of 10th, 11th, and 12th graders might appear to position them as "advanced" relative to their classmates but, as the whole-class conversations indicate, the issues this group faced in making sense of the car value problem very much reflected the issues the entire class was grappling with.

Ms. Sierra is in her 2nd year of full-time teaching at this school, having taught there part-time while completing her doctorate in education. She has a Master's degree in mathematics and was part of the TRU research group during her doctoral studies.

2.2.1 Preview of the Case Study

We will be observing and discussing about 26 minutes of classroom interactions, which begin mid-lesson with the introduction of the car value problem. Here's a quick summary of what happened. We've divided it into five episodes.

In episode 1 Ms. Sierra introduces the problem to the class. We zoom in on the focal group and Ms. Sierra's first interaction with it, where she points out that the formula they have produced to model the car's value needs "tweaking." (See the beginning of Section 2.3 for a more detailed overview.) Overall, this episode sets us up to see the challenges the students will face.

Episode 2 begins with the students grappling with the evaluation from Ms. Sierra: "How did we get the right answer, but the wrong equation?" Over the next five minutes they try various approaches. Their first attempt at a formula provides values that nosedive far too rapidly to be correct, showing the value of the context as a reality check. Toward the end of the episode the students realize that they should compute the car's second year's value based on the value of the car after it has depreciated for one year. But their attempts to find a formula hit a dead end, and they're stymied.

In episode 3, Ms. Sierra pulls the class together to discuss the different approaches that two small groups have taken to the first part of the problem, to find the value of the car after one year. One group's method, subtracting 15% of the car's initial value of $5000, resulted in a value of $4250. A second group's method, dividing the $5000 initial value by 1.15, resulted in a value of $4347. Ms. Sierra validates the first approach and tells the students to resume their work in small groups.

In episode 4, the focal group struggles with the question of how to generate a formula for the car's value. They produce a table of values for the first five years of the car's depreciation, but they can't see how to use the table to generate the formula they need. Ms. Sierra stops by the group and gives them a hint, but the students seem unable to capitalize on it; as they continue without success, their energy flags. They appear to be engaged in a form of guess-and-check (try a formula and see if the resulting numbers match the table) when Ms. Sierra calls the whole class together.

Episode 5 (whole-class discussion) begins with Ms. Sierra focusing on the values the students had produced, year by year, for the car's worth. After some slight confusion, she focuses the students' attention on the approach one student had taken, dividing the value of the car after n years by its value after n + 1 years. The constant ratio of .85 becomes apparent to the class, and a student calls out, "You just multiply 0.85 from everything." Ms. Sierra then helps the students make the connection between the fact that 15% of the car's value was lost each year and the multiplication by .85 to produce the next year's value. She and the students collaboratively produce the equation for the car value.

2.3 What Happened, in Detail (With Some Reflection Questions and Comments Along the Way)

2.3.1 Episode 1 (from 21:12 to 27:48 in the Lesson)

Overview

This episode begins about 20 minutes into the lesson, as Ms. Sierra brings the whole class together to introduce a new problem. She introduces the problem, framing it as an opportunity for sense-making within a real-world context. The focal group calculates the value of the car after one year without difficulty, but they see that their first attempt to find the equation that models the value of the car over time "doesn't make sense" – it says the car will be worth 37 cents after 5 years, a value the students reject. When Ms. Sierra briefly joins the group, she confirms that the students have computed the value of the car after 1 year correctly but that their equation needs "tweaking." She leaves to speak with another group, without pointing the students in any particular mathematical direction.

Narrative

PART 1, WHOLE CLASS

Students are sitting in groups of three or four. The room is full of chattering as the students pass out copies of the activity they're about to engage in (the Car Value Problem).

"3, 2, 1 and 0." Ms. Sierra quiets the room, calling attention to the whiteboard as she turns off the classroom lights. The Car Value Problem is projected on the whiteboard:

Anay buys a car for $5000. The car loses 15% of its value every year.

a. How much is the car worth after 1 year?
b. Write an equation to model the value of the car over time.

Before you go on, find a way to check/justify that your equation is realistic and show your work.

c. How long before the car is worth half of its original value?
d. After owning the car for 10 years, it breaks down. Anay finds out that she will need to replace the clutch to be able to drive the car again. Is it worth it?

Leaning on a desk near the whiteboard, Ms. Sierra gestures toward the board and says, "This situation …" There is a pause as she looks around the room. "… is about buying a car. So, we have done a lot of situations where something is growing by a percentage. Right? You are getting a pay raise, you are having interest. What we haven't done as much …"

She pauses again. "… what we have done a couple times, but really not as much, is when you are losing value by a percentage." She looks around the room and points to the problem on the board.

So, when you buy a car, does the value typically increase or decrease over time for that car?

There is murmuring as multiple students respond to the question. Ms. Sierra hears at least one student say, "increase." She responds,

"OK. If it's a really old car and you take really good care of it, it could go up. But the majority of the time, a typical car is gonna … what?" A student responds, "it's gonna decrease." Ms. Sierra nods and re-states, "It's gonna decrease. OK?" She now stands and points at the whiteboard.

"So, in this situation, Anay buys a car for $5000. The car loses 15 percent of its value each year." She leans against the desk. As she continues, she points to the phrase *1 year* on the board. "I really want you guys to think about… [pause] … You are gonna find out what it is worth after 1 year." Her fingers follow the second prompt as she continues, "You are gonna write an equation to model the value of the car over time."

She then emphasizes her intentions. "This is not one where I'm gonna check and tell you whether the equation is right or wrong. I do that sometimes, but today, I really want you to take a second…" [pointing to the board and reading aloud], *"find a way to check or justify that your equation is realistic."* She continues, "If you think your equation is right, try to prove it to yourself and prove it to your group. Make sure your equation makes sense before you go on."

She continues,

> Okay? Is it gonna be an equation that tells you after one year, after two years? Is it gonna be reasonable? Okay? Then you are gonna go on and answer these two questions. Alright, please get a quick start on this classwork. I'll stamp for your quick start in 30 seconds.

PART 2, FOCAL GROUP

Students begin chatting within their groups. The four students in our focal group are seated at four desks pushed together as shown below. Each student has their classwork and a calculator in front of them. See Figure 2.1.

Blanca says, "750 is what it loses. So do we subtract 750 from 5000?"

Devin quickly responds "I think so," as he shifts his focus to his calculator and starts pushing buttons.

Caleb looks down at his paper as Ayra supports this idea as well, saying "Maybe. So, 5000 minus 750 equals…" She looks at Blanca and says, "4250 dollars." She turns back to her paper and begins to write, saying, "That makes more sense." Blanca agrees, "Yeah."

Having apparently reached a consensus, the four students focus on their individual worksheets. Caleb, Blanca, and Ayra all start writing on their papers. Devin continues pushing buttons on his calculator, every now and then jotting something down. After some seconds of

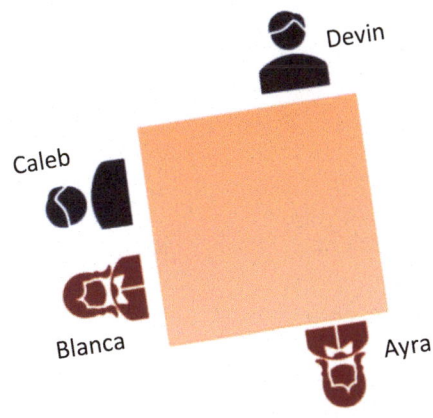

Figure 2.1 Ayra, Blanca, Caleb, and Devin

this independent work, staring at his calculator, Devin shakes his head and says, "I don't think that's how it works. Nope."

Ayra looks up, "Huh?" Blanca erases some writing on her paper. Ayra looks intently at Blanca's paper and then erases what she has written down.

Devin continues, "Because when you plug in 2, it equals 112.25. So that's – you are not subtracting that from 5000. That's how much it's worth." Caleb is now erasing what he originally had written down.

Looking at Devin, Ayra builds on his idea. Both nod in agreement as she says, "So, 750 is actually what it is worth now."

Still staring at his calculator as he begins to push buttons, Devin mumbles, "Unless our equation is wrong."

Blanca responds, "Maybe. Probably." as she continues erasing. Devin stares at his paper. Most of their previous work, subtracting $750 from $5000, has been erased.

Talking mostly to herself, Ayra begins, "Wait, so…" At the same time, Devin begins to chuckle and shake his head: "After 5 years it is worth 37 cents." A big smile appears on his face.

Everyone in the group laughs. Ayra looking down at her calculator, agrees, "Yeah, that can't be realistic."

Devin laughs, "That ain't right." He starts to whistle as he continues to push calculator buttons. Turning his calculator to show Caleb, he continues, "Bruh, it's worth one cent after 8 years." Devin laughs, shakes his head, and puts his calculator back down on the desk.

Ayra, also calculating, confirms Devin's findings. "Wow, after six years… wow … This is weird."

After some laughter, the students continue staring at their work. Blanca looks up from her paper, asking "Does it have to have…" as Devin exclaims, "Oh wait. It says 'write an equation to model the value of the car over time right under it'."

PART 3, TEACHER JOINS FOCAL GROUP

As Devin reads the task prompt, Ms. Sierra walks over to the group and opens up a folder on their desk. While she is leafing through the folder Ayra says, "Wait. This makes … somewhat no sense."

Ms. Sierra stamps a sheet of paper in the folder and then turns toward Ayra. She asks, "What makes somewhat no sense?"

Ayra begins, "Ok, because-" then turns and looks at Blanca. "Do you wanna go?" Blanca takes over. "We made an equation and then we plugged in one and we got 750 and we said it's not realistic."

With a slight nod of her head, Ms. Sierra responds "OK."

Blanca goes on, "And we were like maybe 750 is the percentage it loses. So, we subtracted that from 5000 and we got –"

Ayra finishes Blanca's sentence "– 4250 dollars."

Ms. Sierra again says "OK."

Blanca continues her explanation as Devin and Caleb follow, "But when we kept going, it ended up equaling cents. We plugged in 2 and it wasn't 750."

Ms. Sierra sums up what she understands. "So, you are repeatedly multiplying times 15 and you're getting, it's going down too quickly."

The two girls confirm with "Uhuh" and "Uhmm." Looking at Ms. Sierra, Devin nods his head in agreement and Caleb responds, "Yeah."

Ms. Sierra clarifies, "Ok. What – not times 15, I mean times 15 percent, point 15. OK." The students agree.

Ms. Sierra confirms that the numbers were decreasing too rapidly. "Yeah that's true," and tries to connect this finding with previous work they have done. "Have you ever had that happen before when you were trying to model a pay raise or something?"

The students quickly respond "No" as she continues, "And you accidentally made the equation go down really fast?"

Blanca responds, "I think so," and Ayra says, "Maybe. Yeah."

Ms. Sierra references a previous classroom discussion. "Because I remember when some people did – OK, it's a six percent pay raise and they multiplied by 0.06 and it also went down way too fast." She looks at the students and asks, "Do you guys remember what the problem was there?"

The students respond with head shakes and quiet "No"s.

Ayra begins, "I think…" but then says "I don't remember off the top of my head." Devin, likewise, comes up empty.

Ms. Sierra pulls up short, abandoning this approach. "OK. You're right. You're right that your equation is wrong."

Trying to make sense of what Ms. Sierra just said, Ayra asks, "750 is right, but the equation is wrong?"

Ms. Sierra asks, "750 means what again?"

Ayra responds, "The – how much it's worth after one year," placing her hands on her face.

As the teacher starts to say "No," Blanca corrects Ayra by saying, "It's what it's losing."

Ayra begins to erase what she had written on her paper, as Ms. Sierra confirms that this second interpretation is correct. She says, "750 dollars is the correct amount of what it loses in value in the first year. OK, so how much could it be worth after one year?"

Ayra responds, "4250 dollars."

Ms. Sierra confirms this: "Yes."

With a more upbeat tone, Ayra says, "So, we got it right."

Wrapping up the conversation, Ms. Sierra says, "That is correct. Your equation is gonna need a little tweaking. You're on the right track." She then walks away to check in with another group.

Focus Questions for Episode 1

In these questions, we look at the students' engagement with mathematical content and practices, classroom norms, and issues of formative assessment and cognitive demand.

Please think about them and, if you're working through this book with colleagues, discuss your thoughts collectively before you read what we have to say.

Focus Question 1
What important mathematical ideas and practices are the students invited to engage with in this episode? How are they invited to do so? How do we see/hear the students engaging with important ideas and practices?

Focus Question 2
What norms regarding students' and teachers' roles in generating and working through mathematical ideas do you see in play and/or developing in this episode? In what ways do the norms allocate responsibilities to each of the participants?

Focus Question 3
One thing we see in this case study is Ms. Sierra inferring quickly what the students are thinking, and then commenting in ways that reflect her inferences. That, of course, is a key aspect of formative assessment. Given that we have the luxury of reflecting on what the students said and did, can we infer what individual students were thinking at times? One candidate is when Devin, laughing and shaking his head, said "After 5 years it is worth $0.37." Can you find other places?

Focus Question 4
This is our first pass at an issue we'll return to, since this case study focuses on formative assessment and cognitive demand: Did the context (including the task and classroom norms) and the interactions between students and with the teacher appear to support productive struggle?

Our Commentary

Once you've answered the reflection questions yourself or with colleagues, we invite you to read our commentary. As always, our comments are not meant to represent final answers to these questions. Rather, we offer them as a way to engage with you in conversation.

1. **What important mathematical ideas and practices are the students invited to engage with in this episode? How are they invited to do so? How do we see/hear the students engaging with important ideas and practices?**

As described in Section 2.1, the problem setup offers students the opportunity to expand their understanding of exponential functions and, at least as importantly, to connect different ways of solving the problem.

Because this is a new problem for the students – an extension of work they've done with exponentials, in which quantities are decreasing rather than increasing – it provides an opportunity for the students to build on what they know and to expand the ideas they've learned into new territory. These are key mathematical practices.

Moreover, the fact that this is a context-based modeling problem with plausible "real world" values for the initial car value and the rate of depreciation offers students opportunities for mathematical sense-making – and for checking the plausibility of their answers.

The way the teacher frames the problem,

> I really want you to take a second, [reads from the board] "find a way to check or justify that your equation is realistic."
>
> If you think your equation is right, try to prove it to yourself and prove it to your group mates, your equation makes sense before you go on. Is it gonna be an equation that tells you after one year, after two years? Is it gonna be reasonable? OK?

makes it clear that the point is not simply to answer the questions, but to be thoughtful about the mathematics and how it represents the situation that is being modeled. In short, this is a clear invitation for the students to engage collectively in mathematical sense-making. It reflects Ms. Sierra's description of her reasons for using contextual problems, quoted in the introduction to this chapter:

"Ideally, each context should be

1. Compelling and powerful to many stakeholders. Students should be able to go home and say, "we are learning about fair loan practices" rather than "we are learning about exponential growth by a percent."
2. Familiar to all students and/or worth spending time familiarizing students with.
3. Mathematically productive for cuing student intuitions and/or as an instructional analogy."

In our view, the students are part-way to taking advantage of the context. On the one hand, they reflect on some of their computations, as in this exchange:

Blanca: 750 is what it loses. So do we subtract 750 from 5000?
Devin: I think so.
Ayra: Maybe. 5000–750 = 4250. Yeah, that makes more sense. [presumably, than $750 being the value of the car after one year]

And this:

Devin: After 5 years it is worth $0.37. [Devin is laughing and shaking his head]
Ayra: That cannot be realistic.
Devin: [Laughing] That ain't right ... Bruh [turning calculator toward Caleb] it's worth one cent after 8 years.

In both of these exchanges, the students compare the results of their computations (using calculators) against their sense of what the car's value should be. Those are acts of sense-making.

However, they don't appear to be working collectively through the calculations and linking them to the changes in the car's value. That is, the loss is $750 because it's 15% of the original value of $5000, and you find the next year's value by subtracting the loss from the original value. With this understanding, you then have $4250 as the "new" original value, and you're set up to repeat the chain of operations. The students don't appear to have developed this understanding at this point.

2. **What norms regarding students' and teachers' roles in generating and working through mathematical ideas do you see in play and/or developing in this episode? In what ways do the norms allocate responsibilities to each of the participants?**

Like all classrooms, this classroom has well-established patterns of interactions (sometimes called the "classroom contract") that shape individual and collective responsibilities. Even though we've just "entered" this classroom, a few things are apparent.

As discussed in our response to focal question 1, the framing of the problem makes it clear that this is not a "demonstrate and practice" lesson in which the teacher works on sample problems, and the students are asked to work on similar exercises. Here the students are explicitly asked to develop some of the mathematics themselves. They are the ones doing the math thinking; Ms. Sierra is not doing the thinking for them. This becomes very clear at the end of episode 1, when Ms. Sierra leaves the group, saying "Your equation is gonna need a little tweaking, you're on the right track." It's difficult to know from that one exchange with Ms. Sierra whether this minimal degree of feedback is typical (an issue of formative assessment we pursue later on), but it certainly indicates that some responsibility for idea generation and checking rests on the students' shoulders. The teacher's role is not to give away answers or point to the "right" methods.

It becomes clear just a few seconds into the episode that the students are accustomed to working in groups. Although the computational aspect of the problem sends each student to their calculator, it doesn't take long before the students are exchanging ideas with each other. We say "ideas" because the students, as indicated in our response to Focus Question 1, are not simply "answer-driven." Numbers that don't seem to make sense act as triggers to dig more deeply.

The group dynamics suggest that the students are comfortable working together – that there are norms of sharing and working collaboratively. Three of the four students share their ongoing thoughts aloud and react orally to what each other says. The fourth student, Caleb, tends to be quiet, but he appears to be attending actively to group interactions. The group functions as a team, without clearly defined roles.

One aspect of their collaboration appears to be shared responsibility for understanding. In many groups, one student will take the lead, find the solution method and then tell the other students how to proceed. Of course, they don't have a solution yet, so it's not possible for one student to "teach" the others. But at the same time, the students are actively communicating their ideas with each other, and trying to reconcile the contributions they make. Their sense-making appears to be communal, including a balance of individual accountability and accountability to each other. This is a group effort.

3. One thing we see in this case study is Ms. Sierra inferring quickly what the students are thinking, and then commenting in ways that reflect her inferences. That, of course, is a key aspect of formative assessment. Given that we have the luxury of reflecting on what the students said and did, can we infer what individual students were thinking at times? One candidate is when Devin, laughing and shaking his head, said "After 5 years it is worth $0.37." Can you find other places?

Part of our response to this question was discussed in our response to Focus Question 1. The students do calculate the value of the loss after 1 year, and then on the basis of "reasonableness" – a value of $750 after 1 year is too low – they decide to subtract it from the original value, $750. But, it's clear that this calculation isn't really grounded in a solid understanding of the situation because they don't repeat it for year 2, with $4250 as the starting figure.

Devin's statement

> Because when you plug in 2, it equals 112.25. You are not subtracting that from 5000 that's how much it's worth.

is an indication of his confusion: $112.5 is 15% of $750. It looks like Devin is not subtracting the next year's loss (15% of 4250) but instead is computing the value each year by taking 15 times the value of the previous year. In fact, we can confirm this – if we calculate the supposed value of the car after 5 years, $(5000)(.15)^5$, we get $.3796.

It's worth noting that Ms. Sierra picked up on what the students were doing quite rapidly:

Blanca:	And we were like maybe 750 is the percentage it loses. So we subtract that from 5000 and…
Ayra:	We got $4250.
Ms. Sierra:	Ok.
Blanca:	But we kept going, it ended up with equaling cents. If we plug in 2 it wasn't 750.
Ms. Sierra:	So you are repeatedly multiplying times 15

This is formative assessment in action! Not only that, but in an attempt to help the students make connections, Ms. Sierra asks if the students recall a parallel situation of exponential growth, where the numbers didn't match with their intuitions:

Ms. Sierra:	Have you ever had that happen before when you were trying to model a pay raise or something? … 'Cause I remember when people… Ok… it's a 6% pay raise and they multiplied by 0.06 and it also went down way too fast. Do you guys remember what was the problem there?"

They don't and she doesn't pursue the issue; she's leaving a lot for them to figure out for themselves. This is another judgment involved in formative assessment: when you know there's an issue, how much do you scaffold?

As observers who can see some of what the teacher doesn't, we can also see where Devin begins to get into trouble and carries Ayra along with him. At the beginning of this exchange, Blanca, Devin, and Ayra appear to be on the right track:

Blanca: [working on a calculator] 750 is what it loses. So do we subtract 750 from 5000?
Devin: I think so.
Ayra: Maybe. 5000–750 = 4250. Yeah, that makes more sense.

But what Devin says next indicates that he's not on the same page as Ayra and Blanca. Ayra follows him in that direction:

Devin: I don't think that's how it works. Nope.
Ayra: Huh?
Devin: Because when you plug in 2, it equals 112.25. You are not subtracting that from 5000 that's how much it's worth.
Ayra: So 750 is actually what it is worth now.

In sum, a close look reveals some of the student thinking – and confusion! Our point here is *not* to call for direct instruction or to say anything negative about these students. Rather, it's to highlight the fact that such confusions are natural when students are asked to figure out some things for themselves. Partial understandings and misdirections are typical – the students are juggling a lot! What takes place here indicates the fragile nature of developing understandings. It takes some time before a full, connected grasp of new ideas becomes solid. That reality highlights the complexities of formative assessment. How much room should we allow for confusion – and the possibility that students will resolve things for themselves? (They often do.) How much scaffolding should we offer in the hope of setting the "right" level of cognitive demand? The answers to these questions depend on the students and the context. We highlight these questions here because this case study shows Ms. Sierra actively addressing them.

4. This is our first pass at an issue we'll return to since this case study focuses on cognitive demand: Did the context (including the task and classroom norms) and the interactions between students and with the teacher appear to support productive struggle?

As we see it, the very framing of the problem opened the door for productive struggle. The task was open enough to give students freedom to explore. The students had space and time to work through a number of different methods for solving this problem. They had opportunities to make mistakes to recognize some challenges, and try to address those challenges. As we've noted, that differs substantially from the kind of "demonstrate and practice" lesson that offers little or no opportunity for productive struggle.

There are a few things we can say at this point, and many that we can't – we have to see how things play out. Ms. Sierra's intervention was timely, in that the students were stuck and didn't seem to have a way out of their dilemma. Without that intervention, it's possible that the students would have spun their wheels and given up in frustration.

In her exchange with the students, Ms. Sierra provides minimal feedback. She does let the students know that the value $4250 is correct, which confirms their understanding; and she at least locates the source of difficulty, the fact that their equation [tacitly, $5000(.15)^n$] is incorrect. She does not, at this point, indicate that the equation does not take into account the subtraction that they did to find the value of $4250, or provide any hints that the process of subtraction should be embedded in their general approach.

In short, she leaves them to struggle. Whether that struggle will turn out to be productive remains to be seen. The students have only been working on this problem for a couple of minutes, so Ms. Sierra has decided to give them more time to work things out by themselves. They may continue to be stuck; they may find their way out of their current dilemma; or, she may return later to provide some scaffolding.

Decisions like this occur all the time. There is no "right or wrong" with regard to the choice; such decisions are judgment calls, based on a teacher's knowledge of the students. How and when Ms. Sierra follows up will be a matter of ongoing formative assessment. It will, of course, have implications for the students' individual and collective sense of agency and identity.

2.3.2 Episode 2 (from 27:48 to 33:05 in the Lesson)

Overview

The focal group works to make sense of Ms. Sierra's comment that their calculation of the car's value after 1 year of depreciation is correct but that their equation needs tweaking. Devin suggests that Ms. Sierra might want them to "do the log," but he and others dismiss the suggestion. Individually and collectively they try various approaches without success. After some time Devin says he doesn't think the 0.15 in the expression $(5000)(.15)^n$ is right, but nothing comes of it. Later Blanca says she thinks they have to change the starting value each year. The group calculates the year 2 value of the car using an initial value of $4250 instead of 5000. As the episode comes to a close the focal group has run out of ideas – they've conceptualized the iterative solution, but they "just don't know how to make it into an equation."

Narrative

All four students are writing on their papers as Ms. Sierra walks away. Devin smiles and says, "What the hell? How did we get the right answer, but the wrong equation?" Caleb smiles, shrugging his shoulders and shaking his head. Ayra, still writing, says, "I don't know." Blanca adds, "the equation only works once."

Devin looks over at Blanca's paper and says, "Cause then afterwards it's weird." The students continue working. Devin taps his pencil on his forehead and Caleb punches numbers into his calculator. Blanca has written y= 5000(.15) above y = 750. They're stuck.

Devin asks the group, "Do you think she wants us to do the log?" Ayra picks up her calculator and says, "Maybe log might work ..." but after typing a few numbers into his calculator, Devin rejects his own suggestion: "I doubt that."

Blanca asks, "How do you do log if you don't even know..." and Devin finishes the sentence "the base number." Blanca clarifies, "Yeah. Well. You do know the base number. You don't know what you want it to be equal to."

Ayra, still typing on her calculator, says, "Yeah, you're right. That's true." She puts her calculator down, looks up and says, "I'm not getting that." Devin is still typing on his calculator. Ayra turns to Blanca, who's still looking down at her own paper, and says, "This makes no sense."

The students work individually in silence. After about 15 seconds, Blanca says "Oh, we can do... how long before the car is worth half of its original value?" Devin says "The equation – oh, we can do..." but then trails off uncertainly.

The next few minutes feature individual false starts and collective dead ends, with the students trying things on their calculators and/or writing on their papers. There is no sense of progress. Ayra lays her chin on her folded arms, and the others end up staring at their papers.

Ayra turns to ask Blanca, "Did you get it?" Blanca responds, "I got a negative exponent when I used that equation," flipping the calculator around to show the group. Ayra curses, "Oh ^$%#.

Well that won't work." Returning to Ms. Sierra's parting words Ayra says, "This makes no sense. How do we have the right answer, but the wrong equation?" Devin replies, "That's what I just said" and Blanca, echoing what she had said earlier, adds "The equation only works once." Ayra looks over at Blanca, then puts her chin back atop her folded arms. Ayra and Devin stare at each other, looking confused.

Overhearing conversations between classmates and Ms. Sierra, Ayra says to Devin, "Everybody is using the same equation." Devin, referring to another group, says "No. He said – he said – we're missing …" but stops mid-sentence. He looks over at Devin and adds, "It's probably the most simplest thing." A bit later he says, "It is the 0.15. I don't think the 0.15 is right." Ayra rouses herself a bit and starts to type on a calculator. Meanwhile, Devin picks up his calculator and starts typing once again.

Another fifteen seconds pass in silence. At that point, Blanca looks up and says "Um, I think, I think that each year, we have to change the 500." Sensing that Blanca misspoke, Devin clarifies, "The 5000? to the current value?" Blanca corrects herself quickly and says, "Yeah, I mean, the 5000, yep." Ayra adds on, "To the other current value."

Blanca continues, "So for the second year, it would be 200 … 4250 …" Ayra jumps in, saying, "4250 times 0.15 would give you another answer." Devin says "Yeah, but we have to…what if you …" as Blanca continues, "It gives you 637.5 and then you subtract that. And then you get 3612.50 …"

Ayra adds, "Which gives you the value of the second year. You are supposed to change it every year." Blanca agrees and says, "I just don't know how to make it into an equation."

Despite having found the correct value of the car after 1 and then 2 years of depreciation, the group is deflated. Ayra rests her chin against her folded arms once more. Devin asks, "What if it's not even exponential … the equation is not even exponential?" He smiles and looks over at Caleb's work as Blanca replies, "It has to be." The students go back to working individually. At this point, the camera breaks away to focus on another group, where it remains for about 90 seconds. We do not know what happens in the focal group during those 90 seconds.

Focus Questions for Episode 2

The questions below focus on issues of cognitive demand and sense-making. The goal is to get a sense of what students are grappling with – are they engaged in what might be, or turn out to be, productive struggle, and what kinds of sense-making might they be engaged in?

This is a reminder that our goal in these focus questions is not to evaluate but to clarify what took place and set the stage for a discussion of possible implications. Please think about and/or discuss the following questions before you read our commentary. Where you can, point to evidence in the episode that helps to support what you're thinking.

Focus Question 1

Episode 2 can be challenging to make sense of. It's hard to interpret the starts and stops and hard to see what progress, if any, the focal group makes. This first focal question is an attempt to take stock. Can you summarize what took place and what progress or lack of progress the focal group seems to make in episode 2? What issues does this progress, or lack of it, raise for you?

Focus Question 2

Here are two specific examples of what the students say as they grapple with the question, "How did we get the right answer, but the wrong equation?"

Example 1. Early in Episode 2, after the focal group reconfirms that their previous calculations are in error, Devin asks, "Do you think she wants us to do the log?"

Example 2. Toward the end of the episode Devin guesses that there is something wrong with the .15 in their equation. Shortly afterward, Blanca follows up with "I think that each year, we have to change the... $5000."

In terms of sense-making, how would you characterize the way the students are approaching the problem in examples 1 and 2?

Our Commentary

1. Episode 2 can be challenging to make sense of. It's hard to interpret the starts and stops and hard to see what progress, if any, the focal group makes. This first focal question is an attempt to take stock. Can you summarize what took place and what progress or lack of progress the focal group seems to make in episode 2? What issues does this progress, or lack of it, raise for you?

Here's our interpretation, which includes some reading between the lines. The students arrived at the right answer for the value of the car after 1 year because they *subtracted* the loss in value (15% of $5000) from the initial value of $5000. However, they appeared to lose sight of the subtraction when creating "the equation," which they wrote as $y = (5000)(.15)$. This equation, as written, is a natural for generalization: if you multiplied by .15 the first year, you'd do the same each subsequent year, generating the formula $y = 5000(.15)^n$ for the value of the car after n years. That's what they settled on. Looking for the cause of error in their work, the students rechecked their computations. (This is a perfectly reasonable thing to do.) Once again the numbers shrank too fast, with Ayra concluding "this makes no sense."

A few seconds later, Devin's suggestion "Do you think she wants us to do the log?" seems to come from nowhere. (We'll explore where such a suggestion might have come from in Focus Question 2.) It too is rejected, after some computations – whatever Devin entered into his calculator didn't seem to help – and because the group doesn't have the "base number" (that is, if .15 isn't the number that you multiply by repeatedly, then they don't know enough to build or solve the equation.)

The next few minutes pass without any notable leads, much less progress. If you're not used to seeing students grapple with challenging problems, this can seem frustrating, both for the students and possibly for us as teachers; if we were in the classroom with the focal group, we might feel the temptation to help. Even if you are accustomed to watching students struggle as they grapple with challenging problems, there are any number of questions. For example: Does the level of cognitive demand seem about right? Will the struggle turn out to be productive? How long should I wait before offering some scaffolding? How much scaffolding, of what type, might be appropriate at this point?

About 30 minutes into the class, Devin, who is looking at the expression $5000(.15)^n$, zeros in on a possible cause of difficulty when he says "It is the 0.15. I don't think the 0.15 is right." Soon after, and seemingly out of the blue, Blanca says "Um, I think, I think that each year, we have to change ... the \$5000." Ayra elaborates on the "5000," saying the change has to be to "the other current value."

These two observations contain the seeds of potential progress. The correct value for the exponential equation is .85, not .15, and that value can be used to generate the appropriate equation. It's also the case that you can build on the conversation between Ayra and Blanca to generate the iterative solution to the problem – each year you can compute the next year's initial value by subtracting 15% of this year's initial value. But that's an iterative solution, not an equation. When Blanca ends episode 2 with the statement "I just don't know how to make it into an equation," that challenge is what she's voicing. At that point, the focal group is stuck and feeling somewhat deflated.

Episode 2 is a little over 5 minutes long. In it the students are actively engaged at times, stuck at times, and frustrated at times; they end up stuck. If you're used to observing classrooms in which problem-solving methods have been demonstrated and the students are expected to practice them, seeing students flounder in the ways the focal group does for more than 5 minutes in episode 2 may make you feel uncomfortable. You may wonder if letting students struggle like this is fair to them (why make students go through this?) and if it is a good use of class time (leading students through a solution is faster and brings results). We want to reflect on these questions a bit before proceeding.

In the grand scheme of things, five minutes isn't very long. Even without overt signs of "progress," students may develop a deeper sense of the issues they're working on. Time to think can be useful for generating or incubating ideas, and that's excellent for agency, ownership, and identity. But how much time is enough time, how much is too much? We know that students who are fed a steady diet of short exercises come to believe that "all problems can be solved in just a few minutes" and simply give up when unsuccessful. That's bad for learning. On the other hand, if the work we ask students to do consistently represents too much of a stretch (that is, the level of cognitive demand is too high) the resulting lack of success can be counterproductive: flailing and not making progress isn't fun! The phrase "productive struggle" may sound nice, but it's messy in practice. What's "productive" is a judgment call, and how much struggle students can bear before their work begins to do damage to their AOI depends on the students and the context.

In sum, we want to acknowledge the complexity of making decisions related to cognitive demand, especially given that Ms. Sierra is trying to monitor the progress of multiple student groups working at the same time. We also want to note that this kind of struggle is important and meaningful, no matter how things turn out. If the struggle goes on for "too" long without some sort of payoff, either in deepening understanding or in progress toward a solution, it can be unproductive, with negative implications for AOI. (Obviously, how long is too long and what constitutes a payoff are judgment calls.) If, however, the students do see themselves as making progress or developing deeper understandings, then the struggle can be viewed as productive – and its impact on AOI can be positive.

2. Here are two specific examples of what the students say as they grapple with the question, "How did we get the right answer, but the wrong equation?"

– Example 1. Early in episode 2, after the focal group reconfirms that their previous calculations are in error, Devin asks, "Do you think she wants us to do the log?"
– Example 2. Toward the end of the episode, Devin guesses that there is something wrong with the .15 in their equation. Shortly afterward, Blanca follows up with "I think that each year, we have to change the... \$5000."

- In terms of sense-making, how would you characterize the way the students are approaching the problem in examples 1 and 2?

Before digging into our analysis, we want to recognize that the students really are puzzled – the question "How did we get the right answer, but the wrong equation?" is an honest statement of confusion. With that as context, our view is that the focal group is engaging in two different forms of sense-making during part 1.

EXAMPLE 1. "...DO THE LOG."

One way out of the puzzled state in which the students find themselves involves a classic kind of school-based sense-making – looking for hints in what the class has done recently. In many classrooms, students spend time practicing the techniques and methods that they've learned over the past few days. So, for example, if they've just been taught to solve pairs of simultaneous equations using substitution, they expect to use substitution when they're given a set of simultaneous equations to solve. The same goes for using Boyle's law in chemistry, the inverse square law for gravitational attraction, etc. We don't often go back to first principles in instruction, and students aren't typically expected to do so.

Because of this common classroom pattern, it's quite reasonable for students to ask "what methods have we used recently?" when they are asked to find a method for solving a problem. So, it shouldn't be surprising that Devin asks, "Do you think she wants us to do the log?" The class has been solving exponential equations using logarithms. In a similar vein, doing *something* with the equations – for example dividing by 1.15 instead of multiplying by 1.15, which the students would do if they were modeling exponential growth – makes sense in the context of trying to find a variant of a recently used procedure.

We're not mind readers, of course: "do the log" might have had various meanings. One interpretation is Devin might have thought about starting with $5000 and repeatedly taking its log (He did turn to his calculator right after saying this; we don't know what buttons he pushed.) We do know that he quickly rejected the numbers he got, presumably because they didn't fit with the values he expected. A second interpretation is that "do the log" might mean constructing an equation based on logs instead of exponentials. The third is the one mentioned above – that they should solve the problem, somehow, by taking the log of the exponential equation and capturing the depreciating value. Whichever interpretation represents what was in Devin's mind, Devin and Blanca together realize that they don't know the base, so they can't take logs.

What we see happening in this segment of the episode is a kind of school-based sense-making. The students are using a classic classroom strategy: If you don't know what to try, consider things you've done recently in class. It often works, although it doesn't in this case. It's interesting to note that the idea of using the log is rejected either because the numbers don't work (a productive use of the problem context) or because they don't have the mathematical pieces (in this case the base) to work with. That kind of sense-making is preferable to simply writing down the results of computations, whether they fit or not!

EXAMPLE 2: CHANGING... THE $5000

Toward the end of episode 2, both Devin and Blanca are thinking over their previous work and trying to figure out what went wrong. Devin, assuming that the equation is exponential (a reasonable assumption because the class has been working with exponentials) looks for where their formula, $5000(.15)^n$, might be off:

Devin: It is the 0.15, I don't think the 0.15 is right.

In parallel, Blanca, with Devin attending closely – all the students are listening carefully to each other! – is beginning to think that their starting figure 5000 has to change each year. Here's a condensed version of the dialog:

Blanca: The equation only works once.
...
Blanca: I think that each year, we have to change the ... $5000.
...
Blanca: So for the 2nd year, it would be 200 ... 4250
Ayra: 4250 times 0.15. Would give you another answer.
...
Blanca: It gives you 637.5 and then you subtract that. And then you get 3612.50.
Ayra: Which gives you the value of the second year. You are supposed to change it every year.
Blanca: I just don't know how to make it into an equation.

Although the students wind up feeling stymied, this exchange represents a significant development of understanding. Ayra and Blanca now have just outlined the iterative mechanism for finding the car's value: starting at year 1, you compute 15% of that year's initial value and subtract that number from the initial value. The result is the initial value for the next year.

This iterative approach emerged from their reflections on unsuccessful attempts. They saw that s*omething* had to be changed in the equation, and *something* was wrong with the 5000. Moreover, the iterative solution they outlined followed the mechanism by which the car's value depreciated – that is, the contextual nature of the problem provided a structure that enabled them to correct their previous incorrect approach. Although they may not see it yet – at the moment they are stuck, focusing on their inability to obtain the equation that captures the car value – we see the students' decision to change the initial value in the year 2 calculation from $5000 to $4250 (the result of the first-year depreciation calculation) as an important development, emerging as the result of productive struggle. We want to note, first, that the students had the time to engage with the problem. Second, both aspects of progress in the dialog above come from making use of the context. Devin's statement that the .15 must be wrong came from his observation that his formula produces numbers that get small too rapidly, compared to how a car's value *should* decrease. Similarly, we can't know what triggered Blanca's realization that the initial number in the iterative solution changes every year, but that realization has to be tied to thinking about what happens to the values over time – the car is worth less at the end of year 1, so that's the value you start with for year 2. Making meaningful use of the context, rather than just considering it as "window dressing" and ignoring it once they're enmeshed in writing equations or doing computations, is the students' second form of sense-making.

2.3.3 Episode 3 (from 33:05 to 38:40 in the lesson)

Overview

When Ms. Sierra circulated through the room during episode 2, she saw student groups grappling with a number of challenges. Some, like the focal group, had run into roadblocks. Others had done plausible but incorrect calculations to represent the car's value after 1 year's depreciation, obtaining $4347 rather than $4250 as the value of the car after 1 year.

Ms. Sierra pulls the class together in episode 3 to discuss both answers and the reasoning that led to them. One student explains the reasoning behind the answer of $4347. Ms. Sierra notes that the result is numerically plausible and praises some of the intuitions that led the group to approach it the way they did while indicating that the $4250 figure is correct. After publicly validating the focal group's year-by-year approach to solving the problem, she has the class return to working in small groups, with the goal of finding the equation that models the value of the car over time (part (b) of the original problem).

Narrative

Ms. Sierra has been circulating through the class, checking in with groups regarding their progress with the task and noting some of the challenges they face. She calls for the class's attention. "Alright, pause a second. There's like five groups that need help. Pause a second. [Counting down for quiet] Five, four, three, two, one, and zero."

She picks up a stack of index cards that have students' names on them. As she shuffles the cards, she says, "…This is a little trickier than what we usually do, OK…"

She pulls a card from the deck. "Emily, what did you think? What did you get for part a?" Writing "a" on the board, she quotes part of the problem: "What was the car worth after one year and how did you get it?"

Emily replies, "I got 4347?" her inflection suggesting less than full confidence in her answer. Ms. Sierra writes $4347 on the board. She asks, "Ok. How did you get that?"

Emily says, "I put 5000 parentheses 1.15 close the parentheses and added a negative 1 to it." As Emily speaks, Ms. Sierra transcribes what the student says on the whiteboard as

$$5000(1.15) - 1$$

Ms. Sierra points to the "–1" she has just written and asks, "after closing the parentheses?" and Emily replies, "No, up…", after which the teacher erases the "–1" with her thumb and then re-writes it as an exponent, resulting in the expression

$$5000(1.15)^{-1}.$$

Stepping back from the board, she says, "Ok. And that gave you this number?", pointing to the $4347 she had previously written. She then completes the expression on the board,

$$5000(1.15)^{-1} = \$4347$$

Ms. Sierra again flips through the stack of student name cards, picks another, and says, "Ok, um, Frank, what did you get after one year?" Frank replies, "I got $4250." As she turns to write $4250 on the board she asks, "Ok. How?"

Frank explains, "Cause I multiplied 5000, I mean I put 5000 times 0.15 to the power of 1, which is 750." As he is speaking, the teacher transcribes this as

$$5000(15)^{1} = \$750.$$

Frank continues, "And I subtracted 750 by 5000 and got that" (the $4250).

To clarify, Ms. Sierra points to the $750 she has just written and asks, "So what does this 750 represent?" Frank replies, "The loss in one year."

The teacher writes the phrase "loss after 1 year" under the $750 and says, "This was the loss after one year and [with emphasis] you *subtracted* it." She writes the subtraction on the board as she summarizes what S2 said, "5000 minus 750 and you got 4250."

The whiteboard now looks like this:

$4347
$$5000(1.15)^{-1} = \$4347$$

$4250
$$5000(15)^{1} = \$750$$
loss after 1 year
$$5000 - 750 = 4250$$

Referring to the first computation, Frank asks, "What happens when you put a negative 1?"

Ms. Sierra turns to the class. "Do you guys remember what happens when you put a negative 1 as the exponent?" A few students mumble in response and one student suggests, "It's like a fraction?"

Ms. Sierra circles the expression "1.15^{-1}", saying "It makes it... this ... would become 1 over 1.15." As she draws an arrow from "1.15^{-1}" to write "$1/1.15$," she continues, "and that will make your value go down. OK? This is similar to if you were dividing 1.15."

She looks at the first answer and says "I really like where this group is headed." Then she points to the second answer ($4250) and says "This one is correct."

She clarifies, pointing back to the first equation.

> This one is not correct – *but* I like where this group is headed, because what they were trying to do is make sure the value is going down over time. They are doing something with 15 percent. They saw that if you do your regular 15 percent raise, it would be going up, which we couldn't have. So, they are trying to make it go down over time.

Placing the index cards back on her desk, she addresses the class. "Ok. I want to just clear this up a little bit though at this point, because we've gone in a lot of different directions."

She then steps up to the board with a marker in hand. Drawing a box around the $4250, she says, "This is definitely the amount after one year." She then puts a big X through the calculations involving $4347. Again pointing to the box she has just put around $4250, she says, "This was... This is definitely the amount after one year."

The center of the whiteboard now looks like this:

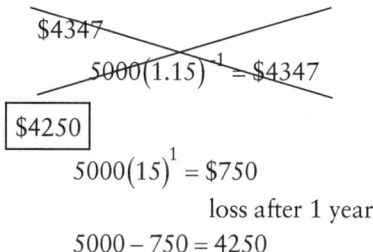

She continues, "We are still in this mode trying to figure out something we could do as an equation, besides dividing, that would give us this correct amount after one year." She points again to the $4250.

Ms. Sierra then starts to make a table. "I'm gonna go ahead and set up a little bit of a table for a second." She turns back to the whiteboard, saying, "years" as she writes the word atop the left-hand column of the table. She says "I don't want to tell you the equation yet" as she begins to fill in the table. "We have years. After zero years, the car was worth 5000." Turning back to the class she asks, "After one year, the car was worth ... what?"

A student replies, "4250." Ms. Sierra writes the value in the table. See Table 2.5.

Table 2.5 The first entries in Ms. Sierra's table

Years	
0	5000
1	4250

Leaning back against a table in the front of the room, she asks, "And why are you guys … why are you guys trying to make … I see people trying to make exponential equations. And why are you guys trying to make exponential equations?"

Blanca, from the focal group, replies, "Because we've always learned that percents usually…" as her voice tails off.

Ms. Sierra continues. "OK. Yes, *and* … is the car going to continue losing $750 every year?" She points to the $750 written on the board. Some students reply, "No." Other students silently shake their heads from side to side. "Why not?" asks the teacher, scanning the classroom.

One student starts to say, "Because it increases…", while another student offers, "Because it loses its value."

Ayra raises her hand and starts to speak in a very quiet voice, "Wait, I think …" The teacher encourages Ayra to continue, pointing to her and saying, "Yeah?" Ayra continues,

> Ok. Well, what we were talking about was that when you, after, ok the $5000 times 0.15 equals 750 that's for the first year, but wouldn't you have to trade out the $5000 for $4250, and then minus it by whatever?

A smile spreads across Ms. Sierra's face as Ayra explains. She nods her head and says, "Yeah… Yeah" affirming Ayra's suggestion. Following up energetically on Ayra's comments, Ms. Sierra raises her voice and asks the class, "Ok, so if I wanted to calculate – in other words, if I wanted to calculate in the second year, how much money was lost… In that second year, in that second year, I wouldn't be calculating 15 percent of the 5000 anymore. I would be calculating 15 percent of the new amount."

While Ms. Sierra is saying this, Ayra looks over at Blanca and says softly, "You were right, you were right."

Ayra asks the teacher, "So how would you write that as an equation?"

Ms. Sierra smiles at Ayra. She addresses the whole class. "Ok. I'm going to give you five to ten minutes to think about what the equation could be." Chuckling to herself, she then says "Think about, think about what this equation should be. It's not 5000 times 0.15 to the x. It's not times 1.15 to the negative x, although those are both good ideas. Keep going."

Focus Questions for Episode 3

Our focus questions explore the character of the conversations that take place in episode 3, in which the teacher allocates time and space to the discussion of an incorrect approach to the problem and provides some but not a great deal of scaffolding for the class as she has them return to small-group work to try to determine the equation that models the car's decreasing value. As always, please think through and discuss the focus questions before reading our comments.

Focus Question 1

In episode 3 Ms. Sierra calls on students to elaborate on two approaches to computing the value of the car after one year. One approach, dividing the initial amount of $5000 by 1.15 each year, is incorrect. The teacher compares the resulting value, $4347, with the value of $4250 obtained by the correct approach. What rationale might a teacher have for giving board space (and classroom time) to an incorrect approach? What might this suggest about the classroom norms this teacher is trying to develop?

Focus Question 2
Near the end of episode 3, Ms. Sierra has an exchange with Ayra regarding the focal group's work. Ayra asks, "but wouldn't you have to trade out the $5000 for $4250, and then be minus it by whatever?" Ms. Sierra brings this statement to the class's attention, saying "In that second year, I wouldn't be calculating 15 percent of the 5000 anymore. I would be calculating 15 percent of the new amount."

What do you see happening in this exchange? Building on focus question 1, what kinds of scaffolding do you see taking place with the clarification above? What do you think the impact on cognitive demand might be? What rationale might a teacher have for laying things out explicitly?

Focus Question 3
At the tail end of episode 3, Ms. Sierra sends the students back to small-group work, saying

> I'm going to give you five to ten minutes to think about what the equation could be. Think about, think about what this equation should be. It's not 5000 times .15 to the x, it's not times 1.15 to the negative x [pointing to equations that had been written on the whiteboard]. Ok? Although those are both good ideas. Keep going.

How much scaffolding do you see taking place in this assignment, and what do you think its impact on cognitive demand might be, as the students settle back into small-group work?

Our Commentary

1. In episode 3 Ms. Sierra calls on students to elaborate on two approaches to computing the value of the car after one year. One approach, dividing the initial amount of $5000 by 1.15 each year, is incorrect. The teacher compares the resulting value, $4347, with the value of $4250 obtained by the correct approach. What rationale might a teacher have for giving board space (and classroom time) to an incorrect approach? What might this suggest about the classroom norms this teacher is trying to develop?

It's worth remembering Ms. Sierra's description of her goals for what she calls "contextual problems," as described in Section 2.1. She made those goals clear when she began the section of the lesson devoted to the car value problem. She said,

> You are gonna write an equation to model the value of the car over time. This is not [a situation] where I'm gonna check and tell you whether the equation is right or wrong. I do that sometimes, but today, I really want you to take a second to ... [pointing to the board and reading]... *"find a way to check or justify that your equation is realistic."* If you think your equation is right, try to prove it to yourself and prove it to your group. Make sure your equation makes sense before you go on.

This kind of sense-making is exactly what took place when, in episode 1, Devin calculated the value of the car to be 37 cents after 7 years. The group knew something had to be wrong.

With that as context, let's look at the discussion of $5000/[(1.15)^n]$ as a possible formula for the value after n years.

From the teacher's point of view, the students who developed the formula $5000/[(1.15)^n]$ *did* engage in a form of sense-making. Over the previous week, the class had studied exponential growth. The students knew that if the car increased 15% in value, they'd be multiplying the value each year by 1.15, yielding $\$5000(1.15)^n$ after n years. Given this, it's reasonable to guess that you'd get a similar decrease by dividing by 1.15 each year.

Unfortunately, then, testing this formula "to see if it's realistic" doesn't help to reject it as a possibility. If someone told you that a car that cost $5000 was worth $4347 after 1 year, that would seem reasonable.

Given her goals for student sense-making, this puts Ms. Sierra in a bind. If her goal were just to make progress on this problem, the most efficient way for her to proceed might be to tell the students that the formula $5000/[(1.15)^n]$ is incorrect and to point them to the correct method. But, doing so could undermine the students' sense-making. Ms. Sierra doesn't want to simply tell the students that they're wrong. So, what can she do to honor their sense-making, while indicating that the answer is wrong?

The first thing she does is to affirm the plausibility of the students' answer: "I really like where this group is headed." But then she clarifies. She points to the second series of equations that produced $4250 and says

> This one is correct. This one [pointing to the $4347] is not correct – *but* I like where this group is headed, because what they were trying to do is make sure the value is going down over time. They are doing something with 15 percent. They saw that if you do your regular 15 percent raise, it would be going up, which we couldn't have. So, they are trying to make it go down over time.

In saying this Ms. Sierra gives the students who had developed the incorrect model credit for making good use of the context and trying to adapt what they have learned. This underscores the view that math class is about sense-making, not simply producing the right 5answer – and that if "near misses" occur as a byproduct of sense-making, it's not a big deal.

Rather than simply saying "this one [$4250] is correct," Ms. Sierra then calls on another student to present the step-by-step calculations for year 1 (take 15% of $5000 = $750 and subtract that number from $5000). Working through this process provides a way of showing the students what the amount left after 1 year *should* be. It also enables her to demonstrate the incorrectness of the first approach using carefully worked-through mathematics (produced by another student, not by herself) as the grounds for judgment rather than by using her authority as a teacher.

In doing things this way, Ms. Sierra comments positively on student sense-making in both the correct and incorrect approaches. Moreover, she models how it is possible to resolve a conflict between two plausible ways to approach a problem by working carefully through the mathematics involved. As the class unfolds in this manner, the students get to see that mistakes are a natural part of the problem-solving process and that they can be overcome by thoughtful reflection.

What does this say about classroom norms? First, Ms. Sierra's intentions are clear. In her discussion she has been very explicit about the fact that what matters is thinking through problems in such a way that things fit together; that math isn't arbitrary, and ideas and formulas should make sense. Moreover, sense-making and explaining one's reasoning are collective endeavors.

We've seen this perspective echoed in the focal group's work. The students have said "that doesn't make sense" a number of times and built an iterative solution that was both logical and grounded in the context. They listen carefully to what each other is saying and build on each other's thinking, with significant investment in individual and collective understanding – note, for example, that toward the very end of episode 3, as Ms. Sierra was working through the iterative approach, Ayra had looked over at Blanca and said softly, "You were right, you were right." These collective sense-making norms ultimately support students both in learning key mathematical practices and in developing a sense of agency.

2. Near the end of episode 3, Ms. Sierra has an exchange with Ayra regarding the focal group's work. Ayra asks, "but wouldn't you have to trade out the $5000 for $4250, and then be minus it by whatever?" The teacher brings this statement to the class's attention, saying "In that second year, I wouldn't be calculating 15 percent of the 5000 anymore. I would be calculating 15 percent of the new amount."

   What do you see happening in this exchange? Building on focus question 1, what kinds of scaffolding do you see taking place with the clarification above? What do you think the impact on cognitive demand might be? What rationale might a teacher have for laying things out explicitly?

We see Ms. Sierra's response to Ayra, "In that second year, I wouldn't be calculating 15 percent of the 5000 anymore. I would be calculating 15 percent of the new amount," as revoicing and clarifying what Ayra said for the class as a whole – that is, paraphrasing Ayra's statement in a way that was more clear, but giving full credit to Ayra for the idea. Ayra's question,

> We were talking about was that when you, after, ok the $5000 times 0.15 equals 750 that's for the first year, but wouldn't you have to trade out the $5000 for $4250, and then be minus it by whatever

is imprecise, although it's not hard for us to figure out what she means to say. When Ms. Sierra addresses the whole class she revoices Ayra's statement clearly:

> Ok, so if I wanted to calculate, in other words, if I wanted to calculate the 2nd year, how much money was lost?...
> In that second year, I wouldn't be calculating 15 percent of the 5000 anymore. I would be calculating 15 percent of the new amount.

The way Ms. Sierra has clarified Ayra's statement does a number of things. First, it's a minor form of scaffolding for Ayra, making sure Ayra's intention is clarified. Second, the way Ms. Sierra rephrases Ayra's comment removes the ambiguity but not in a way that seems like "teaching" new information; it appears to be a casual clarification. That means that the agency and ownership of the mathematics remain with Ayra. While she is providing some scaffolding for Ayra, Ms. Sierra has not taken control of her idea. The student is still in charge of the mathematics – and she knows she's on the right track. In fact, Ayra's comment to Blanca, "You were right, you were right," demonstrates that the ownership remains with the students.

Third, Ms. Sierra's announcement sets the class on a clearer path to finding the value of the car year after year. This too is a deliberate teaching move: all the students will be able to extend the table that contains the yearly values, as either a precursor to finding the equation or to check the values that their equations produce. By clarifying and validating

Ayra's comment, Ms. Sierra reduces uncertainty and points the students in the direction of the next big task. That lowers the cognitive demand to some degree, but not a tremendous amount. The aim is to keep the ultimate task (finding the equation) within reach, for as many students as possible.

3. At the tail end of episode 3, Ms. Sierra sends the students back to small-group work, saying

> I'm going to give you five to ten minutes to think about what the equation could be. Think about, think about what this equation should be. It's not 5000 times .15 to the x, it's not times 1.15 to the negative x [pointing to equations that had been written on the whiteboard]. Ok? Although those are both good ideas. Keep going.

How much scaffolding do you see taking place in this assignment, and what do you think its impact on cognitive demand might be, as the students settle back into small group work?

Ms. Sierra's clarifications set all the students on the same path, computing the value of the car year after year. In doing this, she has kept many of the students from pursuing unprofitable directions. That's a strong move. At the same time, when she tells them "I'm going to give you five to ten minutes to think about what the equation could be" she provides no hints regarding how the students should derive the equation; she just reminds them that two possible equations have been ruled out.

It's worth thinking about the degree of challenge from this point on. As we saw in episode 2, the focal group appeared to have no idea about how to obtain an equation after having outlined an iterative solution. Thus, although the whole class has been set on the path of building a table with the values obtained from the iterative approach, the challenge that remains is substantial. Will the students find a solution? If not, what might the consequences be? Time will tell, but it's quite possible that the level of cognitive demand is very high, even though Ms. Sierra has kept some students from pursuing what she knows to be dead ends.

2.3.4 Episode 4 (from 38:40 to 42:55 in the Lesson)

Overview

Episode 4 begins as the focal group re-convenes, working on parts b and c of the car value task. The students begin by calculating the value of the car for the first 5 years, noting that the car would be worth half its original $5000 value between 4 and 5 years. They are unable to find the equation, however, and report this when the teacher checks in with them. She then asks if they remember the method they'd previously used to derive an equation from a table. They don't recall it, and she leaves the group with a somewhat vague hint. After her departure, the students try out candidate equations by doing computations on their calculators, but the numbers don't work; they're stymied. At this point, the teacher calls the whole class together.

Here are the relevant parts of the task:

Anay buys a car for $5000. The car loses 15% of its value every year.

a. How much is the car worth after 1 year?
b. Write an equation to model the value of the car over time.
   Before you go on, find a way to check/justify that your equation is realistic and show your work.
c. How long before the car is worth half of its original value?

Narrative

Blanca is speaking when the camera returns to the focal group. She has done some calculations. Devin and Ayra are both looking at her while she speaks and Caleb is looking at his worksheet.

Blanca says, "$2610.03 which is almost almost half" –

Caleb utters, "almost." Ayra now is looking forward and Devin has turned to look down at his worksheet while holding his calculator with his left hand.

Blanca continues, "I might, I'm gonna do 5 to see if it even...."

Devin jumps in, "It's going to be less than it, but ..." and Blanca overlaps, saying, "Yes, just to make sure."

They look down for a moment while Ayra turns to Blanca while Caleb continues to look down at his paper.

Then Devin says, "When we get the actual equation we can do the log and answer it quickly. We need to find the equation first and then we can do the number C 'cause..."

[He interrupts himself, laughing, and shaking his head because he has just called "C" a number]

"... I mean the number C ... the letter C. [Ayra laughs as well, and Devin collects himself.] Then we can do the letter C."

Ayra has been calculating and speaking softly to herself: "Oh god, I think we have...", at which point she addresses the group: "I got $2218.53." Blanca and Devin alternate looking at Ayra and her worksheet. Caleb continues looking down at his worksheet and starts annotating.

Ayra asks, "the loss?" as Blanca continues: "So after 5 years it is less than half and after 4 years it is more than half." At this point, the whole group is looking at Blanca's worksheet.

Devin asks, "What equation were you using? Was it the same one?" to which Blanca says what she'd been doing: "No, I just kept umm ... whatever this was, I multiplied it by fifteen and subtracted that by it and we got that."

Reacting to this, Devin says "I thought the equation is exponential." Blanca responds, "It is. She just said it was. It can't be linear because it's not gonna decrease by the same amount."

Somewhat dismayed, Devin says "I feel like ....ugghh." Blanca says "I feel like we could find the equation from the table, but I don't remember how to do that." Devin starts to say "maybe you divide the..." as Ayra, pointing to the values written on Blanca's worksheet (Table 2.6),

says "Don't you see ... like... what the... from there... from there to there is and then from there and then from there to there ... "

Breaking off, Blanca says, "Oh, minus 15 percent. Ohhh! So maybe this is..." but Devin says, "I already tried that though." As he enters numbers into his calculator, she starts erasing the equation she had begun to write. He continues: "It gives you... the one with the weird... the weird numbers. Oh, that actually gives you the same thing." He shows the calculator screen to Blanca and says, "Oh no, the one that gave me the weird number was the..." as he starts entering data again.

Table 2.6 Blanca's worksheet

x	y
0	5000
1	4250
2	3612.5
3	3070.63
4	2610.03
5	2218.53

Trying to clarify, Ayra asks, "You got the negative exponent or the one that gave you the really weird number?" Blanca begins to say, "Maybe the…" as Devin turns the calculator screen toward Blanca again and says, "Yeah, this." Blanca replies, "Maybe this is negative." Ayra jumps in and says, "No, the 15 percent…" but then trails off. Blanca asks, "You tried that?" Devin begins to respond as Blanca, looking over at Ayra, says, "He already tried it." Devin types into his calculator while shaking his head, smiling, and saying, "Oh wait, that's wrong." They pause as the teacher approaches.

In response to her question "How are we doing?" the group collectively replies, "Uhhhhhhhhh." After a slight pause, Blanca takes the initiative and shares, "I made a table that goes up to 5 years."

"OK," says Ms. Sierra, giving Blanca time to continue. Blanca says, "But… I can't. And like I was thinking … I know we can find a table, the thing … the equation from the graph, but I don't remember how."

After pausing for about five seconds, Ms. Sierra asks, "Ok, you guys remember something that we did with dividing in tables to make equations?" All four students shake their heads. Blanca replies, "I do not remember." Ayra echoes, "I don't remember," adding "That is what I'm saying." Devin looks over at Caleb shaking his head and smiling.

Ms. Sierra responds by summarizing what the students have done to this point "So you're trying to find what can be a pattern in this table then." Blanca replies, "It goes down by 15 percent, the y." After a pause, Devin asks, "Wait, what was the bottom?" He reaches across the table to take a closer look at Blanca's worksheet and begins typing into his calculator.

As Ms. Sierra begins to walk away from the group, she replies to Blanca's comment, "Ok. It goes down by 15 percent. But is there anything that is being constantly added, subtracted, multiplied, or divided within this table? That could definitely help you make an equation. Ok that's one hint."

After Ms. Sierra leaves, Ayra says "The thing that's always multiplied is the 15 percent."

"Yeah," Blanca agrees. Ayra turns to Devin, who is keying numbers into his calculator and asks, "What are you doing?" [Ayra laughs as she asks him] Devin passes the worksheet back to Blanca, smiles and says, "I'm trying to unleash my algebra."

With her paperback in hand, Blanca asks, "What if we flip these two?" She uses her pencil to circle two numbers and shows them to her group mates.

"You wanna…. Ohhh wait, that… let me check it," says Devin. He begins typing on his calculator again. "Flip it?" says Ayra, picking up her calculator. Blanca continues, "Ok, um. You do 1.5 times…"

Devin jumps in, "No, that did not work" as Ayra asks, "Why'd you put 1.5?"

Laughing and turning his calculator to Blanca, Devin says, "That's 3 million 750…" "Let me see!" says Ayra. He holds up his calculator. She leans in, grins, and says, "Damn." Turning to Blanca and smiling, she says, "Yeah, the car would probably not be that much."

Devin, looking down at his calculator, asks the group, "What if we make it negative?" "Yeah, yeah, yeah," replies Ayra. She picks up her calculator and inputs some numbers. A few seconds later she puts the calculator back down and looks at Devin. He can tell it didn't work. "Never mind," he says. Laughing, Ayra repeats. "Never mind." The camera pans away as the teacher begins to call the class together.

Focus Questions for Episode 4

Episode 4 is complex, there's a lot going on. We've done what we can to tell the narrative in straightforward ways, but there's a lot to unpack. In what follows we're going to look at what happened in close detail, taking advantage of the opportunity for "instant replay." Our motivation is that making sense of student thinking – trying to figure out what students understand, and what they're struggling with – is a key aspect of formative assessment and helps to make decisions regarding cognitive demand.

> Again, our goal in these focus questions is not to evaluate, but to clarify what took place and set the stage for a discussion of possible implications. Please think about and/or discuss the following questions before you read our commentary. Where you can, point to evidence in the episode that helps to support what you're thinking.

Focus Question 1
Can you trace the group's mathematical understanding up to the point when Ms. Sierra approaches their table?

It may be helpful to focus on the major strands of discussion that emerge, specifically (a) the group's pursuit of the year-by-year approach, and the progress they are able to make when using that approach; and (b) the group's attempt to express the equation for the car's depreciating value in the form $y = A \cdot B^x$, and the progress they are able to make when using that approach.

Focus Question 2
When Ms. Sierra joins the focal group, Blanca shows her their data table and explains their dilemma: "I know we can find ... the equation from the graph, but I don't remember how."

In this brief interaction, the teacher provides two hints. First, in direct response to Blanca's statement, she says "You guys remember something that we did with dividing in tables to make equations?" The students say no. Then, as she's about to leave the group, she says "OK. It goes down by 15 percent. But is there anything that is being constantly added, subtracted, multiplied, divided within this table? That could definitely help you make an equation."

The decisions any teacher makes in moments like these are acts of formative assessment, based on the teacher's sense of what the students understand and how much progress they seem to be making. What did the students reveal, and how might that have served as the grounds for Ms. Sierra's changing the hint (and, as a result, changing the level of cognitive demand)?

Focus Question 3
In the final part of this episode, we see the focal group trying various equations on the calculator – with results that get as large as 3 million! Think back to the students' work in episode 1, when we saw them plugging values into their calculators. How do you compare their work in the two episodes? What does this suggest about the level of cognitive demand and productive struggle at this point?

Our Commentary

1. Can you trace the group's mathematical understanding up to the point when Ms. Sierra approaches their table?

    It may be helpful to focus on the major strands of discussion that emerge, specifically (a) the group's pursuit of the year-by-year approach, and the progress they are able to make when using that approach; and (b) the group's attempt to express the equation for the car's depreciating value in the form y = A·Bx, and the progress they are able to make when using that approach.

Episode 4 begins with Blanca telling the group about her work on the approach suggested by Ms. Sierra at the end of episode 3, calculating the value of the car for the first five years of depreciation. The value of the car after 4 years is $2610.03, slightly more than half of the car's original value. The value after 5 years is $2218.53, which is below half. So, she says that the answer to part (c) of the task is between 4 and 5 years. The others, following the change in yearly values, are on the same page. So far, so good.

But Ms. Sierra has asked the class to produce the equation that generates those yearly values. As becomes clear later in the episode, the class has produced equations from tables of values that document exponential growth. Unfortunately, none of the students remembers how. Blanca says, "I think we can find the equation from the table, but I don't remember how to do that." Ayra tries to remember, saying "Don't you see what from there to there is and then from there to there?" as she points to the yearly values, with other group members following. But things don't click. The students have made progress but hit a roadblock.

In parallel and overlapping with these efforts, Devin has been trying to generate the general equation for the car's value. When he says "We need to find the equation first and then we can do part C", he's talking about finding the exact amount of time when the car is worth half of its original value. (This was part C of the original problem; using the table of values, the group had observed that the value is between 4 and 5 years.)

In saying that they need to find the equation, Devin is referring to the fact that the class knows how to solve equations of the form A Bx = C by taking logs:

$$\text{if } A \cdot B^x = C, \text{ then } B^x = C/A, \text{ so } x = \log_B(C/A).$$

In that equation, A would be $5000 and C would be $2500; the problem is that the group doesn't know what the value of B should be. Devin has tried B = .15, but that hasn't worked; the values of $5000(.15)^x$ got small too fast. For now, at least, the attempt to make use of an equation of the form $y = 5000(something)^x$ to generate the right values has run aground.

There is some interplay between the approaches, with some checking of basic assumptions. After Blanca explains that her solution was year by year, involving multiplying by 15% and then subtracting, Devin reacts by saying "I thought the equation is exponential." Blanca's response confirms Devin's statement. She says, "It is [exponential]. She just said it was. It can't be linear because it's not gonna decrease by the same amount."

At this point, the well has run dry. The group has produced a table of values but can't remember the procedure they'd used to generate an equation from a table. They have tried without success to find the right value for B in the equation $y = A B^x$. Various members of the group have suggested ideas in the hope something might work. Devin mentioned division in the context of looking at the table of values, but he trailed off; Blanca and then Ayra suggested trying various values in the calculator, but Devin said that doing so produced "weird numbers", and so on. While not making suggestions, Caleb has been leaning in to follow and sometimes checking this on his calculator. In sum, the group has run out of ideas as the teacher approaches –the reason for the collective "Uhhhhhhhhhh" when Ms. Sierra comes up to the table and asks how they're doing.

2. When Ms. Sierra joins the focal group, Blanca shows her their data table and explains their dilemma: "I know we can find ... the equation from the graph, but I don't remember how."

   In this brief interaction, Ms. Sierra provides two hints. First, in direct response to Blanca's statement, she says "You guys remember something that we did with dividing in tables to make equations?" The students say no. Then, as she's about to leave the group, she says "OK. It goes down by 15 percent. But is there anything that is being constantly added, subtracted, multiplied, divided within this table? That could definitely help you make an equation."

   The decisions any teacher makes in moments like these are acts of formative assessment, based on the teacher's sense of what the students understand and how much progress they seem to be making. What did the students reveal, and how might that have served as the grounds for Ms. Sierra's changing the hint (and, as a result, changing the level of cognitive demand)?

Ms. Sierra's first hint could have solved the problem – *if* the students remembered what she was suggesting. When they had been working with exponential growth, they had seen that you can find the exponential growth factor from the table by dividing the value of the quantity at any year by its value the previous year. Looking at the entries in the table of values, the growth factor is 4250/5000 = .85 (or 3612.50/4250 = .85, etc., or the ratio of any two consecutive entries in the table, since all those ratios are the same). With an initial value of 5000 and a yearly growth/shrinking factor of .85, the car's value after x years would be $y = 5000 \cdot .85^x$.

We can't know what was in Ms. Sierra's mind, but it's possible that when she gave them the hint "You guys remember something that we did with dividing in tables to make equations?" she thought that the hint would cause them to remember the way that dividing produces the growth factor. If they did, then this group (and the rest of the class) would wrap up the problem quickly. Under these circumstances, her question about division would have served as a stimulus to direct recall. That's low cognitive demand, but finding the growth factor quickly would have advanced the lesson and allowed Ms. Sierra to move to the big picture, where she could make all of the relevant connections between:

- the table of values arrived at year by year,
- the base (.85) as derived from the table,
- the equation that the students could create once they had derived the base from the table, and
- the alternative way to find the base (subtracting 15% from 100%).

Given that there is so much to do, providing the hint can be seen as a way of expediting progress in an already long lesson. Unfortunately, the hint didn't work: none of the students showed any indication of remembering the division. At that point, this teacher – any teacher – faces the challenge of what, if anything, to suggest next.

One option is to tell the students about the division or lead them directly to it. Doing so is efficient in terms of the overall lesson plan, but it removes the opportunity for productive struggle and the benefits of that struggle. If the teacher provides the solution path, then the students haven't made the connection themselves. That has implications regarding the students' agency, ownership of the content, and their identities, a point we'll return to in our concluding discussion.

A second option is to suggest that the group return to work, while not providing further support or information. If the students succeed, then they've done so on their own; that's a plus. If, however, they continue to struggle without seeing themselves as making progress, it could be disheartening. The students' lack of progress up to that point (remember the "Uhhhhhhhhh" when Ms. Sierra had asked how they were doing, and their inability to remember what they'd done with the table) suggests that the level of cognitive demand might be too great at this point and that further struggle without support may not be productive.

Options between these two extremes include various ways of narrowing the students' focus while not giving them the answer. Up to the point where Ms. Sierra joined them, the students had focused on two things: the iterative nature of the solution (subtracting 15% each year) and the search for the equation. In saying, "OK. It goes down by 15 percent," the teacher acknowledges the first approach. Then she advises the students not to focus on it, but to focus instead on the values in the table. She says, "*But* is there anything that is being constantly added, subtracted, multiplied, divided within this table? That could definitely help you make an equation."

In saying this, Ms. Sierra directs the students to focus their attention on the numbers in the table, while not specifically saying which operation to use. Since she hasn't told them what to do, she hasn't removed opportunities for agency; there are still decisions to be made and a fair amount of work for the students to do. At the same time, what Ms. Sierra has just said focuses the students' attention in a direction that she hopes will be more profitable for them. This downward adjustment of cognitive demand may enable the students to make visible progress, in which case some of their struggles will have produced results.

It remains to be seen, of course, how this kind of decision will play out. Making adjustments in cognitive demand is a balancing act; we never know how easy or hard a particular suggestion or direction will turn out to be for students. And, it's worth noting that the perfect hint for one group may be too much of a hint for a second group and not enough of a hint for a third! Our intention is not to suggest that there is an "ideal" level of cognitive demand or that there are the "right" moves for any situation. It's to highlight the tensions involved in formative assessment and to point to the fact that knowing your students and monitoring their ongoing work are essential. Consider the focal group's situation at the moment. On the one hand, something will need to be done if they have an insight and move quickly to a solution; on the other hand, something may need to be done if they remain stuck in a way that seems unproductive.

3. In the final part of this episode, we see the focal group trying various equations on the calculator – with results that get as large as 3 million! Think back to the students' work in episode 1, when we saw them plugging values into their calculators. How do you compare their work in the two episodes? What does this suggest about the level of cognitive demand and productive struggle at this point?

In episode 1 the students appeared to be engaged in a somewhat purposeful attempt to use what they knew. Devin had assumed that the equation was exponential, and he used the calculator to test values of $5000(.15)^x$. The values his calculations produced were unrealistic, and he concluded as a result that the equation must be incorrect. As we discussed, doing these computations at that time, in the way Devin did, was a reasonable act of context-based sense-making. Likewise, Blanca was moving toward a year-by-year solution in subtracting 15% of the car's initial value to arrive at its value at the end of year 1. So, although the students were just beginning their work on the problem, they were making sense of what it demanded of them – reasonable progress, indicating an appropriate level of cognitive demand at that point.

In this episode, however, the students find themselves in a different situation. Their teacher has offered them a hint, but they didn't see how it was useful. The focal group sees no new avenues for exploration, and that puts them in a bind.

What do you do when you're stuck like this? You might go back and try some of the things you'd tried before. That's what they do. They're still engaged, as their smiles and banter reveal.

But the previous methods are still unhelpful. Given what they're able to bring to mind, there are no further productive directions to pursue. At the moment, then, the level of cognitive demand is too high. They've reached a point where, absent some sort of inspiration or intervention, further struggle is not likely to be productive.

This state of affairs set the stage for episode 5.

2.3.5 Episode 5 (from 42:55 to 46:30 in the Lesson)

Overview

Seeing that many student groups are struggling, Ms. Sierra calls the class together. Pointing to the table that contains the year 0 and year 1 car values, she highlights the fact that a particular student had divided those two numbers. After some confusion, the group collectively realizes that each time they divide, the result (starting with 4250/5000) will be .85. When Ms. Sierra asks what number they could multiply by every time, a number of students call out ".85". The idea catches on, as students "see" the ratio in the table. Ms. Sierra then asks what the common ratio, .85, has to do with the 15%. The students don't see the connection, so the teacher focuses on the fact that subtracting 15% of something requires two operations, and she'd like to get the result in one. A student calls out "divide [by] 85 percent," which another amends to "multiply by .85." Ms. Sierra unpacks why this makes sense and then, in interaction with the students, formulates the exponential equation $y = 5000(.85)^x$.

Narrative

Ms. Sierra walks up to the front of the class and says, "Ok, you guys ... We are still pretty stuck. I'm gonna give you a bigger hint."

A student loudly responds "Nah" as the teacher continues, saying, "majority of groups are still pretty stuck." She points to the small table written on the whiteboard (Table 2.7), saying, "So, um, [student name] was doing something that I want to really highlight. She was looking at these two numbers. Looking for a pattern, and she divided them." She uses a marker to link the two numbers with a downward arrow (Table 2.8), and writes "5000/4250 =" on the board. The class is abuzz, with some laughter, an audible "Oh wow," and some murmuring.

Table 2.7 The small table on the whiteboard

Years	
0	5000
1	4250

Table 2.8 Ms. Sierra links the two numbers

Years	
0	5000 ⟩ .85
1	4250

She asks "What do you guys get when you divide these?", and then appears nonplussed when a student calls out "1.17! It's 1.17!" The same student repeats, "It is 1.17!"

Ms. Sierra stops and says "Did you guys divide 5000…" as she points to the whiteboard. Some students say "yeah" and the student who had called out 1.17 says "The calculator don't lie." The source of confusion is revealed when Ms. Sierra looks at the board, sees that she had written 5000/4250, and says "oh, that."

As she goes to the board to erase the fraction 5000/4250 she asks, "What did I mean to do? Thank you for catching my mistake. What did I mean to do? Why? Why should I do it the other way?"

She begins to write the correct fraction, 4250/5000. As she is doing so a student says "So you can see what…" As other students begin talking Ms. Sierra continues, "I want to know what number could be multiplied every time." She faces the students for a second and turns her back to the whiteboard as she continues writing 4250/5000.

Multiple students respond with "0.85." Ms. Sierra replies "Thank you." She finishes writing

$$4250/5000 = .85$$

on the board. There is some chattering as she writes this, and numerous students respond with an excited "ohhhhh!!!" She moves to capitalize on this:

> OK! OK! What did you guys get after two years? Stick with me, stick with me, I know it's hot, I hear that. I hear an ohhh. Can three people explain to me what you said ohhh about?

A student responds, "You just multiply 0.85 from everything" and another says "huh." Ms. Sierra revoices, saying "OK, you're saying I can just continue multiplying [by] 0.8…" As she writes on the board, she says "I'm going to keep multiplying 0.85 times everything to get the future one."

The board now contains the following table (Table 2.9).

Looking to make connections, Ms. Sierra asks, "What does 0.85 have to do with 15 percent?" The class is silent.

A student raises her hand, and Ms. Sierra calls on her. The student says, "4250 is 85 percent of 5000." Ms. Sierra repeats this word for word as she writes it on the board: "4250 is 85 percent of 5000".

There are varied reactions. Focal group student Ayra exclaims, "Wowwww! Wow." But another student says "I still don't get it."

Ms. Sierra works to keep the class together.

> OK, so. Wait, wait, wait. Just stick with me. I see a lot of side conversations start to come up. I know this is more whole class than we usually do. But stick with me as a whole class for a second.

"If we have a 15 percent, we are losing 15 percent of the value. What percent of the value is left?" Various students respond, "85 percent." Ms. Sierra smiles. Working to bring things to closure, she says, with some emphasis: "OK. So if I want to get directly from the value to the new value every year, what is *one single operation* I could do every year?"

Table 2.9 The expanded table

Years	
0	5000 .85
1	4250 .85
2	

One student utters, "Subtract 15" while another says "divide." Responding to the first, Ms. Sierra says, "OK, subtracting is gonna take two operations. I would have first to find what is 15 percent and then subtract it."

One student suggests, "Divide 85 percent." With a clearly questioning tone, Ms. Sierra responds "Divide ... Divide 85 percent?" Another student quickly says "No, multiply by .85."

Ms. Sierra builds on this. "OK, it is a smaller number ... OK, I'm trying to make the number smaller, but I can multiply by a number smaller than one." She moves to the whiteboard.

"OK, sooo ... OK, so, what can my equation be then?" she asks. As students say "5000," she writes it on the board and turns to the class, and says, "I have an initial value 5000. You guys saw that it is exponential. Everybody, pretty much every group, was trying to do something where you're multiplying or dividing every time. OK?"

She then writes the next part of the equation, so that

$$5000(.85)$$

is written on the board. She points to the ".85" and says, "It's gonna be multiplied times 85 percent..." and continues, "To the power of x, t, whatever you want to call it." She finishes writing the equation, so that

$$5000(.85)^x = y$$

Is written on the board. We hear Ayra say "wow" once again. Ms. Sierra brings this part of the lesson to a close, saying "OK. All right. Using this equation... using this equation you are going to solve the rest of the problem."

Focus Questions for Episode 5

As always, please think through and discuss the focus questions before reading our comments.

Focus Question 1
A significant number of mathematical ideas emerge and get explored in episode 5. How do those ideas compare to the mathematical ideas discussed in Section 2.1?

Focus Question 2
What stands out to you in what the students appeared to find easy or difficult in episode 5? What issues might this raise about student understanding and how to address it?

Focus Question 3
We'd like to reflect on the difference in the level of scaffolding the teacher provides in episodes 1 and 5. In episode 1, she provided the focal group with almost no scaffolding, telling them that their equation [$y = 5000(.15)^x$] "is gonna need a little tweaking" before walking away from them. In episode 5 she asked leading questions to support the class in making connections. What reasons might the teacher have had for acting so differently in those two situations?

Focus Question 4
We'd like to explore the phrase "productive struggle" a bit more. Toward the end of episode 4, it was clear that the focal group (and, we think, some other groups) were running out of steam. They hadn't generated new ideas for some time. Does that indicate that their struggle was not productive? Please explain your thoughts, taking into account the events that unfolded in episode 5.

Our Commentary

1. A significant number of mathematical ideas emerge and get explored in episode 5. How do those ideas compare to the mathematical ideas discussed in Section 2.1?

Much of the mathematical content described in Section 2.1 comes together in this final episode. Let's consider the various pieces.

The iterative (year-by-year) solution. We can't know if all of the groups had successfully calculated the value of the car after 1, 2, 3, or more years, but the table of values on the board was accepted without question. It's probably safe to say that the students understood how to compute the values of the car year after year, and accepted the values written on the board. It's also worth noting that the contextual nature of the problem had helped (early on, the students had rejected the results of their calculations because they were too small – they "didn't make sense.")

Determining the base of the exponential equation. The focal group knew soon after beginning to work on the problem that they wanted to find the base B in the equation $y = 5000B^x$, but neither they nor their fellow students recalled how to do so from previous work with exponential growth. In episode 4, they ran out of steam. Here in episode 5, Ms. Sierra led a somewhat directive discussion of the annual ratio of the year's ending value to its beginning value. Although there was a slight hiccup when she wrote 5000/4250 on the board rather than 4250/5000, the scattered "oh"s throughout the classroom when students calculated the ratios of the car's value over successive years indicated that the students were making the relevant connection – that the ratio of annual car values was .85. To put it another way, .85 was the constant multiplier in the table, meaning that the equation was exponential and of the form $y = 5000(.85)^x$.

Connecting the "15% off" to the common ratio/multiplicative factor of .85. It did take some work, but with consistent prompting by Ms. Sierra, it seems the students did make the connection that "15% off" can also be seen as "85% of." This observation "connected the dots" mathematically, allowing the students to see that (a) the year-by-year calculations produced a table whose values can be used to compute the common ratio/multiplicative factor and (b) thinking of how much was left after the subtraction, 85%, also gives the multiplicative factor.

In sum, all the pieces for making the connections described in Section 2.1 are present and the choruses of "oh"s at times suggest that at least some of the students made the connections and were excited by them. How much sense any particular student made of those connections at the moment is impossible to say, of course; many students found the problem challenging and some partial understandings were aired in the discussions. But the ideas were publicly tied together by the teacher.

It's also worth thinking about sense-making in this episode and in the lesson as a whole. In this episode the resolution seems primarily mathematical – we don't see recourse to the context, as in previous episodes – but, tying things together mathematically and justifying the same result two different ways is an important aspect of mathematical sense-making. Looking back at episodes 1–4, we see that the implausibility of results produced by plugging into the formula $5000(.15)^n$ played an important role in rejecting that formula; and in episode 5, some students rejected the idea of the multiplier being 1.17 because the value of the car should be decreasing over time. In addition, part (d) of the problem, which asks if it's worth replacing the clutch after 10 years, will once again demand meaningful attention to the context. Thus, in the conversations overall, we see students encouraged both to make use of the context and to make mathematical connections.

2. **What stands out to you in what the students appeared to find easy or difficult in episode 5? What issues might this raise about student understanding and how to address it?**

When the authors worked on the problem, our first thought was that "15% off" meant "85% of." This gave us the multiplicative factor of .85 and allowed us to write the exponential equation. That's what seemed natural to us. Of course, we could have calculated a string of year-by-year values – but we didn't see, right off, what good that would do us in looking for the multiplicative factor.

This points out the importance of instructional history. Because the class had been working with tables when they studied exponential growth, working with tables seemed natural for the students. Given that they had this recent experience, we were a bit surprised that determining the ratios in the tables didn't occur to more students. We were also surprised that the class (which is pretty vocal) didn't have an answer to Ms. Sierra's question, "What does 0.85 have to do with 15 percent?" One would think that students in an Algebra II class would have this information at their fingertips.

It seemed Ms. Sierra was a bit surprised at this too. In addition, although we can't know for sure, it's a reasonable guess that when she set the groups to work on the tables at the end of episode 3, she expected them to find the common ratios in the tables; yet, after observing the challenges they faced during episode 4, she felt the need to call the class together to work on the question.

What do we take from this? Here are a few things. First, what students know – or, at least, are able to bring to mind – is a function of their instructional history. Such history includes not only what they're *supposed to* have as part of their background, but what they've done recently. (This is an important point for all of us when we watch videos of teaching. They provide glimpses of classroom practice without context, so our understanding of teaching and learning possibilities is limited.) Second, context can make a difference. On another day, Ms. Sierra might have asked the students "If I have a 15% discount, how much is left?" and gotten an immediate response of "85%"; "And what can I multiply by to get 85%?" ".85?" and so on.

We'll never know why the students didn't make the connection that day. Were they tired? Were they fixated on subtraction because that's how they'd built the table of values? It's hard to say. But, for whatever reason, they didn't have the desired knowledge at their fingertips. And that's the point. What matters isn't what the students should know, but what they actually know in the moment. The only way to find out is through formative assessment – providing opportunities for the students to reveal what they know, and then adjusting instruction in response.

3. **We'd like to reflect on the difference in the level of scaffolding the teacher provides in episodes 1 and 5. In episode 1 she provided the focal group almost no scaffolding, telling them that their equation [$y = 5000(.15)^x$] "is gonna need a little tweaking" before walking away from them. In Episode 5 she asked leading questions to support the class in making connections. What reasons might the teacher have had for acting so differently in those two situations?**

There are always issues of balance when we think about providing support for students as they struggle. On the one hand, students don't learn to persevere or to dig into problems unless they have opportunities to engage deeply, and (at least some of the time) to see the fruits of their labors. On the other hand, long periods of frustration can be alienating and have a negative impact on people's sense of agency and their mathematical identities. The way we act in any situation depends on the context – what do the students seem to know, how long have they been working, how fatigued or frustrated do they seem to be, what do we know about their past experiences with mathematics?

In episode 1, the students were just getting started on the car value problem. They had made some tangible progress in modeling the car's loss of value over year 1 and were just beginning to play with equations. There was time and space for them to explore, and they were still "fresh." We don't want to delve into the specific support the teacher gave them – one never knows how things will turn out when we make decisions, and second-guessing any teacher's decisions isn't fair – but we will note that a relatively low level of scaffolding seems appropriate at this point. The students haven't been at it for that long, and who knows what they'll come up with? The teacher has determined what they seem to know and how they're progressing at this point. That's the first part of formative assessment. Adjusting the demands on the students is the second half of formative assessment. The teacher's decision is to tell them what they need to work on ("your equation is gonna need a little tweaking") but not to provide guidance as to what to do. That leaves the level of cognitive demand on the students pretty high. Will it turn out to be too high? Possibly, but it's early in the lesson, the students are fresh, and they've been making progress. If the teacher keeps tabs on how things are going, she can adjust later.

The situation is very different at the beginning of episode 5. As we saw in episode 4, the nature of the focal group's work is different from what it was early in the lesson. Not only had they run out of new ideas to try – a situation that can always change – but they'd run out of energy, and were halfheartedly rehashing things they'd tried before. As a result, spending more time without additional support might be frustrating.

From Ms. Sierra's comments at the beginning of episode 5, it appears that the focal group's experience was mirrored by other groups: Ms. Sierra said that a lot of groups were stuck. Given that, more direct support may well have been called for.

When to intervene and how much support to provide is always a judgment call. The issue is, do students seem positioned to profit from further engagement, and if so, what kind of engagement? This depends on what they seem to know, how much energy they seem to have, and our personal judgment of whether the level of demand at this particular point seems to be reasonable.

The phrase "do students seem positioned to profit from further engagement" raises a further question, which we pose as focus question 4.

4. **We'd like to explore the phrase "productive struggle" a bit more. Toward the end of episode 4, it was clear that the focal group (and, we think, some other groups) were running out of steam. They hadn't generated new ideas for some time. Does that indicate that their struggle was not productive? Please explain your thoughts, taking into account the events that unfolded in episode 5.**

"Productive struggle" isn't an easy thing to get our heads around. There are times when it's obvious that the efforts students have put into grappling with a problem have paid off – the times when you can see the progress they've made. For the contextual problem explored in this case study, there is clear evidence of progress when (for example) the students build the year-by-year model or put together the equation characterizing the decrease in car value. In such cases, it's easy to say that the struggle that produced such results was productive. But what can we say when they don't make such obvious progress?

Some of the productive outcomes from engaging deeply with a problem are not obvious or easily measurable. They may involve deeper understanding, seeing more connections, or being in a position to learn more – a parallel to making the ground more fertile for further learning. In episodes 4 and 5 we heard members of the focal group say "wow" when Ms. Sierra and others explained or summarized how key mathematical ideas fit together. Our sense of those exchanges was that the students wouldn't have found the new information that meaningful if they hadn't themselves struggled to see the connections. If that's right, then there was a payoff to their struggles. They may not have produced the "right" final solution, but their struggle may well have prepared them to understand and appreciate that solution. In that sense, some of the students' struggle may well have been productive, even though it may not have seemed so at the end of episode 4.

2.4 Reflecting on "The Car Value Problem"

Now that we've worked through this case in detail, we'd like to step back and look at the big picture. Overall, we'll take the student perspective as reflected in Figure 2.2.

We'll start with the mathematics. Then, because the focus of this case study is on issues related to cognitive demand, we'll turn to Dimension 2. Since formative assessment is critical for setting the level of cognitive demand, we'll look at Dimension 5 in conjunction with Dimension 2. We'll wrap things up with a discussion of Dimensions 3 and 4, Equitable Access and AOI. They too are closely linked.

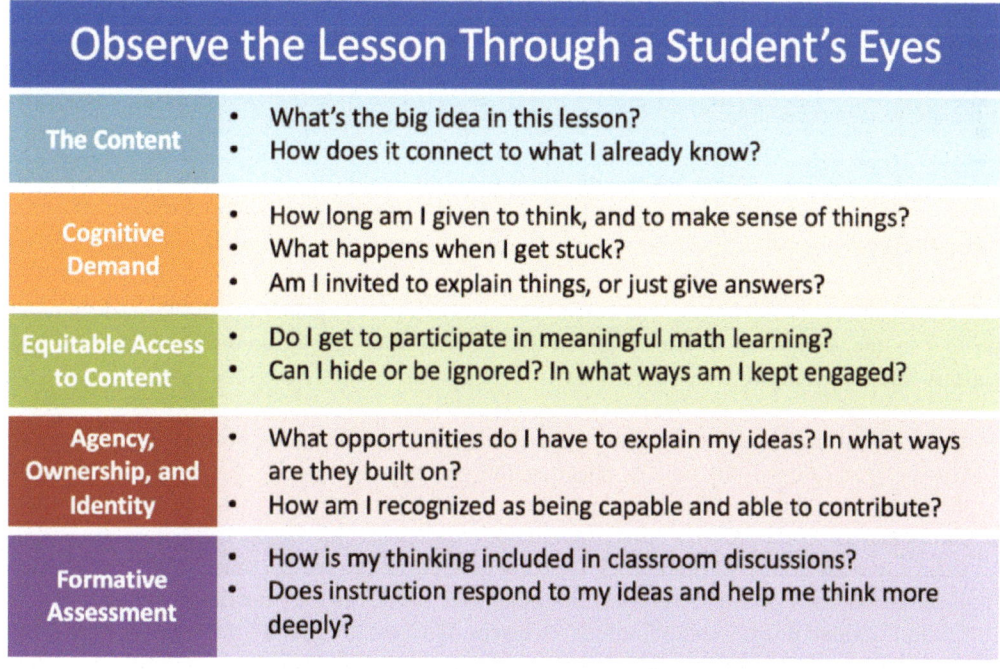

Figure 2.2 What's important, from the student perspective

Dimension 1, the Mathematics

As we saw in Section 2.1, the car value problem provides opportunities for students to engage with some big mathematical ideas:

-   Contextual problems provide opportunities to do meaningful mathematical sense-making.
-   Exponential growth and decay are effectively the "same thing."
-   Two different approaches to deriving the equation for an exponential function (from a table of values, and from the growth factor) are mathematically linked.

There is, of course, a lot of structure supporting these ideas, for instance, that:

-   A person's understanding of "real world" contexts can help them to think about how the mathematics fits together, and whether possible answers are plausible.
-   The key link between exponential growth and decay is that the underlying equation, $y = AB^x$, models both situations; B > 1 in the case of exponential growth and B < 1 in the case of exponential decay.
-   Decay (or growth) can be modeled iteratively.
-   Each year's depreciation (or growth) can be calculated in two steps, by subtracting (or adding) the given % of that year's initial value.
-   If you have constructed a table of values, then finding the common ratio.
     [(year $n+1$ value)/(year n value)] from the table provides the multiplicative factor for the exponential equation. This allows you to derive the exponential equation.
-   Each year's depreciation can be calculated in 1 step, multiplying the initial value by the percentage that *remains* after depreciation. That remaining percentage is the multiplicative factor in the exponential equation.

All these ideas are connected. The opportunity to make those connections, along with the fact that they provide different ways to get started on the problem, is what makes the problem (at least potentially) "groupworthy." The question from the student perspective is how much of all of the above was accessible.

| **The Mathematics** | • What are the big ideas in this lesson? |
| | • How do they connect to what I already know? |

What we did see in this lesson was a significant amount of engagement with the mathematical specifics in the second set of bullets given above, with very strong connections to what the students knew. Specifically,

-   The students made use of the contextual character of the problem, both in developing the year-by-year calculations and in rejecting the values produced by the calculator because they "didn't make sense." More than once, the students worked to make sure that their computations fit the situation described in the problem setup, "losing 15% of its value each year." Over the course of their work the focal group came to realize that the value of the car at the end of each year had to be less than the value of the car the previous year and that as a result, the value it lost would diminish each year. Without the real-world situation to anchor their thinking, it's quite possible that they'd have subtracted $750 each year. It's clear that student sense-making was facilitated by the fact that they could check their intuitions about the problem scenario against the numbers and equations they produced.

- They did, ultimately, see the equation as $y = 5000(.85)^x$, which had the same form as exponential growth equations, where the growth ratio had been larger than 1. Throughout the lesson, the students worked hard to make the connection to the exponential form of the equation.
- They did, with the teacher's guidance, find the common ratio in the table, and use it to construct the exponential equation. Their "oh"s indicated that they were making connections to their previous understandings.
- They did, with the teacher's guidance, connect the iterative procedure (subtracting a fixed % of the value each year) with the exponential equation, using the % of the value that remained each year.

In sum, it appeared that many students made many if not most of the connections listed in the second set of bullets above, connecting to and building on prior knowledge. (We say "appeared" because we can't know for sure.)

How much the students understood of the big ideas in the first set of bullets is hard to say – building understandings of big ideas takes time. What we can say is that the students made good use of the contextual nature of the problem and that the teacher emphasized contextual sense-making; that the students had tried to use the equation for exponential growth in attempting to solve this problem, and that the solution they arrived at, with the growth factor less than 1, elicited "oh"s when they did solve the problem; and that they had experienced the connection between the iterative and exponential forms of the solution.

Dimensions 2 and 5, Cognitive Demand and Formative Assessment

Some reminders:

> The key phrase with regard to *cognitive demand* is "productive struggle." There's not much to be learned if students aren't challenged. At the same time, if the challenge is too great, then learning may be out of reach! Here's a weightlifting analogy. If you give someone a 1-pound weight to lift (even with multiple repetitions), they're unlikely to get appreciably stronger. But the same is the case if you give them 500-pound weights! They'd struggle without results. What matters is using weights that are within people's current capacity – and adjusting when they get stronger.
>
> Figuring out students' current capacity and making adjustments when it seems appropriate is what *formative assessment* is all about. As students work on tasks, individually and collectively, the teacher can get a sense of what they know. Then, what happens is a judgment call. If it appears that the students aren't challenged at all, does the teacher have a more challenging task up their sleeve? If the students seem (at least momentarily) to be stuck, what might be a reasonable choice? Sometimes "keep working, I'll be back" may be enough – the students may simply need more time to sort things out, or to build on each other's thinking. Sometimes, a reminder of key information or a question of the type "have you thought about X?" is enough to orient the students and support their progress. Sometimes a more direct suggestion may be helpful. The goal, however, is to avoid being *too* helpful. If a suggestion removes the challenge, then it removes the potential for productive struggle and consequently, mathematical sensemaking.

With these issues in mind, let's take a pass through the whole case. Cognitive demand begins with the problem setup, and what is made salient to the students. In episode 1, Ms. Sierra worked to make sure that the students saw the problem as a real-world problem. She talked about car values decreasing, then set the students to work on the task. The fact that the focal group first arrived at a value of 37 cents after 5 years, and didn't know how to amend their approach, makes it clear that the task wasn't too easy!

When Ms. Sierra joined the focal group and they said their answer didn't make sense, she figured out that they had computed $5000(.15)^n$. This is formative assessment in action! To be more precise, this is the first part of formative assessment, creating an environment in which the students feel comfortable voicing their understandings and working to make sense of them. The second part of formative assessment consists of making instructional decisions based on that sense-making. In this case, Ms. Sierra confirmed that the loss in the first year was $750. Then, when Ayra said that the amount the car was worth after 1 year was $4250, she left the group with what they found to be a somewhat mysterious comment: "That is correct. Your equation is gonna need a little tweaking. You're on the right track."

With this decision, Ms. Sierra solidified the ground on which the students stood and clarified what they needed to work on. Her comment left the students with a fair amount to struggle with (and they did – remember Caleb's statement, "What the hell? How did we get the right answer, but the wrong equation?"). Her move was intended to keep the students' struggle productive; time would tell if it would turn out to be. We note that this is precisely what formative assessment is about: the teacher makes a judgment about how much support will enable the students to engage in productive struggle, while at the same time not giving away too much. All such decisions are judgment calls. It's not for us to say whether it's a good or bad decision. We can, however, look at how the decision played out on this day with this group of students.

In episode 2, the focal group made some progress, conceptualizing the iterative solution to the problem. In the meantime, Ms. Sierra observed that some groups of students arrived at differing values for the value of the car after 1 year – some obtaining $4250, some $4347. Her intervention at this point is more deliberate: while providing encouragement for the sense-making involved in the second answer, she also (by means of calling on students and having them explain their thinking) makes it clear that the first answer is the correct one and that the students should not pursue the line of thinking that yielded $4347.

This more overt action, in comparison with the previous example, demonstrates the contingent nature of formative assessment. Again, without judging the decisions but focusing on what may have been the thinking that produced them: At the end of episode 1, Ms. Sierra appeared to decide that the students in the focal group might make progress with the small amount of scaffolding she had given them. The case was different in episode 2. The value $4347 is plausible. Because of that, the kind of sense-making that alerted the focal group to the fact that something was wrong with their initial calculations wouldn't get triggered here.

Moreover, the method that produced $4347 (dividing by 1.15) can be iterated. Thus, the students might continue dividing by 1.15 without there being any signals from the context that what they were doing was wrong. That possibility, we imagine, was the reason for Ms. Sierra's intervention. Her intention was to maintain a productive level of cognitive demand, as the students worked in directions that were either productive or from which they could recover naturally. It's healthy to learn from one's mistakes if those mistakes don't cost too much. For example, in episode 1 the focal group quickly saw that their exponential equation was wrong – the car couldn't possibly be worth 37 cents after 5 years! In this case, however, the students' calculations wouldn't reveal that they were heading in the wrong direction.

In episode 3, the focal group implemented their iterative solution. Then they hit a wall when they tried to find an equation to generate the values they had computed. Ms. Sierra asked if they remembered a method they had used when they had worked on exponential growth, but they didn't. At that point, she left them with a vague hint, "Ok. It goes down by 15 percent. But is there anything that is being constantly added, subtracted, multiplied, divided within this table? That could definitely help you make an equation." As it turns out the hint wasn't sufficient, and (outside the teacher's view) the students were unable to make progress. But, it reflects an attempt to give the students room to find the solution. Had they found the solution on their own, it might have contributed to their sense of agency and ownership of the result.

Other groups also failed to make progress in episode 4, and in episode 5 Ms. Sierra called the class together for a discussion. There she led them interactively to consider the ratio $f(n+1)/f(n)$, finding the value of .85 as the annual multiplier. She then moved to pull things together, leading the students to the observation that when you subtract 15% of the value of an object, what's left is 85% of the object.

Here again, we focus on how formative assessment informs cognitive demand. From her statement "Wait, wait, wait. Just stick with me. I see a lot of side conversations start to come up. I know this is more whole class than we usually do. But stick with me as a whole class for a second" it's clear that Ms. Sierra saw the students as tired and somewhat lost; her judgment was that continued explorations were unlikely to be productive and she acted accordingly. Unlike the previous cases of formative assessment, where Ms. Sierra sent small groups back to work, she engaged here with the whole class. In addition, the level of scaffolding has just jumped. Depending on your perspective, this action might be seen as "too much," or, perhaps it might be seen as "finally doing what the students needed earlier." Once again, those are judgment calls, which depend on both the students and the teacher. (And the teacher is there, and we're not; she knows her students, and we don't.)

Let us turn to the issue of cognitive demand in episode 5. The incorrect answers called out by students make it clear that they were significantly challenged at this point – and they may have been for some time. Does that mean that an earlier intervention might have been appropriate? It's not for us to say. We will note that the periodic "Oh"s of recognition from students as they put things together indicate their readiness for the revelations. Their efforts, even if unsuccessful in terms of reaching a solution on their own, lay the groundwork in ways that made the connections meaningful when they did make them.

With this recap, let's turn to the student view.

Cognitive Demand	• How long am I given to think, and to make sense of things? • What happens when I get stuck? • Am I invited to explain things, or just give answers?

Formative Assessment	• How is my thinking included in classroom discussions? • Does instruction respond to my ideas and help me think more deeply?

We can think about these issues with regard to the focal group and with regard to the whole class.

To start with cognitive demand, it's clear that the students were stretched – we saw the focal group working hard each time the camera focused on them. Although we can't know what happened in the other groups, Ms. Sierra's summary comment in episode 5, that most groups were stuck, suggests that the focal group's experience was at least somewhat typical. It certainly appears that Ms. Sierra didn't rush the class through the problem. The students were given time to struggle.

Different things happened when the students got stuck. Early on, Ms. Sierra gave the focal group vague hints; the idea may have been to nudge them a bit in the right directions but not to provide much guidance since it was early in the process. Later, her hints to the focal group were more pointed. And in whole-class discussions, Ms. Sierra made somewhat pointed comments to rule out possible dead ends (episode 3) and led interactive discussions aimed at pulling together the threads of a solution (episode 5). That is, what happened when students got stuck varied quite a bit, depending on Ms. Sierra's sense of how stuck they were and how much energy they had.

Finally, with regard to cognitive demand, this lesson was replete with student explanations. Students expected explanations from each other in the focal group, and in whole-class discussions, the teacher asked students to explain their thinking. This was critical, given the fact that the students did find the problem challenging. At the same time, there is the question of which students get to explain. This is a fundamental question related to issues of AOI; see the discussion in the next section.

Let's turn to formative assessment. In conversations with the focal group, Ms. Sierra mostly listened and then tried to nudge the students in profitable directions, building on what they'd done. This is important: the goal of formative instruction is not to channel students into predetermined directions but to support their productive thinking.

In whole-class discussions, Ms. Sierra started by putting examples of student thinking on the board and then working through them. Here too, the discussions were grounded in student thinking, which served as the base for classroom conversations. Only at the end, when the students seemed to be running out of steam, did she take a more direct lead.

To reflect:

> Maintaining a workable level of cognitive demand is a balancing act. Different students (and groups of students) have different levels of knowledge, agency, and tolerance for struggling with a lack of progress. At times a slight nudge may seem to be all that is needed to keep the students working productively; at times, more significant interventions may seem warranted. It goes without saying that making such decisions is a real challenge. We make them on the fly, with whatever information we can gather; and, of course, different students or different groups of students may want or need different levels of support and scaffolding. Too little scaffolding may leave the students unsupported; too much may deprive them of significant opportunities to put things together for themselves. The most we can hope for is to be "close" a reasonable percent of the time.

The way we get close is through Formative Assessment. FA begins when student thinking is made public. Prior to class, teachers may assign tasks that can help to reveal some things that students are likely to find challenging. (See the Formative Assessment Lessons at https://www.map.mathshell.org/lessons.php for examples.) Those, plus our experience, give us some idea of what to expect. When the class is working in small groups, teachers can circulate through the class and observe. On the one hand, this may lead to scaffolding individual groups in case of difficulty or offering more challenging tasks if a group finds the current work too easy. On the other hand, it may well open up discussions in interesting mathematical ways. Students often come up with ideas that, if pursued with an open mind, can enrich the mathematical space. (We've seen students come up with new approaches to problems that we've been using for more than a decade!)

Whole-class discussions occur for many reasons, e.g., to kick off the consideration of a new topic, to bring the class together after a prescribed amount of small-group work, or (as happened in episode 2, when Ms. Sierra worked through the reason that $4347 was not the value of the car after 1 year) to alert students to possibly problematic issues that have emerged during small-group discussions. In whole-class discussions, having the students voice their own ideas provides opportunities for us, and other students, to hear and work with student thinking. It also does so much more, of course – that's what Dimensions 3 (equitable access) and 4 (AOI), discussed below, are all about.

Finally, it's good to remember that students are powerful resources for each other and that student-to-student conversations are a major component of formative assessment. In this "car value" case study – and in the following two case studies as well – significant learning occurs when students listen to each other's ideas, probe each other's understandings, and build on each other's thinking.

The teacher's responsibility with regard to formative assessment is not to address the challenges each student faces individually. It's to set up the learning environment so that those challenges are addressed – and all of the students in the class are partners in helping each other learn.

Dimensions 3 and 4, Equitable Access and Agency, Ownership, and Identity

Our focus in this case study was on Dimensions 1, 2, and 5, so our comments on dimensions 3 and 4 will be relatively brief.

The heart of dimension 3, equitable access, is that it focuses on the degree to which every student has opportunities to engage in meaningful ways with the core content of the lesson. This is impossible to monitor fully in any classroom, and drawing inferences about any one student is especially challenging in just one lesson: a particular student may be seemingly engaged on some days and not others, for example. Some patterns raise red flags, e.g., when a teacher calls on the same few students every day. But, even given the challenges of "sampling" participation in one classroom for one day, there are things we can look for. Does the environment seem safe, in that many students appear comfortable venturing ideas? Which students have opportunities to contribute, to get their ideas on the table? How are their contributions received, and whether those contributions are correct or not? Who explains, who evaluates, and what is the spirit of the evaluations? How are students positioned (e.g., as competent or not) by fellow students and by the teacher?

Issues of access matter for Dimension 4, AOI. Students who are left out of the conversation or who are positioned by others as having little to contribute have little opportunity to develop a sense of agency (the willingness to engage and contribute), ownership over the content (it's someone else's, not theirs), or positive mathematical identities. In contrast, those who do venture ideas and see them validated, built on, or amended in productive ways can come to see themselves as participants in the mathematical enterprise.

A quick scan of this lesson suggests that (to the degree one can tell in any such sampling) many students feel comfortable venturing ideas in this classroom, whether or not those ideas ultimately turn out to be correct. We hear numerous students call out ideas in the whole-class discussions and in the focal group we see students dig in without hesitation, bouncing ideas off each other. Indeed, when discussing the incorrect value for 1 year's depreciation, $5000 \times (1.15)^{-1} = \$4347$, Ms. Sierra stresses its reasonableness:

> This one is not correct – *but* I like where this group is headed, because what they were trying to do is make sure the value is going down over time. They are doing something with 15 percent. They saw that if you do your regular 15 percent raise, it would be going up, which we couldn't have. So, they are trying to make it go down over time.

In whole-class activities, Ms. Sierra solicits contributions broadly; at times she chooses students' names randomly from a collection of index cards that have students' names on them. Using randomizing devices such as index cards or "equity sticks" is, if done with appropriate sensitivity, one way to democratize student participation. In this lesson, it indicates a deliberate attempt to distribute opportunities across the classroom.

Before turning to the student perspective, let's reflect on the way this lesson provided opportunities for AOI. Looking at depreciating car values is a common textbook activity. Typically students experience them through "demonstrate and practice" instruction: the teacher models an example of the form $y = a(1-r)^x$ and students work through similar examples. When this happens students don't have much of an opportunity to develop and take ownership of the general function form. The problem context and the way this teacher gave the students room to explore it pushed for sense-making to be at the heart of the day's activities. That is, the activities of relating the formal mathematics to the context, and checking that everything they did fit together, was *authentic*.

As we discussed in Section 2.1, connections to real-world problems often seem arbitrary. Here, the connections were used in meaningful ways. The students had to come up with the formula on their own from the table. Moreover, the chorus of "oh"s when the students realized that 15% *off* meant 85% *of*, meant that the resulting formula was meaningful, not simply memorized. The way the problem was presented, and the ways in which students engaged in sense-making, opened up opportunities for the development of AOI.

Now let's turn to the student perspective.

| Equitable Access to Content | • Do I get to participate in meaningful math learning? |
| | • Can I hide or be ignored? In what ways am I kept engaged? |

| Agency, Ownership, and Identity | • What opportunities do I have to explain my ideas? In what ways are they built on? |
| | • How am I recognized as being capable and able to contribute? |

Participation varies in the focal group. Devin and Blanca are quite vocal; it seems at times that they carry the action. What does that mean in terms of equitable access? There's only so much we can say, especially when one or more participants are quiet. A close look at the focal group's interactions indicates signs of engagement from all four of the students for the duration of the task. They all attend to the conversations, making eye contact with each during discussions and, even when not saying anything, respond to others' comments by doing calculations or writing on their notes. Ayra is actively invested in the group's work: we saw her lean over to Blanca at the end of episode 2 to say "you were right, you were right" and then ask Ms. Sierra, "So how would you write that as an equation?" There's less to say about Caleb, the quietest member of the group. He seemed to be listening to his group's conversation, using his calculator and leaning over periodically to check the results he was getting with those obtained by his group mates. From their body language (which, of course, you can't see) he was welcomed by the others – not as someone "copying answers" but as someone tracking the conversation and contributing. (In episode 1, for example, Devin had turned his calculator to show Caleb an implausible result: "Bruh, it's worth one cent after 8 years." Caleb smiled in acknowledgment.) So, Caleb didn't hide and he wasn't ignored.

Before proceeding, we should note that *equal* participation is probably impossible and it may not be ideal. At the same time, the degree and kind of participation does have implications for AOI, to which we now turn.

Overall, many of the students we were able to observe appeared to have meaningful opportunities for the development of AOI. This class was about sense-making, and all such efforts were greeted positively, even when they didn't pan out. This is supportive of the development of agency and ownership. For some, ownership of results was clear and strong. Within the focal group, Devin was consistently engaged vocally in developing the mathematics. Toward the end of episode 2, both Ayra and Blanca contributed to the whole-class discussions. And, ownership was shared and attributed: toward the end of episode 2, when Ms. Sierra was discussing how to extend the computation of the car's year 1 value to year 2, Ayra looked over at Blanca and said softly, "You were right, you were right." Again, we are less certain about what Caleb took from and contributed to the focal group's exchanges. As discussed above, he seemed to be meaningfully engaged in the focal group's conversations. Yet, how much this day's conversations supported Caleb's growth of AOI is open to question. We should note, however, this is just one lesson on one particular day. Student participation varies depending on a number of factors including a student's mental and physical health, the assigned task, and group composition. The goal is to provide a variety of recurring opportunities for participation over time.

The development of agency, ownership over content, and positive mathematical identity is a long-term process. Each of these changes slowly as the result of ongoing experiences. What that means is that nobody can make claims about the growth and change of AOI in this one lesson. But, we can note the general climate of opportunity and some experiences that could contribute to positive change. Over time, such opportunities could make a difference.

Concluding Comments

In these final comments we step back to look at the big picture. This is our opportunity to build on the case study to think about our own teaching.

Our goal in discussing any case is not to evaluate, but to "problematize." We hope to have identified some of the interesting issues in this classroom, using the TRU Framework. But we want to leave you, and ourselves (we continue to think about these issues!) with the following (very big!) questions. They're things to carry with you – the kinds of things we'd like to be part of our thinking, all of the time.

Looking at the whole task posed by Ms. Sierra (parts a through d) at the beginning of the case, are there extensions you might think of? Are there connections to related mathematics we might explore, in our own classrooms?

A lot of what happens regarding cognitive demand depends on what the students know. What do your students know about the tasks at hand? As we've seen in this case, what students are *supposed* to know may not be what they can bring to mind on any particular day. How do you find out what the students do know? How might their approaches to any particular task be different from one another's? Where might they run into difficulty and how might you adjust, or be ready for what they say and do?

Can you think about things you might do, in whole-class discussions and when observing small groups and facilitating their interactions, that might shift participation patterns in productive ways?

Similarly, how can you work with students to construct norms that encourage student contributions and support the growth of student AOI?

Finally, besides listening carefully, can you think about actions you might take to make student thinking more public? These are discussion items for you, and us, for as long as we are working to support students' meaningful engagement with rich mathematics.

Notes

1 All names in the case studies are pseudonyms.
2 Note the connections to the mathematics itself (Dimension 1) and how we observe and respond to student thinking (formative assessment, Dimension 5). A focus on cognitive demand necessarily entails attention to Dimensions 2 and 5.

Chapter 3

Where is the Ten?

A Focus on Equitable Access and Agency, Ownership, and Identity

3.0 Introduction

This case study takes place in an introductory Algebra I lesson. It focuses on three students, Alicia, Bernardo, and Carla, as they work on a problem that deals with the algebraic representation of a geometric figure. The teacher, Ms. Davis, has posed a challenging question, which requires a deep understanding of the representation to address. When we join the group, Alicia has been given the responsibility of explaining the group's answer to the teacher. The group works collectively to support Alicia's understanding, but when the teacher joins them she is expected to explain their work without assistance from the other students.

This is a Complex Instruction (CI) classroom, which features challenging problems and clearly defined norms aimed at supporting ambitious and equitable instruction.[1] There is a lot of give-and-take in the discussions between the students, and a lot of frustration. In such conversations, students' mathematical identities can be at risk. Being told the answer to a problem may position a student as not being able to figure things out, with negative consequences for their identity; but struggling unsuccessfully can also be damaging to a student's identity. The student-to-student exchanges, which include some digs the students take at each other, and the exchanges with the teacher, raise fascinating questions about how to support all of the students in the group to engage with the core content in meaningful ways (Dimension 3, equitable access) and on the possible impacts of the discussions, both with and without the teacher present, on the identity development of all three students (Dimension 4). For that reason, the primary focus of this case study is on equitable access and AOI.

A quick summary reminder:

> *Equitable Access to Rich Mathematics* focuses on the opportunities that each and every student has to engage with mathematically central content and practices – the opportunities students have to become knowledgeable, flexible, and resourceful thinkers and problem solvers. This includes equitably distributed opportunities to essential practices such as reasoning and explaining, as well as engaging with mathematically rich ideas.
>
> Students can develop a sense of *agency* with mathematics when learning environments provide opportunities for them to jump in, make important decisions, and talk about math –

DOI: 10.4324/9781003375197-5

"In this classroom, I get to do mathematics. When I think, talk, or act, math happens." Students who have mathematical agency take initiative on their own.

Students can *own* mathematical ideas when learning environments support them to make their own sense of mathematics and share their thinking with others – "I figured this out. It makes sense. It's not simply what 'they' told me is true. I told them why it's true."

Students build positive mathematical *identities* when they participate in mathematical communities that see them and help them see themselves as mathematically powerful – "I am a member of this mathematical community. I am figuring out mathematical things and making sense of them with other people."

Agency, ownership, and identity develop over time, mostly through interactions between teachers, students, and content, as the product of students' mathematical opportunities and classroom experiences. Yes, interactions outside the classroom matter as well. However, it's mostly classroom experiences that shape people's mathematics identities, because when someone says they're good or bad at math, they almost always mean "school math."

We begin this lesson discussion, as we begin all three lesson analyses in this book, with a detailed look at the mathematics involved and its potential for classroom discussion. The math is rich and the students do get deeply involved – and challenged. They work very hard to make sense of, and explain, the problem at hand. Because of the challenge, the level of cognitive demand is important. What's *too* hard? When might a teacher consider intervening, in what ways, and to what effect? Figuring out what the students are understanding and considering options are fundamentally issues of formative assessment.

It's important to remember that all five dimensions of TRU are not simply present; they interact. The Mathematics (Dimension 1) needs to be worth engaging with. If the math that students experience is consistently rote or mechanical, students are not going to get invested in it or excited about having done it. If there's no challenge (Dimension 2), students won't experience a sense of growth or ownership; but if the task is simply beyond reach, students will be frustrated and disheartened. Equitable access (Dimension 3) means that *every* student has the opportunity to engage with the central mathematical ideas not only in ways that contribute to their understanding, but also in ways that provide opportunities for the development of AOI (Dimension 4). And as always, formative assessment (Dimension 5) is the mechanism that makes things work. When teachers attend to what students are thinking and make adjustments when needed, they can provide every student with opportunities to take agency, own mathematical ideas, and build a more positive mathematical identity. Thus, although equity-related issues are the focus of this case study, we'll also highlight how the other dimensions play significant roles.

You can read this chapter and think about our questions on your own. Better, you can read and discuss the questions with one or more colleagues. As you move through the chapter, we encourage you to address the questions yourself before you read our commentary. Do the task, reflect on what happened, and collect your own conclusions and questions. Our commentary is part of a conversation we're having with you and that we hope you are having with others.

3.1 The Mathematical Task

This chapter is named "Where is the Ten?" because that question becomes the central focus of the students' mathematical work as they work on a task dealing with the algebraic representation of a geometric figure. At the start of the lesson the students were asked to find the perimeter of a figure composed of algebra tiles. (We describe these in the next section.) The perimeter turned out to be $10x + 10$. To probe the students' understanding, the teacher followed up with a question about part of their answer: "Where is the ten?" This is not an easy question to answer.

Before we see how students took up this question and how the teacher supported them, we want to explore how this question had the potential to support students to engage with important mathematics. We will also explore how the task and follow-up question had the potential to support students' agency, ownership, and identity. We refer to the *potential* of the task and follow-up question because as any teacher knows, plans and tasks are only part of what makes a math classroom work. What students and teachers do with plans and tasks is equally as important to what students learn. Nonetheless, coming to a full understanding of the task and the teacher's "on paper" plans for it will help us make sense of what took place as the lesson unfolded. As we explore the task, we'll pay particular attention to features that support both students' engagement with important mathematical ideas and their developing senses of agency, ownership, and identity.

This case study takes place in an Algebra I class. The teacher, Ms. Davis, uses a type of pedagogy called Complex Instruction (CI). The goal of CI is the creation of classrooms in which students make sense of rich mathematical ideas together, with norms that support all students to see themselves as smart. (One idea behind CI is that there are multiple ways to be smart beyond getting correct answers rapidly.)

A major component of CI involves the use of "group-worthy" tasks. They are open-ended, mathematically rich, collaborative tasks that provide multiple ways for students to show to themselves, their classmates, and their teacher that they can do important mathematics (Cohen & Lotan 2014, Cohen et al. 1999). Students work on the problems in small groups.

Working with group-worthy tasks involves a number of factors. First, there's the task itself – a problem or set of problems chosen to be approachable in different ways, allowing for multiple solution paths. (This kind of openness supports rich mathematical conversations and explanations, with students comparing and contrasting ideas.) Then, there are follow-up questions that the teacher can use to push and probe student thinking. Sometimes these follow-up questions come in the lesson plan provided by the curriculum. Other times, teachers plan questions of their own. Finally, CI specifies roles and structures for how students and teachers interact. We will talk more about these interaction structures later in this chapter, once we dive into the case.

When supportive classroom norms have been established, group-worthy tasks have the potential for supporting every student's engagement with important mathematics, in ways that can help them to develop a sense of mathematical agency, ownership, and identity. Rich tasks provide opportunities for students to address complex problems, make connections, and explain their thinking. Thus they create opportunities for students to see themselves and be seen as mathematically smart (contributing ideas, reflecting, challenging, organizing, summarizing, explaining), often in ways that challenge traditional classroom status hierarchies. In the discussion that follows we'll explore how the initial task and follow-up question, along with the classroom norms, provided such opportunities for the three students in the focal group.

The students in this class were exploring different ways of representing the perimeter of an unusual shape that was composed of algebra tiles. Algebra tiles are flat plastic tiles that play a role similar to the role that base-10 blocks play in arithmetic. Just as base-10 blocks help students represent arithmetical expressions visually, algebra tiles help students represent algebraic expressions visually. They give students a hands-on way to make sense of an abstract concept, that of unknown quantities.

The students were using a set of algebra tiles that included multiple copies of the two tiles illustrated in Figure 3.1. The first tile is a square, one unit on each side. It has a perimeter of four units and an area of one square unit. The second tile is a rectangle. Two of its sides are one unit long, and two of its sides are x units long. It has a perimeter of $2x + 2$ units and an area of x square units.

Where is the Ten?

During this lesson, students were asked to calculate the areas or perimeters of various figures built from algebra tiles. They were also asked to build figures out of algebra tiles that had particular areas and perimeters. Each problem was written on a task card. Each group of students had one task card and a bin of algebra tiles. The students we will observe had recently finished solving one of these problems together. We will see them working on answering the teacher's follow-up question about the problem. First, though, let's work on the problem itself.

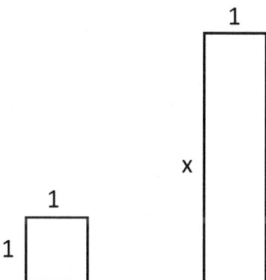

Figure 3.1 1-by-1 and 1-by-*x* algebra tiles

The Initial Task

Here is the task that the students in this case study had just completed.
Now it's your turn.

Figure 3.2 What is the perimeter of this object? Explain

Please work through this task – possibly by yourself and, if possible, in discussion with your colleagues. Build (if you have algebra tiles) or sketch the figure first, label the sides, and show work that supports your final answer.

For those who are new to algebra (as the students in this lesson were), the mathematics in this task is complex. There are numerous ways to think about and calculate the perimeter. Try to experience this task from the perspective of a student.

 *After you have completed the task, answer these questions:*

- How many ways can you confirm and justify the perimeter you found? What are they?
- What connections can you make between the figure made from algebra tiles and the algebraic expression you found for the figure's perimeter?
- With what mathematical concepts and/or practices are you engaging?
- What potential mathematical challenges does this task present?
- If you're working with colleagues, how are your ways of thinking about this problem similar or different? What opportunities do you have to build on each other's ideas?

The perimeter of this figure is $10x + 10$. If you did not get this answer, take a few minutes to go back over your work. A full solution to this problem involves much more than simply writing the expression. A full solution includes explaining how you got this answer and making connections between the concrete algebra tile representation and the abstract algebra expression. In what follows we'll show three solution methods.

Solutions

In how many different ways did you solve for the perimeter of the shape in the initial task? This task has a single final answer – the perimeter of the figure – but there are numerous ways to explain and justify it. Would you be surprised to learn that we've seen at least a dozen different ways to find this answer? Let's look at a few possible solution methods.

All of the solutions hinge on noticing a key feature of the figure. No matter how you wind up finding the perimeter, the key point to observe is that inside the figure, the length of the uncovered part of each partly covered rectangle is $(x - 1)$. See Figure 3.3.

Method 1

One way we've seen students find the perimeter is to add the lengths of the exposed sides of the object individually as they make their way around the object, one side at a time. Then, students combine like terms to arrive at the simplified perimeter. If you follow this method, starting on the left with the full x and proceeding clockwise, you get:

$$x + 1 + (x - 1) + 3 + (x - 1) + 1 + (x - 1) + 3 + (x - 1) + 1 + x + 2 + (x - 1) + 1 + (x - 1).$$

When you combine like terms, you end up with

$$10x + 10.$$

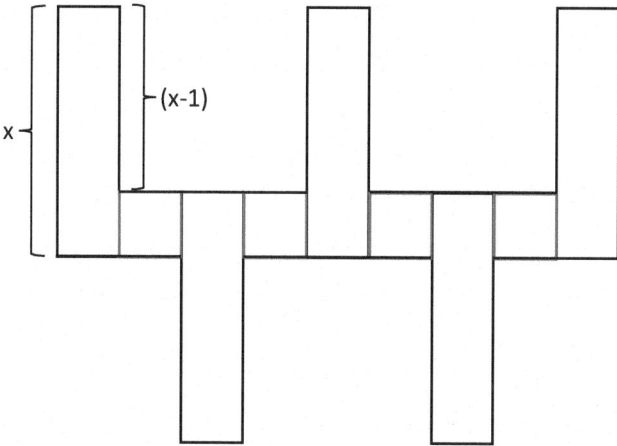

Figure 3.3 The "inside" parts of the rectangles

Method 2

A second way that we've seen students solve this problem is to count the "long," "medium," and "short" sides separately and then add their lengths together. The "long" sides have length x. The "medium" sides have length $(x - 1)$. The "short" sides have length 1.

If you tally up the sides in this way, you'll find:

Two long sides of length $x : x + x = 2x$

Eight medium sides of length $(x - 1) : 8(x - 1) = 8x - 8$

Eighteen short sides of length 1:18

Add the sorted lengths together to find the total perimeter,

$2x + 8x - 8 + 18,$

and combine like terms to get the final answer,

$10x + 10.$

Method 3

A third way that we've seen students solve this problem is to, in essence, take the figure apart. Students find the perimeters of each tile individually, add those perimeters together, and then subtract the lengths of the sides that overlap when the figure is put back together. This way is tricky! You have to remember to subtract each overlapping length twice. Two tiles meet in each overlap, so your taken-apart-perimeter has overcounted each overlapping length two times. But, this way of finding the perimeter often makes sense to students who have algebra tiles in front of them, tiles that are easily slid around and reconfigured.

When you take the figure apart, you find that each of the five rectangular tiles has a perimeter of $2x + 2$, and each of the four square tiles has a perimeter of 4. Adding the perimeters of the tiles together, you get:

$5(2x + 2) + 16$
$= 10x + 26$

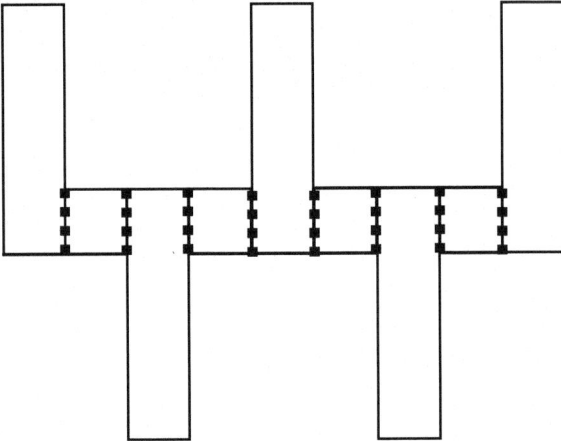

Figure 3.4 The overlaps are each 1 unit long

But don't forget about the overlaps! The tiles overlap in eight places. Each overlap is one unit long. See Figure 3.4.

You have to subtract the overlapping lengths twice, since every overlap results in losing a length of 1 from *both* overlapping tiles. So, students arrive at the final answer by subtracting 16 from their previous expression:

$$10x + 26 - 16$$
$$= 10x + 10.$$

There are other ways to arrive at $10x + 10$. (In fact, we used this problem with a class of pre-service teachers as we were writing this book. One of the students in the class found a solution we'd never seen before!) We have shared just a few solutions. If your method was different from these, take a moment to reflect on how it compared. We do not know for sure how the students in the lesson we're discussing found their solution because we joined their small group after they had arrived at their answer. However, we do know that they found the perimeter of the figure to be $10x + 10$, and that they justified their solution algebraically.

Showing and explaining any one of these solutions to the task demonstrates a great deal of mathematical understanding. But, this teacher wanted to push her students' mathematical thinking further. She did this by asking a follow-up question. Let's see where that leads us.

The Follow-up Question

The follow-up question, "Show me where the ten is," was not part of the original task. It was part of the teacher's lesson plan, intended to probe student thinking. The teacher asked the follow-up question after students shared the correct answer for the perimeter of the figure. She pointed to the figure and asked, "Where's the ten?"

The algebraic expression for the perimeter, $10x + 10$ has two parts – the "x's part" and the "1's part." If you interpret the expression for the perimeter of the figure geometrically, by comparing it to the figure itself, it says that there are ten lengths of x and ten lengths of 1 in the perimeter. But are there ten lengths of x and ten lengths of 1 *in the figure*? Look back at the figure and count the number of lengths in the perimeter that are one unit long. There are *eighteen* 1s, not ten. Can you explain, pointing to specific parts of the figure, why there are only ten units of 1 in the final expression for the perimeter of the figure? In the shorthand used by the teacher and students, "Can you show me where the 10 is?"

Compare the algebraic expression you found for the perimeter of Figure 3.5 to the figure itself. The expression for the perimeter, 10x + 10 has a 10 in it. Can you show where the 10 is in the figure?

Figure 3.5 Where is the 10?

Work on this follow-up question by yourself or with colleagues. Explaining why you can see eighteen unit lengths in the picture, but why there are only ten in the final computation of the perimeter can be challenging. As you'll see, it's a real challenge for the students in this class.

Solutions

Could you explain how to find the 10 in Figure 3.5? It's not easy. Just as with finding the perimeter, there are numerous ways to show the 10. We will share a few.

Students often find it easier to "show where the 10x is" than to "show where the 10 is." As a result, we've seen many students start by finding the 10x and then adapt their approach to help them answer the more difficult question, finding the 10. Sometimes students start with the 10x accidentally! But the work they do finding the 10x still helps them find the 10.

Here is one example. Although there are two complete x 's in the figure, eight of the other interior pieces are partly covered and only contribute (x–1) to the perimeter. What if we were to "extend" their contribution by 1, by borrowing one of the unit lengths to make up the difference? That would make eight more pieces of length x, giving us ten x's. In the process we've "used up" eight of the eighteen 1's, leaving ten. So the re-formed perimeter consists of two x's, eight two-part pieces whose length is x, and ten pieces of length 1.

Making this argument verbally, without referring closely to the figure, may not be clear or convincing. So, let's take a closer look at how some students have used this approach.

Sometimes students notice that they can make an L-shape that has a length of x by adding an inside (x – 1) side with the adjacent side length of 1, so (x – 1) + 1 = x, as in Figure 3.6.

The students then realize that they can make L-shaped pieces for *every* inside length of (x – 1), making a figure that looks like Figure 3.7. In that figure, you can "see" the 10 x's and the 10 remaining 1's.

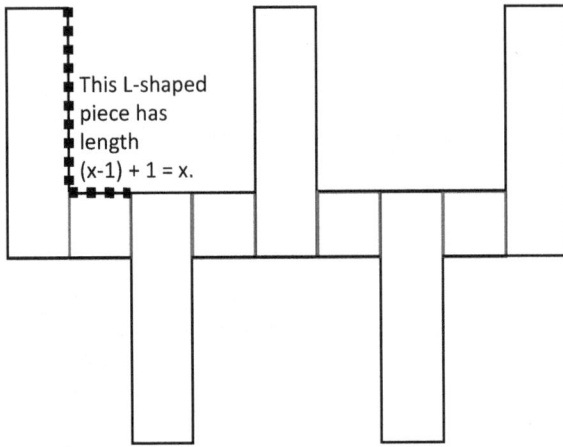

Figure 3.6 The L-shaped piece has length *x*

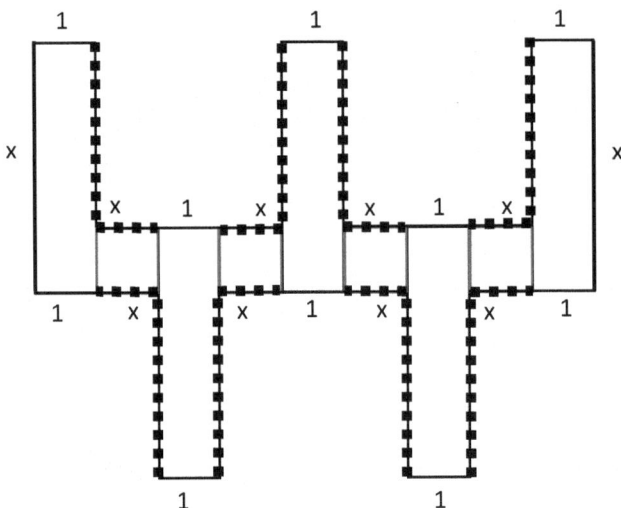

Figure 3.7 With 8 L-shaped pieces there are 10 pieces of length X and 10 pieces of length 1

Other students have used a different strategy to find and show the 10. They start by drawing the outline of the figure, as shown in Figure 3.8a.

They then "move" the segment of length 1 circled in Figure 3.8b to a new position in Figure 3.8c.

A similar sequence of moves, this time repositioning the segment circled in Figure 3.8d to the new position indicated in Figure 3.8e, leaves a 1-by-*x* rectangle isolated by itself:

With six more repositionings, the students arrive at the figure shown in Figure 3.8f.

When you label the lengths of the five 1-by-*x* rectangles in Figure 3.8f, you get Figure 3.8g – in which both the ten 1's and the ten x's are clearly apparent.

We know we've devoted a lot of time to this. Here's why: (1) part of "thinking mathematically" involves making mathematical connections; and, (2) our students have come up with all of these, so (3) it's good to be prepared for the approaches that your students might come up with!

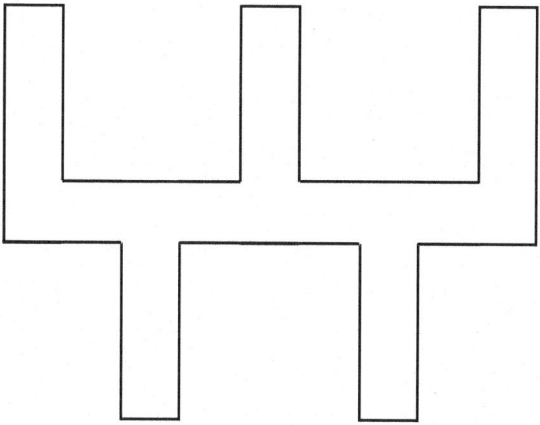

Figure 3.8a The outline of the figure

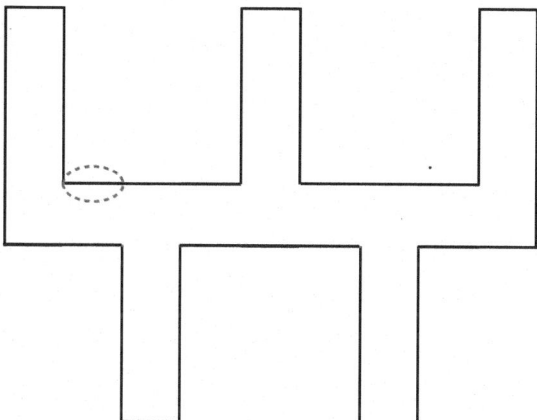

Figure 3.8b A piece of unit length about to be repositioned

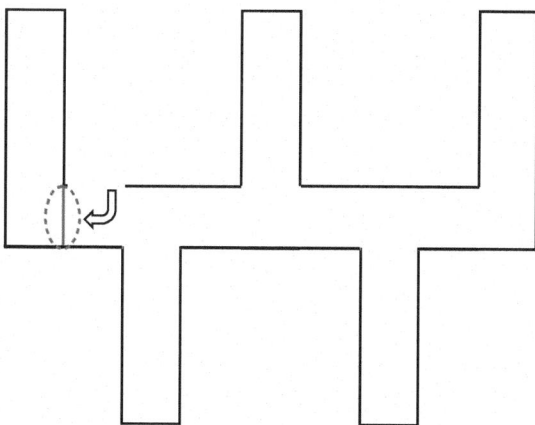

Figure 3.8c The outline after the piece of unit length has been repositioned

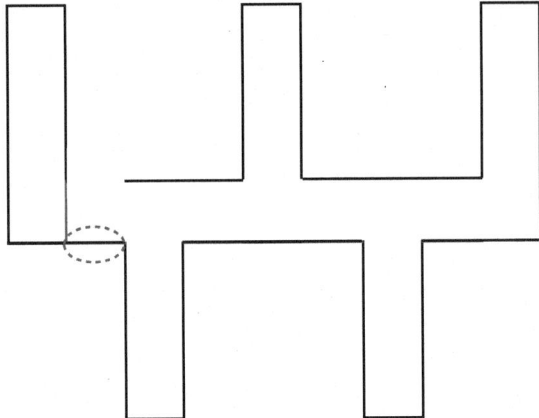

Figure 3.8d A second piece of unit length about to be repositioned

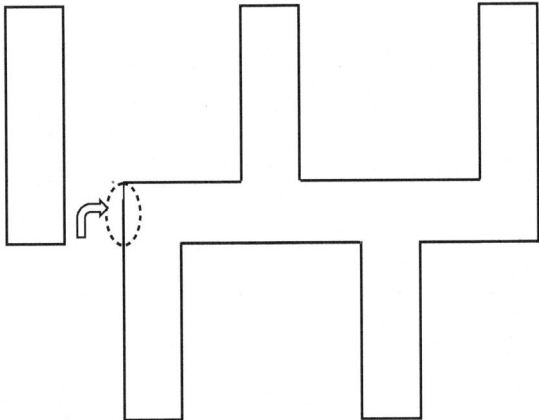

Figure 3.8e The outline after the second piece has been repositioned. A 1-by-x rectangle has been "set free" on the left

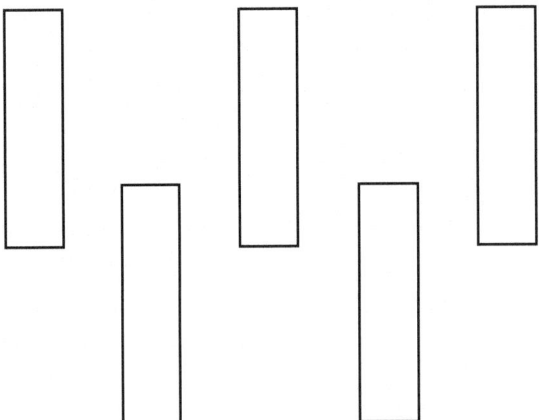

Figure 3.8f The figure that remains after all of the repositionings

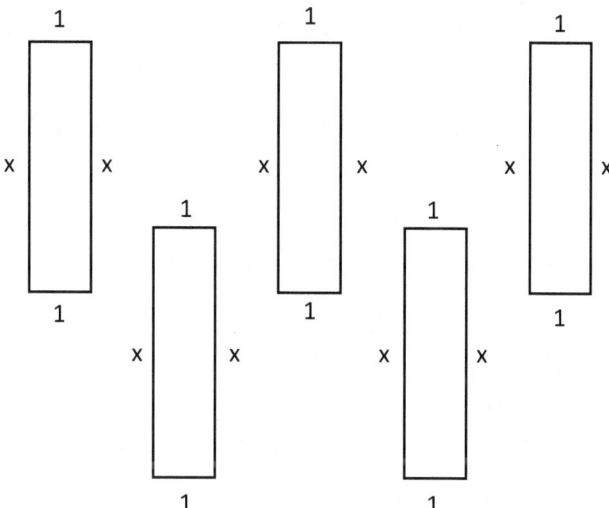

Figure 3.8g You can see the 10 1's and 10 x's in the perimeter!

 Now that you've worked on the on-paper task and follow-up question, we'd like to explore how the elements of this lesson have the potential to support students. Before you read on, take some time to think and, if possible, talk with colleagues about the following questions.

1. What mathematical understandings might students develop from engaging in this lesson? How are those understandings important?
2. How can challenging lessons like this one support rich mathematical learning environments that help build students' agency, ownership, and identity with math?

Focusing on these two questions will help you reflect on how the task and follow-up question have the potential to support students' engagement with important mathematics and development of agency, ownership, and identity, our focus for this chapter.

Our Commentary

Once you've answered the reflection questions yourself or with colleagues, we invite you to read our commentary. As always, our comments are not meant to represent final answers to these questions. Rather, we offer them as a way to engage with you in conversation.

1. **What mathematical understandings might students develop from engaging in this lesson? How are those understandings important?**

We think that the original on-paper task and the teacher's follow-up question have the potential to help beginning algebra students:

- Make connections between algebraic and geometric representations,
- Learn how to combine like terms and do algebraic manipulation by applying new algebraic knowledge in a geometric context, and
- Explain their thinking about algebra, both as they're engaged in sense-making and in summary, to their classmates and peers.

The goals for students' learning to work with algebraic symbols are complex. Part of the power of algebraic language involves its ability to represent "real world" phenomena in meaningful ways – all of "mathematical modeling" is based on the power of symbolic manipulation. So, there are many times we want symbols to be seen as representing objects in meaningful ways. At the same time, part of the power of algebraic symbolism is that we can combine terms, solve equations, and perform other symbolic manipulations, without paying attention to the meanings of the terms when manipulating the symbols. This is a tricky balance! If we focus too much on the procedural aspects of algebra without connecting symbolic operations to the objects being represented, the result can be what's been called "meaningless symbol manipulation."

Algebra tiles can be useful in helping students learn fundamental algebra skills, such as combining like terms and simplifying expressions, while also reasoning about connections between algebraic expressions, perimeter, and area. The on-paper task helps students do both things by asking them to find the perimeter of a figure with whole-number and unknown-length sides. While the perimeter of the figure is concrete in the geometric model, the algebraic expression for perimeter that results from combining like terms is an abstraction, not easily identifiable with any of the model's quantifiable features. By resolving this apparent tension, students can experience how algebra represents concrete situations but also allows for flexible interpretation. The algebra tiles help highlight the differences between numbers and unknowns by representing them as different lengths. At the same time, the algebra tiles help highlight the similarities between numbers and unknowns by representing them as sides of a quadrilateral, sides that can be measured and whose lengths can be summed to compute a perimeter. When students group side-lengths with similar features and simplify the expression they find for the perimeter, they make sense of how to do arithmetic with unknowns, just like they already know how to do arithmetic with whole numbers. The tangible things they do with algebra tiles – physically counting, gesturing at, and moving them – support them in making sense of abstract ideas.

We think that the teacher's follow-up question, "Where's the ten?", pushes students to make deeper connections. That's because answering the question calls for *explaining* the connection between the expression they found for the perimeter, $10x + 10$, and the figure itself. The students may have explored this connection while finding the perimeter – but in our opinion, going from the geometric figure to the algebraic expression is more straightforward than the "Where's the ten?" task. To find the ten the students have to go from the simplified algebraic expression back to the much messier geometric figure. Where *did* that ten come from after all? It came from the figure, yes, but also from combining like terms. And while algebra treats all x-terms and all whole numbers as like terms, the x's and whole numbers in the simplified perimeter expression do not correspond to sides of the same length in the figure. Remember, the figure has sides of length x, 1, and $x - 1$. Combining like terms is an algebra thing, not a geometry thing. That's what makes the follow-up question so challenging and mathematically important. While answering the follow-up question, students explain their thinking about important algebra skills and connections between algebra and geometry. And they learn a lot as they explain.

We can imagine other possible follow-up questions the teacher might have asked that would have been less mathematically challenging. For instance, she could have asked them to explain their reasoning, which the students might have interpreted as a prompt to merely reiterate the algebra steps they followed. But the follow-up question "Where's the ten?" was carefully designed to encourage students to explain their thinking *and* to push their understanding of algebra to a deeper level. How challenging to make follow-up questions, and how much to scaffold, is always a judgment call. We want to stress that the teacher in this case study knows her students; we don't. So, we're in the position of watching her make a choice based on her best judgment and seeing how it plays out.

We think that students who work together to answer the follow-up question will develop a deeper understanding of fundamental algebra skills and connections between algebra and perimeter than students who just complete the on-paper task. What do you think?

2. **How can challenging lessons like this one support rich mathematical learning environments that help build every student's agency, ownership, and identity with math?**

The question is, what opportunities are there for interactions with the task that allow students to contribute meaningfully to their own and to collective mathematical sense-making? We think that this task and follow-up question can provide opportunities for *all* students to see themselves and each other as having productive and interesting mathematical ideas, making meaningful contributions to problem-solving, and being smart math students.

We consider this task to hold potential for supporting the development of students' agency, ownership, and identity – and more! – because:

- It provides students with opportunities to connect prior knowledge with new, essential, and interesting concepts. Doing some of this work may be a challenge – we'll see it as the case unfolds – but the level of challenge can be kept within reach. That's a matter for formative assessment (Dimension 5) and cognitive demand (Dimension 2).
- There are multiple ways to reach a solution. This contributes to the students' developing sense of what mathematics is, and what it means to do mathematics (Dimension 1) and provides multiple points of access (Dimension 3).
- Similarly with regard to access, there are various ways for students to contribute meaningfully to the solving process.

Returning to issues of AOI, it's helpful to put yourself in the shoes of a beginning algebra student. Unknown quantities are scary new ideas. Coming to grips with the meaning of an unknown quantity and learning how to talk about and do arithmetic with unknowns and whole numbers is a big step for beginners. The cognitive demand is steep. Taking big steps can be scary for students. Knowing this about students requires thinking about how they might experience a task through the lens of cognitive demand (Dimension 2). What challenges will students face? How can we help them face those challenges without diminishing their learning power? We want to support them to take these challenging steps with agency, to take ownership over the new ideas, and to build the mathematical confidence essential in positive mathematical identities.

The algebra tiles task is designed to help students take that big step with agency and ownership by combining a familiar problem – finding perimeter – with a new concept, dealing with unknown quantities. Finding perimeter is a mathematical concept over which we *hope* beginning algebra students already have ownership. (Of course, it requires some formative assessment to know this for sure.) Even familiar problems can hold unexpected challenges for students, so it's good to continue learning about what they know. Pairing a familiar concept, perimeter, with a new and challenging concept, working with unknowns, has the potential to support students to take ownership over the new concept. The familiar question about perimeter supports students in building on prior knowledge. To support students in making these connections, the task must have the appropriate level of cognitive demand. If the task is too challenging, students may be unable to access it entirely, but with too much support, students will not have the opportunity to make those connections on their own. This is an example of how attending to cognitive demand (Dimension 2) is needed to create rich opportunities for AOI (Dimension 4).

This task and follow-up question also provide opportunities for building students' agency, ownership, and identity by inviting students to reach a solution in a variety of different ways. Remember how many different ways there were to find $10x + 10$? And how many ways were there to explain where to find 10 in the figure? That's important. If there's only one way to solve a problem, then students who fail to see that one way may be at a loss – and if they're simply told what to do, there isn't much opportunity for developing agency and initiative. In contrast, this task and follow-up have multiple entry points. When it's possible to get started in various ways, a problem is open to more students – giving them access to rich mathematical ideas and agency-building opportunities. Also, the mathematical enterprise has changed in an important way. The students' goal is not to find the one "right" way of solving the problem. They are searching for one or more ways that make sense to them, rather than focusing on the single "approved" method validated by the teacher or a textbook.

Again, we see important connections to other dimensions. If the mathematics involved in the task was less deep or more procedural, students might be more constrained in their choices of approach, denying them opportunities to exercise mathematical agency. Here, engaging with the task includes making sense of the figure, representing it algebraically, combining like terms, and providing explanations to a challenging question. All of this relates to the mathematics (Dimension 1). If the task was less cognitively demanding (Dimension 2), students' choices and sense-making would feel less mathematically important. And by providing multiple entry points, the task is made more accessible to students with different mathematical backgrounds and prior knowledge. The more opportunities students have to engage with a task in different ways, the greater access students have to the task (Dimension 3). And when students can make important decisions, they can come to see themselves and their classmates as mathematically powerful and smart. They can take ownership over their problem-solving methods and the explanations they develop (Dimension 4).

It's important to note that both dimensions 3 and 4, equitable access and opportunities for AOI, are never guaranteed by task construction. What kinds of opportunities each student has to engage will depend on how an activity is set up and what kinds of norms exist for classroom discussions. Will an idea that is partially correct be rejected because it's "wrong," or built on because it has potential? If only correct answers are approved, the environment doesn't feel safe for many students and they may ultimately shy away from the conversations, effectively being shut out of access. On the other hand, if their ideas get built on, they can see themselves being welcomed into the mathematical conversation. This has implications for access and AOI.

> [You'll notice we haven't discussed Dimension 5, formative assessment, in the previous paragraph. That's because formative assessment comes alive as the students *engage* with the task! Seeing what the students understand, how the task might be modified and how the students might be positioned as resources for each other may turn out to be crucial decisions in-the-moment.]

Throughout this section, we've been saying that the task and follow-up question *have the potential to...* We've used that phrase on purpose. Opportunities to make sense of important mathematics and develop agency, ownership, and identity don't live in the task or lesson plan; they live in the ways that the students interact with the mathematics, with each other, and with the teacher.

You have now experienced the task first-hand as a learner of mathematics. In the following case study we share an example of how this task came to life in one particular classroom, with one teacher and one group of students, on one particular day.

3.2 Some Background about the Classroom We're Visiting

It's important to know some things about this specific school and classroom before we dig into details. The mathematics department in this school has a very strong equity focus, and a strong sense of community. Department members work collectively on lesson and curriculum planning. Student teachers and new teachers will frequently shadow experienced teachers before they teach a particular lesson, and then teach that lesson the next day. Ms. Davis is in her second year of teaching, and she is well integrated into the department's teaching community.

This is a "sheltered algebra" classroom, in which English language and content instruction are integrated. The idea is to support students in developing fluency with English while learning important mathematical ideas. A lot of the student-to-student dialogue in class is multilingual. You can learn more about sheltered English classrooms in Freeman, Freeman, and Gonzales (1987), Katznelson and Bernstein (2017), and Krashen (1985).

Most of the students in this classroom are fluent Spanish speakers. Their English language proficiency varies. Our focal students, Alicia, Bernardo, and Carla, are all native Spanish speakers. Alicia and Carla appear to be comfortable speaking English, switching between English and Spanish while they discuss the problems. When they talk to Bernardo, they speak Spanish; he speaks Spanish exclusively when the three students are working together. The few words of English Bernardo speaks during this lesson segment are in direct response to a question asked by Ms. Davis. Ms. Davis does not speak Spanish, and all of her conversations with the group are conducted in English.

In reproducing what took place we transcribe the students' Spanish-language conversations and provide English translations following them, in parentheses. In our discussions, we'll consider the impact of the students' ability to converse in Spanish. Is there an upside, in terms of access? Is there a downside, in terms of less fluency with academic language? These are critical issues when thinking about Dimensions 3 and 4.

As we've noted, this is a CI classroom (Cohen & Lotan 1997, Cohen et al. 1999). CI aims to support students in making sense of important mathematics by engaging them in carefully structured collaborative work on group-worthy problems. CI classrooms build norms for how students interact with one another and co-construct knowledge. They address problematic and inequitable patterns of interaction that can arise in classrooms and make some students feel less smart or worthy of speaking, taking risks, or learning than other students. In what follows, we elaborate on some of the CI norms that we observed playing a significant role in this classroom.

One norm in this classroom is that students working in a group are responsible for each other's mathematical understanding. Students are supposed to make sure that everyone in their group makes sense of the mathematics they are addressing, and to work together until it happens. The class knows that this is expected of them, and the teacher helps them understand why. She uses a range of routines to maintain this norm. Specifically, we'll see her use the following routine: When students in a group say they have solved a problem, the teacher chooses one student at random to explain the group's work. Each time the students in that group call the teacher over to discuss their work on that problem, that student serves as the group's representative. If the student's explanation is clear and compelling, the teacher certifies the group's work. If there are difficulties with the explanation, the teacher will ask the students to work together to resolve them. It will be the representative's responsibility to explain how they resolved the issue when the teacher returns – but everyone knows that the representative speaks for the group, not just for themselves. By choosing a student in a visibly random way, the teacher makes sure that the students don't think that that particular student was chosen because the teacher felt she knew more or less than the other students. The random selection helps sustain the norm that all students are responsible for each other's understanding, and no one student is more worthy of sharing that understanding than the others.

3.2.1 Preview of the Case Study

Here's a quick summary of what happened during the 12 minutes of class time that are the focus of this case study. We've divided this case study into four episodes – four places at which there's a break or switch in what's taking place (typically when the teacher joins the group or leaves them to work by themselves). We also describe a brief episode ("epilogue") following the students' work on the problem, where we see the impact of the discussions.

In episode 1 Alicia, Bernardo, and Carla are working on the "Where is the ten?" follow-up question discussed in Section 1, to show where the "ten" in the perimeter $10x + 10$ can be seen. Alicia is the group representative, so it is her job to present the group's work to Ms. Davis when Ms. Davis joins the group. When Ms. Davis does, it becomes apparent that Alicia had been focusing on the "$10x$" part of the perimeter rather than the "10." Ms. Davis says "I didn't ask for ten x. I asked for ten." She reminds the group that they have to work together to help Alicia understand and explain where the 10 comes from. She leaves, saying she'll be back when they're ready.

In episode 2 the three students dig back into their work. Bernardo jumps into an explanation, which is only partly understood by Alicia and Carla. This starts a series of exchanges, in which all of the students (mostly in Spanish) work hard to clarify their understanding. There are several rounds of back and forth, after which each of the students appears to feel that they understand where the ten comes from. They call Ms. Davis over.

In episode 3, it becomes clear that Alicia has once again focused on the "$10x$" part of the figure. When Ms. Davis stops her explanation, saying that she wants Alicia to explain where the *ten* comes from, Bernardo and Carla try to suggest what Alicia should say. But, this violates the "classroom contract" – it's Alicia's job to explain, and Bernardo and Carla's job to make sure she can explain, before they call Ms. Davis over. Gently but firmly, Ms. Davis reminds the students of this and sends them back to work.

In episode 4, the three students dig back into the problem, with a clearer understanding of what they need to do. This time they develop a clear and correct explanation of where the "10" can be found in the perimeter. All three students understand it and Alicia explains it clearly when Ms. Davis joins the group. The students, somewhat relieved, begin work on the next problem.

In the Epilogue Ms. Davis joins the group as they set up a new problem. In their exchanges, Alicia displays a degree of confidence that was not apparent during the group's work on the "Where is the ten?" problem.

After presenting each of these episodes we'll ask a few focus questions to help you collect evidence about what happened from the perspective of the students. By the time we finish the case we'll have looked at numerous issues related to equitable access, agency, ownership, and identity, and how the other three dimensions (the math, cognitive demand, and formative assessment) have played a role in supporting them.

3.3 What Happened, in Detail (with Some Reflection Questions and Comments Along the Way)

3.3.1 Episode 1 (from 0:00 to 0:56 in the Lesson Segment)

Overview

The class has been working in small groups to find the perimeter of Figure 3.2. Alicia, Bernardo, and Carla call Ms. Davis over to validate the answer they have produced. Alicia is the group spokesperson, and it's her responsibility to answer the question "Where is the 10?" It turns out that her/the group's answer isn't correct; they have identified $10x$ in the figure rather than 10. When it becomes clear that the group hasn't figured out why there are ten 1's in the figure, Ms. Davis sets them back to work and says she'll be back when they think they've got it.

Narrative

It is about ten minutes into class. Students are seated in groups of three or four, working together on area and perimeter problems with algebra tiles. The room is full of lively mathematical conversation. Some students are writing, others are moving tiles or pointing to sketches they have drawn on their papers.

Ms. Davis circulates through the class, listening and watching as students work. Groups of students call her over to ask questions or to show their solutions to the problems. When she certifies a solution as correct, she stamps it and the students move on to the next problem.

Alicia, Bernardo, and Carla call Ms. Davis over to their group. She had visited the group earlier in the lesson. The students, with Alicia as spokesperson, had correctly found the perimeter to be *10x + 10*. They had been working on the much more challenging follow-up question, explaining "where the 10 is."

Pointing to Alicia, Ms. Davis asks, "So, if I ask her to show me where 10 is, she'd be able to show me?"

Alicia quickly responds. "Yes!"

Ms. Davis positions herself at the fourth side of their table. "OK," she says. "So show me where 10 is." See Figure 3.9.

Alicia confidently grabs her pencil and points to the labeled parts of Figure 3.10, counting the long and medium sides aloud:

It's 1, 2, 3, 4, 5, 6, 7, 8, 9, 10!

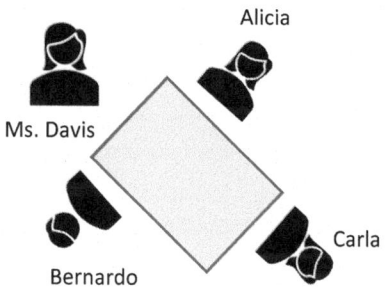

Figure 3.9 Alicia, Bernardo, Carla, and Ms. Davis at the table

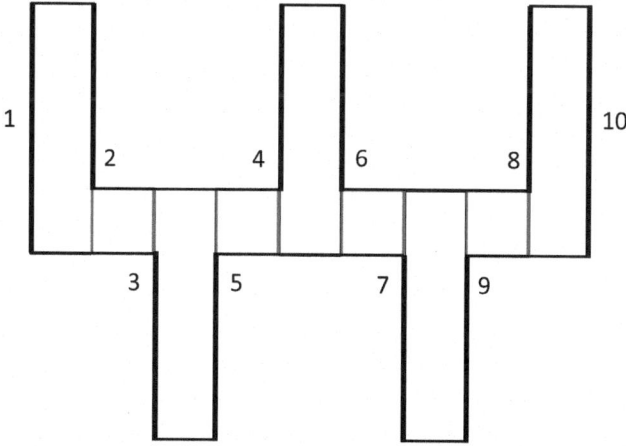

Figure 3.10 The sequence Alicia marks off as she counts

She stops. Then she frowns. She knows something isn't right.

Ms. Davis steps in. She points to the unit length at the top of the first rectangle. "But isn't this a one here?"

Alicia barely lets her finish her question. "Oh no, wait!" she interjects, shaking her head. "$10x$!" She starts to count again, this time using her pencil more intentionally to trace up and down the long and medium segments of the figure, the sides of length x and $(x - 1)$. "1, 2, 3, ..." See Figure 3.11.

But Ms. Davis interrupts her. "I didn't ask for ten x. I asked for ten."

Alicia whispers through the end of the count and puts her head in her hands.

As Alicia folds into nervous laughter, Carla speaks up. "Ten!" she demands, echoing Ms. Davis. "Where do you get 10?"

"Where is the ten?" Ms. Davis repeats.

Bernardo begins to offer an explanation. "No. You..."

But Ms. Davis stops him. "No no no" she says gently and smiles, looking from Bernardo to Alicia and back again. "Alicia... [to Bernardo and Carla] You told me that Alicia was ready. Right?"

Alicia smiles through her fingers.

Bernardo defends himself. "She said she was ready!"

Alicia reacts, pointing at Bernardo. "You...!"

Also responding to Ms. Davis's question about Alicia's being ready, Carla says. "She is not! She didn't even understand the question."

Alicia once again puts her head in her hands.

"So," Ms. Davis regroups. Smiling and looking down at the figure on the table, not at any student in particular, she asks again. "Show me where ten is."

Carla grabs Alicia's arm. "OK," she begins, switching to Spanish to talk to Alicia, "te acuerdas del dieciocho? (Ok, remember the 18?)"

"Aja (Yeah)."

"Te acuerdas que le quitabamos... (Remember that we take away...)"

"No, Carla." Ms. Davis stops her. She taps Carla's notebook to get her attention.

"No, she's explaining to me," Alicia clarifies. "She's telling me, 'You remember'..."

Ms. Davis nods. "OK then, I'll need to come back." She moves away from the table, leaving the three students the task of working things through together.

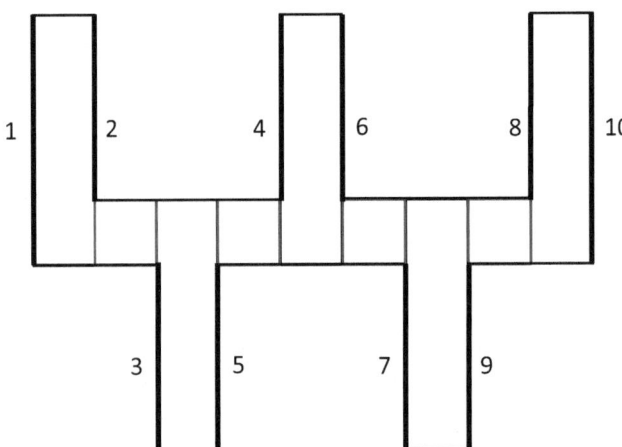

Figure 3.11 The sequence Alicia marks the second time she counts

Focus Questions for Episode 1

 Episode 1 is short (it lasts slightly less than a minute), serving for the most part to set up what follows. Our first set of discussion questions starts with reflecting on the ways in which the students make sense of the mathematics. It also focuses on how we see the norms we described in the background to this case study being established and developed. Please think about or discuss the following questions before you read our commentary. Where you can, point to evidence in the episode that helps to support what you're thinking.

Focus Question 1

What important mathematical ideas and practices are the students invited to engage with in this episode? How are they invited to do so? How do we see/hear the students engaging with important ideas and practices?

Focus Question 2

What norms regarding students' and teachers' roles in generating and working through mathematical ideas do you see in play and/or developing in this episode? In what ways do the norms allocate responsibilities to each of the participants?

Focus Question 3

How do the interactions in this episode appear to position each of the participants with regard to their mathematical proficiency? Although it's very early in the vignette and we want to be careful about drawing conclusions, are there points to note about the interactions that might play out with regard to AOI?

Our Commentary

1. What important mathematical ideas and practices are the students invited to engage with in this episode? How are they invited to do so? How do we see/hear the students engaging with important ideas and practices?

When we discussed the task and follow-up question in Section 3.1, we explored the mathematical ideas and practices that students had the potential to engage with as they worked on them. Students had the potential – there are never guarantees! – to build understandings of unknown quantities and how to manipulate them by making connections between algebraic expressions and the concrete geometric figure. The task and follow-up question invited students to dig deeply into the relationship between the terms "10" and "$10x$" in the algebraic expression $10x + 10$ and to figure out where they can be "seen" in the perimeter of the shape made from algebra tiles. We also found that the question "where is the 10"? may seem simple, but it's challenging!

Episode 1 shows that the question *is* challenging for these students. The cognitive demand (Dimension 2) is high. But is it so high that the students were unable to engage productively with the important mathematics? We don't think so. We see the students engaging with the mathematical ideas and practices identified in Section 3.1. In this episode, Ms. Davis invited Alicia to share her group's reasoning. Alicia correctly explained where the $10x$ is in the figure. She did this by counting and carefully pointing to the sides of length x and $(x - 1)$ in the figure, which do contribute to the $10x$ in the algebraic expression for the perimeter.

But... Ms. Davis was asking for the 10, not the $10x$. Ms. Davis listened to Alicia's explanation of the $10x$, but then pointed out the gap between her question and Alicia and her group-mates' answer. We know, from working on the problem ourselves, that finding the $10x$ can be a productive (even if accidental) first step toward finding the 10. In this episode, we sense that Alicia, Bernardo, and Carla recognized that they had counted the $10x$ instead of the 10. Ms. Davis seemed to sense that, too, and used what she had learned when she decided (using formative assessment, Dimension 5) to leave them so that they could continue to work on the problem by themselves.

How productive their engagement will be remains to be seen. If Alicia, Bernardo, and Carla can dig themselves out of the situation they find themselves in, this will have been a surmountable challenge. If they get stuck... then the question will be, what steps might be taken via formative assessment to adjust the level of cognitive demand? But for now, they've taken up the invitation to engage with some mathematics that is worth engaging with.

2. **What norms regarding students' and teachers' roles in generating and working through mathematical ideas do you see in play and/or developing in this episode? In what ways do the norms allocate responsibilities to each of the participants?**

In this episode we see Ms. Davis reinforcing the classroom norms discussed in the overview – norms that are meant in part to support students to collectively take ownership over mathematical ideas and develop individual and collective identities as capable learners and mathematical problem solvers. The students are accountable not only to the teacher, and to the mathematics, but to each other: it is their collective responsibility to ensure that all of them understand the mathematics and can explain it. Ms. Davis does not scaffold away the challenge in the task, at least at this point.

To make it easier to follow the discussion, we're going to name three norms here so that we can see them operate as this lesson plays out. There are more norms shaping what happens, of course; but these three are central. They are:

- **Norm 1:** When a group calls the teacher over to explain, this certifies that the group representative is ready to present the group's explanations on their own and without help. The teacher will stop attempts from other students to scaffold a group representative's efforts during the explanation.
- **Norm 2:** All group members are responsible both for understanding the mathematics themselves and for making sure the group representative understands it and is ready to explain it.
- **Norm 3:** The mathematical standards of the original task are high, and they are maintained. Students are expected to address the question as posed, and to provide clear, coherent, and detailed explanations.

With her questions and body language, Ms. Davis made space for Alicia to explain where the 10 is, offering her room for agency – and at the same time reinforcing the norm that it was Alicia's responsibility not only to demonstrate understanding, but to speak for the group.

When Ms. Davis joined the group, her question, "So, if I ask her [Alicia] to show me where 10 is, she'd be able to show me?" prompted the students to confirm that Alicia was ready to share the students' collective thinking (norm 1). Throughout the episode, Ms. Davis directed her questions and comments to Alicia. She deflected Bernardo and Carla's attempts to provide assistance to Alicia, as in this exchange:

> But Ms. Davis interrupts her. "I didn't ask for ten x. I asked for ten."
> Alicia whispers through the end of the count and puts her head in her hands.
> As Alicia folds into nervous laughter, Carla speaks up. "Ten!" she demands, echoing Ms. Davis. "Where do you get 10?"
> "Where is the ten?" Ms. Davis repeats.
> Bernardo begins to offer an explanation. "No. You…"
> But Ms. Davis stops him. "No no no" she says gently and smiles, looking from Bernardo to Alicia and back again. "Alicia… [to Bernardo and Carla] You told me that Alicia was ready. Right?"

This last comment makes it clear that the explanation Alicia provided is supposed to be the group's explanation, not Alicia's alone (norm 2). When Alicia did not give a correct explanation, all three students owned the incomplete but in-progress response she had shared.

Similarly, Bernardo and Carla's disavowals of responsibility for what Alicia said – Bernardo saying "She said she was ready!" and Carla saying. "She is not! She did not even understand the question" – were not aligned with the norm of collective responsibility. Ms. Davis addresses this twice, though she does this indirectly.

First, Ms. Davis prevents Bernardo and Carla from answering for Alicia. Second, when Carla starts explaining the mathematics to Alicia, Ms. Davis says "OK then, I'll need to come back." That is: by virtue of her having been chosen as group spokesperson this time, Alicia is responsible for giving the group's explanation to Ms. Davis. The others are not supposed to pitch in at this point; the group should have agreed on an explanation and made sure that Alicia was able to provide it. In this way, Ms. Davis is reinforcing the collaborative norms. By not allowing Bernardo or Carla to explain for Alicia or to scaffold her in the moment, Ms. Davis is holding Alicia and the group to a very high standard. Bernardo and Carla are supposed to work with Alicia, the group's designated spokesperson, until they all have a solid grasp of the mathematics and she can explain it by herself, on their behalf (norm 3).

3. **How do the interactions in this episode appear to position each of the participants with regard to their mathematical proficiency? Although it's very early in the episode and we want to be careful about drawing conclusions, are there points to note about the interactions that might play out with regard to AOI?**

It's early, so we don't want to jump to conclusions. And, it's hard to read between the lines of the students' comments. But, it's certainly worth noting some of the back and forth between the students. Consider the dialogue that follows where we left off, above:

> But Ms. Davis stops him. "No no no" she says gently and smiles, looking from Bernardo to Alicia and back again. "Alicia… [to Bernardo and Carla] You told me that Alicia was ready. Right?"
> Alicia smiles through her fingers.
> Bernardo defends himself. "She said she was ready!"
> Alicia reacts, pointing at Bernardo. "You…!"
> Also responding to Ms. Davis's question about Alicia's being ready, Carla says, "She is not! She didn't even understand the question."
> Alicia once again puts her head in her hands.

All this goes by very fast. But, Bernardo's comment "She said she was ready!" places the blame for Alicia's incorrect response squarely on Alicia's shoulders – and results in a somewhat indignant response from Alicia, "You…"

Carla's response, "She is not [ready]! She didn't even understand the question" clearly positions Alicia as the weak link in the group.

We note that this is a CI classroom in which issues of student positioning and status typically get significant attention. In the preceding exchanges, Bernardo and Carla have clearly positioned themselves as more knowledgeable than Alicia, and we have to wonder at this point about the impact of this exchange on Alicia's mathematical sense of self. It's reasonable to wonder whether one might be tempted, as a teacher, to say something at this point – either about the way Bernardo and Carla are treating Alicia or about whether to provide some scaffolding that would ease the burden on Alicia's shoulders. Surely Ms. Davis noticed the ways in which Alicia was positioned.

That's where things stand at present. Let's see what happens after Ms. Davis leaves the group.

3.3.2 Episode 2 (from 10:56 to 3:48 in the Lesson Segment)

Overview

The three students work collaboratively on their explanation, with a lot of back and forth until Alicia thinks she can provide an explanation of where the 10 (ten 1's) in the expression $10x + 10$ comes from. At the end of the episode, they call Ms. Davis over.

Narrative

As soon as Ms. Davis walks away, the three students lean in toward the table. Bernardo points to the tiles and begins to explain.

"'ira, aqui esta. Solo dile (Look, here it is. Just tell her …)"

"Shhh!" Alicia interrupts. "Que me lo explique Carla! (Let Carla explain it to me!)"

Carla begins her explanation. "Acuerdate del 10. Ok, 'ira. (Remember about the 10. Ok, look.) 1, 2, 3.…"

Carla trails off. She turns to Bernardo.

"De donde sacaste el 8? (Where did you get the 8?)"

Alicia nods along. "Son 18, aja, (There are 18, yeah,)" she adds, pointing out that there are 18 sides of length one exposed in the figure, not ten.

Bernardo has an answer ready. "Son 8 equis menos… (There are eight *x*-minus…)"

"Oh. Son 8 equis menos (There are eight *x* minus)," Carla repeats.

"Menos una (minus one)," Bernardo interjects, clarifying that there are eight lengths of "*x* – 1" in the figure.

"Menos una. (Minus one.)"

Corrected, Carla continues her explanation. Alicia stares at the page as she talks, a bemused smile on her face.

"So, a esta (these)" Carla points to the 1-by-1 tiles that overlap with the x-by-1 tiles, "se la quitaste a este man – (you took from this guy –)," she continues, pointing to the *x*-by-1 tiles. See Figure 3.12.

Alicia

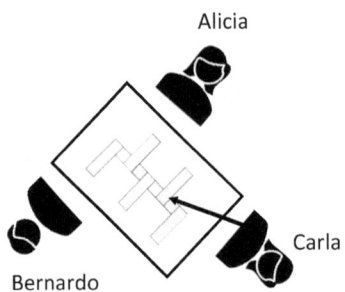

Carla

Bernardo

Figure 3.12 Carla pointing at the tiles

Bernardo interjects, finishing her thought. "Para ponerselas a este. (To put them on this.)"

Alicia pushes back from the table, exasperated. "No me hablen los dos a la misma vez! (Don't talk to me at the same time!)" She fixes her hair and puts her pencil down, giggling self-consciously.

"Mira no!" Bernardo gently persists. ("No, look!") "Son 8. Son 8 menos 18. (It's 8. It's 8 less than 18). Cuanto son? (How much is that?)" he quizzes Alicia.

"Diez (Ten)."

"Ay esta. Y porque 8 menos 18? (OK. And why 18 minus 8?) Porque (Because)…"

But Alicia has a question of her own.

"Y de donde agarraste 8? (And where did you get the 8 from?)" she demands, frustration in her voice.

"De estas (From here)." Bernardo gestures to the x-lengths in the figure that Carla had just pointed out and counts aloud. "1,2,3,4,5,6,7,8."

"No, man!" Carla exclaims. "Pero no cuentas estas de aquí! (But don't count these!)" She points to the outermost x-length on the left side of the figure. Alicia glares at Bernardo as Carla corrects him, shaking her head. "Que las de afuera son x! (Because the outside ones are x's!)" See Figure 3.13.

"Las de afuera no (I'm not counting the outside ones)," Bernardo agrees, pointing to the same outermost x-length. "Si pero yo no mas estoy contando las de adentro (I'm only counting the inside ones)," he adds, indicating by pointing to an inner $x - 1$ length.

Alicia points to the side of length 1 on an x-by-1 tile. "Y estas? (And these?)"

"Estas cuentan como uno (These count as one)," Carla explains.

Bernardo counts aloud, pointing to sides of length 1. But Alicia impatiently swats his hand away.

"Y estas? (And these?)" she urges, giggling, pointing to an exposed side of one of the 1-by-1 squares.

"Estas ya están contadas con el 18! (These were already counted with the 18!)" Bernardo exclaims. Alicia hides her face in her arms and laughs.

Carla talks over him, more patience in her tone. "Estas son 18 juntas. (These are 18 all together.)"

"Oh. Aha." Says Alicia. "Estas son 10 mas 5. (These are 10 plus 5.)"

Carla points to one of the sides of length 1 on an x-by-1 tile.

"Tienes que agarrar una de estas (You have to take one of these)" she explains, and then shifts to point to one of the sides of a 1-by-1 tile, "para sumarsela a esta (to add it on to one of these). So, that is where you get…" She trails off in English, nodding and glancing at Bernardo, her explanation of the 18 unit lengths complete.

Alicia ventures another question. "Y de donde agarraste menos… (And where did you get minus…)" She stops and shakes her head, struggling to remember how much is taken away from the 18 unit lengths they had just accounted for. "menos … 8?"

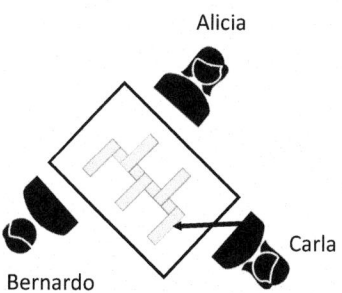

Figure 3.13 Carla pointing at the end tiles

Carla drops her head into her hands. "Ugh."

"Esta es x menos una (This is x minus 1)," Bernardo gently reminds Alicia. He points to the x-minus-1 lengths. "Son 8 x-menos-una. (There are 8 x-minus-1's.)" He begins to count the x-minus-1 lengths again, the second time in several minutes. See Figure 3.14.

Carla clicks her tongue at Bernardo as he talks. She has had enough. They are going in circles. She takes a deep breath.

"Ay, no," Carla sighs, "y aquien le fue a preguntar? (and who did she ask?)"

Alicia glaring at her, Carla launches her own explanation. This time, she starts from the very beginning. And she speaks in English.

"So," begins Carla. "10x. So where do you get 10x? So, right here." Carla leans in to point to the tiles.

She pauses. She cannot remember where to find the 10x in the diagram. She turns to Bernardo.

"Where do you get 10x?!" she demands.

Bernardo quickly answers. "Habia 2x mas 8x son 10. (There was 2x plus 8x, makes 10.)"

"Oh, yeah, yeah, yeah. It's true. It's true. So..."

"Ya?" Bernardo looks at Alicia to see if she's following.

"No." Alicia shakes her head and looks away.

"Are you sure?" Carla asks Bernardo. She is not.

"Si!" Bernardo exclaims. "Ay dos x! (There are two x's!)" He taps the two x-lengths on the far left and right of the diagram. "Una, dos x! Mas 8x. (One, two x's! Plus 8x.)" See Figure 3.15.

"Oh! It's true!" Carla remembers. "Yeah, yeah, yeah! Tell her," she says, pointing to Alicia. Then she yells for Ms. Davis. "Miss!"

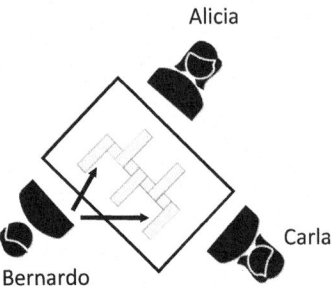

Figure 3.14 Bernardo counts the inside tiles

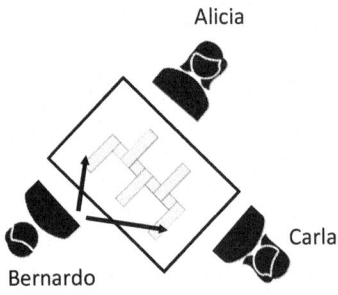

Figure 3.15 Bernardo points to the outside tiles

Alicia joins in. "Son $8x$ verdad? (There's $8x$, right?)" she asks Bernardo.

"Aqui son $8x$ y luego (Here's there's $8x$, then...)"

"$8x$. Aha."

Alicia writes "$8x$" as Carla again calls for the teacher. "Ms. Davis!"

Bernardo continues. "Mas $2x$, y ... ya. (Plus $2x$, and ... there.)"

Alicia writes "+ 2x" and nods.

"Ya..." Alicia stares at the page. Her gaze flips to Bernardo. "Eso es todo? (That's it?)" She does not seem to believe how simple it is.

"Ms. Davis!" Carla yells, louder.

Ms. Davis calls from across the room. "Yeah? Are you guys ready?"

"Yeah – uh..." Carla pulls Alicia's sleeve.

"Son igual a 10 (They equal 10)," Bernardo replies.

"Como sabes... (How do you know...)" Alicia trails off.

Suddenly, Alicia exclaims, "Oh!" A huge smile breaks across her face as she throws her hands into the air with joy. She points at Bernardo as if to say, "You were right all along!", nodding and bouncing in her chair.

Ms. Davis has reached their table. "Are you sure?" she asks, observing the celebration.

"Yeah!" They look relieved and excited.

But something catches Ms. Davis's eye across the room. "Oh," she says, turning to walk away from them. "I'll give you one more minute. I'll be right back."

Alicia seizes the extra minute to make sure she understands.

"Pero, (But,)" she begins, waiting until Bernardo and Carla are paying attention again. "En la piece, estos son los verdaderos 8? (In this figure, these are the real 8?)" She points to the eight interior x-lengths as she asks.

"Yes, sweetheart," Carla replies. "It is."

"Y, (And,)" Alicia points to the two x-lengths on the far left and right of the figure, "los dos x's son estos dos? (the two x's are these two?)"

"Uh-huh!" Carla cheers.

Bernardo offers Alicia some advice for when Ms. Davis returns.

"Si te pregunta porque son ocho x-menos-1, (If she asks you why there are eight x-minus-1's,)" he suggests, pointing to the places in the figure where the 1-by-1 tiles overlap with the x-by-1 tiles, "es porque hay uno, uno que le estorba. (it's because there's one, one that's in the way.)"

"Uh-huh." Alicia nods.

Carla squints at Bernardo and teases him. "Tu como sabes que 'le estorba'? (How do you know it's 'in the way'?) Pobrecito. A lo mejor está allí para acompañarlo. (Poor thing. Maybe the only reason it's there is to keep the ones company.)"

"Ay," Alicia giggles, rolling her eyes at Carla, who hides her face. "Ya a va salir con sus chistes. (Here you come with your jokes.)"

"No es para menos (No wonder)" Carla giggles back.

They lean back, relaxing into their laughter and starting to tidy up their workspace. Alicia stacks stray papers while Carla pulls two x-by-1 tiles from the bin of algebra tiles on their table, preparing to move on to the next question on the sheet.

Suddenly Carla asks again, "$2x$? Which ones are the $2x$?"

Alicia and Bernardo jump into action, pointing to the x lengths on the far right and left of the figure.

"Ok," Carla acknowledges, remembering. "It's true."

As Alicia sits up and gives Carla a side-eye glare, Ms. Davis arrives at the table, ready to hear what they have to offer.

Focus Questions for Episode 2

 In episode 2, we get a close look at the students' conversation without Ms. Davis. This is where we see how the students dig into the mathematics and how they interpret and implement classroom norms. Please think about or discuss the following questions before you read our commentary. Where you can, point to evidence in the episode that helps to support what you're thinking.

Focus Question 1
In episode 1, we invited you to think about the students' engagement with important mathematical ideas and practices. What can we say at this point, having watched them work together for some time without the teacher present?

Focus Question 2
Similarly, we considered the norms in play during episode 1. Is there more to say at this point?

Focus Question 3
There was some concern in episode 1 about the ways in which Alicia was being positioned by Bernardo and Carla. How do we see all of the students positioning each other in this episode?

Focus Question 4
In this episode, the students switch back and forth between Spanish and English. What role does the use of both languages play? How does it appear to affect opportunities for mathematical sense-making? Moreover, we're privileged to see some things the teacher doesn't, in terms of the ways Alicia, Bernardo, and Carla interact. There's some banter. Is it just "time off task?" Does it play a role in any potentially productive ways?

Our Commentary

1. In episode 1, we invited you to think about the students' engagement with important mathematical ideas and practices. What can we say at this point, having watched them work together for some time without the teacher present?

   We consider three issues: how the students engaged with mathematical content and practices, the mathematics that they engaged with, and the progress they made.

   ### 1a. The way Alicia, Bernardo, and Carla engaged with mathematical content and practices

   When Ms. Davis leaves the group, Alicia, Bernardo, and Carla jump into explaining where the ten is. Without hesitation, they take on the goals of understanding the mathematics, explaining to each other, and being able to explain why the representation works the way it does.

All three students participate actively. They may at times lack confidence in specific answers or lines of reasoning, but they show no lack of confidence in the eventual success of their problem-solving activity. Alicia, who will need to provide the group's explanation to Ms. Davis, is assertive in her demands for mathematical justification for every step of the reasoning that Bernardo and Carla provide – in her attempts to understand the mathematics she is unwilling to merely repeat what the others say or to take the correct answer on faith. Bernardo and Carla pitch in fully to meet the group's responsibilities (see our response to Question 2). All the students demonstrate perseverance (one of the Common Core State Standards for Mathematical Practices) in the service of making and explaining the relevant mathematical connections (two more practice standards).

1b. The mathematics they engaged with

Alicia, Bernardo, and Carla work hard at making connections between the physical representation and the symbolic expressions, repeatedly pointing to different parts of the figure in front of them as they explain and ask questions of each other. Alicia points to different lengths in the figure and asks how they contribute to the 10 and $10x$; all three students interact with lengths of 1, $x - 1$, and x, grappling with how those lengths contribute to the $10x$ and 10. This is challenging to get right.

1c. The progress they made

This episode shows how challenging these ideas are for beginners! The students start out focused on Ms. Davis's question "Where is the ten?," with Bernardo pointing out that there are 18 exposed 1's, but that there are 8 lengths of $(x - 1)$, so you have to subtract 8 from 18. This is hard to see in the figure, and when Bernardo points to the tiles where the unit lengths (the minus 1's) are covered, there is potential confusion – the tiles are of length x.

When Carla picks up the conversation, she focuses on the x's:

"So," begins Carla. "$10x$. So where do you get $10x$? So, right here." Carla leans in to point to the tiles.
She pauses. She cannot remember where to find the $10x$ in the diagram. She turns to Bernardo. "Where do you get $10x$?!" she demands.

From this point on, the group – quite possibly without realizing that they have shifted gears – is focused on explaining the "$10x$" part of the answer. The question of 10 has receded into the background. The three students do converge on a robust explanation of why the perimeter of the object contains $10x$, and they call Ms. Davis over to explain. But, they have not yet made collective sense of where the 10 is.

2. Similarly, we considered the norms in play during episode 1. Is there more to say at this point?

From their discussions, we see that norms 1, 2, and 3, which we highlighted in the discussion of episode 1, have been well internalized.
Here they are again:

- **Norm 1:** When a group calls the teacher over to explain, this certifies that the group representative is ready to present the group's explanations on their own and without help. The teacher will stop attempts from other students to scaffold a group representative's efforts during the explanation.
- **Norm 2:** All group members are responsible both for understanding the mathematics themselves and for making sure the group representative understands it and is ready to explain it.

- **Norm 3:** The mathematical standards of the original task are high, and they are maintained. Students are expected to address the question as posed, and to provide clear, coherent, and detailed explanations.

Alicia, Bernardo, and Carla clearly understand that while Alicia has been selected to explain their thinking to Ms. Davis, the group has collective responsibility for constructing and understanding the explanation that Alicia will present on their behalf (norm 1). The mathematical standards are high, and Ms. Davis is present even though she is absent: all three students know that they will need to provide Ms. Davis with a clear explanation (norm 3). They all work to make sense of the question, and to understand one another's thinking. Bernardo and Carla both act as though Alicia's understanding of the mathematics is important to them (norm 2).

The conversation is open and flowing, with all three students fully engaged throughout the episode and all three accountable to each other. Alicia is an active and assertive listener, who advocates for herself by interrupting with questions when she does not understand something Bernardo or Carla has said. Carla initially takes an explaining role at the request of Alicia herself, but later alternates between revoicing Bernardo's ideas ("There are 8 x minus"), asking questions ("Where did you get the 8?"), and challenging Bernardo's reasoning ("But don't count these!"). She is the one who recognizes when the discussion has stagnated: toward the end of the episode she prompts Bernardo to start from the beginning and recap his thinking. In these ways Carla works toward individual and collective understandings. Bernardo takes his role as a group member seriously, not simply explaining the mathematics but inserting gentle quiz questions into his explanations ("How much is that?").

3. **There was some concern in episode 1 about the ways in which Alicia was being positioned by Bernardo and Carla. How do we see all of the students positioning each other in this episode?**

Bernardo, Carla, and Alicia positioned each other very differently with respect to their knowledge of the content and the associated roles of "explainers" and "learners" within the group. It is worth considering how various acts of positioning in this episode may have opened up or closed down opportunities for each student to engage with the mathematics.

Throughout this episode, Bernardo is consistently positioned by himself and his peers as mathematically knowledgeable – both understanding the mathematics and being ready to explain it. Just after Ms. Davis leaves the group, Bernardo positions himself as the group's lead by jumping right in to explain to Alicia what she needs to know... " 'ira, aqui esta. Solo dile (Look, here it is. Just tell her ...)." Interestingly, Alicia initially contests this positioning by shushing Bernardo and expressing her preference for having Carla explain the math to her instead. As the episode continues, Alicia begins to accept Bernardo's offers to help. Carla, too, looks to Bernardo to answer her questions when she gets stuck.

In addition, Bernardo solidifies his position by not asking any questions, being eager and ready to jump in with an explanation whenever Alicia or Carla poses a question, by advising Alicia of what she should say to Ms. Davis, and by exuding confidence that his explanations are correct. There is no clear evidence that Bernardo's mathematical understanding changed at all throughout this episode, although Alicia's and Carla's questions prompted him to explain in more detail than he had previously.

In a continuation of the positioning from episode 1, Carla positions herself (and is positioned by Alicia) as an explainer. Carla's explainer status is less robust than Bernardo's, in that she has moments of doubt and turns to Bernardo for clarification.

Carla readily responds to Alicia's repeated requests to explain the math to her, looking to Bernardo periodically to fill in the gaps when she gets stuck. After receiving help from Bernardo,

Carla is quick to indicate that she now understands. She verbalizes her agreement with Bernardo and signals her newly established expertise with her exclamations of, "It's true, it's true!" Thus, although Carla asks for help occasionally herself, her primary role is to help and support Alicia by responding to Alicia's questions and offering her own explanations for where the ten is.

Throughout this episode, Alicia is consistently positioned, by herself and her peers, as a less-knowledgeable "learner" within the group. At the same time, Alicia is positioned, again by herself and her peers, as one who is capable of understanding the mathematics (and eager to do so).

Alicia is the group representative, so she will be the one who is put on the spot when Ms. Davis returns. Therefore, the group's focus is on preparing Alicia. All three students acknowledge that Alicia needs to learn more since she was unable to offer a complete and accurate explanation to Ms. Davis on her first attempt. Alicia is not content with the other students merely telling her what to say – she demonstrates a strong sense of agency by asking clarifying questions and probing for deeper understandings. Although Alicia is positioned as "not yet understanding," her line of questioning when receiving Bernardo's explanation actually reveals a lot of mathematical sophistication – carefully accounting for each geometric unit ("And where did you get the 8 from?"; "What about this one (1x1 unit)?" She is very much an active mathematics learner and sees herself as such (as do Bernardo and Carla).

While at times Alicia exhibits frustration and seems uncomfortable with her need for help (e.g., giggling, smiling, giving head shakes), all three students remain committed to making sure Alicia understands. When joking and teasing occurs, the students refocus their attention quickly on the task at hand. Without knowing anything about the personal relationships of these students, we simply can't know how some of these teasing comments were intended and how they were received (For example, when Carla calls Alicia "sweetheart," was Alicia offended or amused?).

4. In this episode the students switch back and forth between Spanish and English. What role does the use of both languages play? How does it appear to affect opportunities for mathematical sense-making? Moreover, we're privileged to see some things the teacher doesn't, in terms of the ways Alicia, Bernardo, and Carla interact. There's some banter. Is it just "time off task?" Does it play a role in any potentially productive way?

The Role of Both Languages

The fluid back and forth between Spanish and English signals that students use whatever language seems most comfortable for them at any given time. Speaking Spanish provides Alicia, Bernardo, and Carla an opportunity for students to voice mathematical ideas in their own language – and thus take ownership of the mathematics. Further, using Spanish allows them to make sense of the problem together. For example, it appears that Bernardo is comfortable and fluent in Spanish, which allows him to bring his understanding to bear. If he were limited to speaking English, he'd be less helpful to the group's progress.

Some might argue that the students compromise their development of English language proficiency moving between two languages. However, Ms. Davis requires that the group present their work in English, which does appear to have an impact. At one point in the conversation, Carla breaks from their conversation in Spanish and speaks mostly English – perhaps her way of taking the next step in formalizing their understanding and preparing for the teacher. Another way to interpret Carla's switching from Spanish to English may help to illustrate a potentially harmful consequence rising from the hierarchical nature of language use in the classroom. For example, when Carla "takes over" for Bernardo midway through the episode, she switches to English, which might be read as a possible power or positioning move, signaling that English is the language of serious work.

The Impact on Mathematical Sense-Making

Having two languages to draw from most likely created opportunities for Alicia, Bernardo, and Carla to express their ideas in the ways that made sense to them and allowed them to focus on the mathematical concepts without worrying about additional language barriers. The mathematical exchange might have been less robust or fluid if the students could only communicate in English; in this episode, it appears that at times the students can get more deeply into the math when speaking in Spanish. Additionally, the heavy use of Spanish indicates that the students were really trying to understand the mathematics for themselves and not yet worrying about what to say to the teacher.

We do note that there are some difficulties. In Spanish, *"Son 8 x-menos-una."* can be heard as "there are $8x - 1$" or "there are $8(x - 1)$'s." We think (but there is no way of knowing for sure) that Bernardo intended the latter, but Alicia heard the former. That plus the ambiguity of pointing and saying "this" or "these" makes it challenging for us as observers, and possibly for the other students, to follow.

Bantering: Potentially Productive?

That the students can banter in Spanish contributes to a sense of the group being in a "safe mathematical space." The social exchanges help, and support the math exchanges – because they're comfortable, the students can exchange ideas. Bernardo's fluency in Spanish clearly contributes to the mathematical discussions. In addition, being able to speak in Spanish provides room for the group to lower the tension.

Consider this exchange.

> Bernardo offers Alicia some advice for when Ms. Davis returns.
>
> "Si te pregunta porque son ocho x-menos-1, (If she asks you why there are eight x-minus-1's,)" he suggests, pointing to the places in the figure where the 1-by-1 tiles overlap with the x-by-1 tiles, "es porque hay uno, uno que le estorba. (it's because there's one, one that's in the way.)"
>
> "Uh-huh." Alicia nods.
>
> Carla squints at Bernardo and teases him. "Tu como sabes que 'le estorba'? (How do you know it's 'in the way'?) Pobrecito. A lo mejor está allí para acompañarlo. (Poor thing. Maybe the only reason it's there is to keep the ones company.)"
>
> "Ay," Alicia giggles, rolling her eyes at Carla, who hides her face. "Ya a va salir con sus chistes. (Here you come with your jokes.)"

The first three lines are about a mathematically challenging point, and the moment is somewhat intense. Carla's joke about the tile *estorbando* and being a *pobrecito* injects a moment of levity and defuses the tension, without taking away from the mathematics. This is much easier to do in one's first language!

3.3.3 Episode 3 (from 3:48 to 5:27 in the Lesson Segment)

Overview

When Ms. Davis returns to the group, Alicia, confused, shows her where the $10x$ is in the figure. Ms. Davis stops her, clarifying that Alicia is showing $10x$ and that Alicia's task is to find the 10. Bernardo and Carla try to pitch in with explanations and support for Alicia, but Ms. Davis stops them, saying that their job is to make sure Alicia can provide the explanation by herself. When Alicia, somewhat upset, notes that Bernardo can explain where the ten comes from, Ms. Davis reinforces the norms by saying "So listen to him." She then leaves the group.

Narrative

Ms. Davis approaches the group and stands in between Alicia and Bernardo. She kneels down on the floor next to the group's table and rests her chin on her folded arms. She is at the same level as the sitting students.

She begins, "Ok, then. Let's see. Where's the ten?"

Alicia grabs her pencil and explains, "It's these. 1, 2, 3, 4, 5, 6, 7, no wait. It's 8, right? So then we add $2x$, so then it would equal 10. 8 plus 2, 10. I did it." Alicia smiles and moves her hands with a flourish.

Ms. Davis, now standing, continues to stare down at the figure as she speaks. "8? Wait. I'm confused. Did you show me the 10 or did you show me the $10x$?"

Both Alicia and Carla start talking, but Alicia quiets her group-mates quickly with a finger wag and a "Shhhh!"

Alicia continues explaining on her own. "I'm showing you how I got the 10, because since these are x, it was 1, 2, 3, 4, 5, 6, 7, 8,... 9, 10. And that's how we got 10."

As Alicia speaks, she points to different parts of the figure, indicated in Figure 3.16.

"When you go 9, 10, what are you pointing at?" Ms. Davis asks.

With her pencil, Alicia draws lines next to the outer edges of the figure (as shown by the red lines in Figure 3) and says, "this part right here." Ms. Davis continues to probe, "What is this part?" and she points to the outer edges of the figure.

Alicia responds, "x."

Ms. Davis responds confused. "x. But then I asked you for... I thought you were showing me the 10?"

Alicia leans in and continues, "I am, but-"

Ms. Davis quickly responds, "So, you counted 1, 2, 3, 4, 5, 6, 7, 8... 9, 10." As Ms. Davis counts, she repeats Alicia's pointing as indicated in Figure 3.16.

She continues, "But that is 8 plus ... you told me these were x's," pointing to the two sides that were counted as 9 and 10.

Carla jumps in, "No, no, no. She is confusing."

Alicia quickly responds, "No, I am not."

Ms. Davis says, "I'm confused... maybe I'm confused."

Bernardo jumps in and makes a circular motion with his hand, "Cuenta todas las de la orilla. (Count all the edge ones.)"

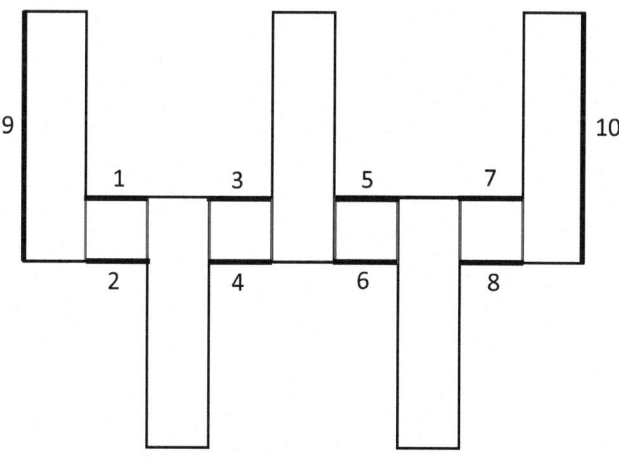

Figure 3.16 Alicia's counting

Talking over Bernardo, Carla speaks to Alicia. "You're not going to get a simple, 'ah, yeah. The 10 is right here.'"

There is tension in the air.

Alicia responds with a tight voice. "I know that, but let me explain it, then argue with me."

Carla interjects matter-of-factly: "Ok, well, you're just confused."

Alicia, clearly annoyed, continues. "Well, why don't you all explain it then?"

Carla responds, "Because she wants you to explain it."

Alicia, now a bit more relaxed, says, "And what if I want you to help me explain it?"

Ms. Davis jumps back into the conversation and says, "That's right. You help her and I'll be back."

Alicia puts her head down on the desk and protests, "No. You stay here so I can explain it to you."

Ms. Davis doesn't give in to Alicia's plea. "Nooooo. You need to be able to-"

Alicia tries again, "Why are you asking me? Why don't you ask them?" She points to Bernardo and says, "He's the one who can explain it."

Ms. Davis builds on Alicia's observation, saying, "Yeah. So listen to him," as she walks away from the group.

Focus Questions for Episode 3

In episode 3, there are some significant tensions when Alicia is not successful in responding to Ms. Davis's request to show her the ten. Our questions are aimed at clarifying what happened. As always, think about the evidence in the episode that helps to substantiate the claims you're making.

Focus Question 1
In episode 2, we saw Alicia, Bernardo, and Carla begin to grapple seriously with the relationships between the 1's and x's. What happens here, with regard to mathematical content and practices?

Focus Question 2
Much of the "action" in episode 3 centers around the classroom norms. What evidence is there about the norms we've been focusing on and how they're being reinforced?

Focus Question 3
Things are a bit tense as the students are positioned (either tacitly or explicitly) by each other when they interact with Ms. Davis. Is there anything to note about what the students say, or about how Ms. Davis deals with the situation?

Our Commentary

1. In episode 2, we saw Alicia, Bernardo, and Carla begin to grapple seriously with the relationships between the 1's and x's. What happens here, with regard to mathematical content and practices?

The content issue at the heart of this episode involves the confusion between unlike terms – specifically, between x-length units (the long side of the algebra tile) and the 1-length units (the short side). In the previous episode, the students had slipped into counting x's, despite the fact that they were asked to find the "10" – that is, to count 10 unit lengths in the diagram. Ms. Davis calls Alicia's attention to this when she points out that, although she had asked for the "10," Alicia's explanation focuses on the "x's": "Wait. I'm confused. Did you show me the 10 or did you show me the $10x$?"

Without more information, it is difficult to say precisely what Alicia is thinking at this point. She may, for example, be treating the unit lengths and the x lengths as being fundamentally of the same kind (that is, as like terms). She may not have realized that finding the "10" referred specifically to unit lengths, or may be confused by the difference between the x and $x - 1$ lengths. Or, she may have understood the underlying mathematics well but became distracted from her original goal in the course of the difficult cognitive work of solving the problem.

Bernardo gives some evidence that he perceives the difficulty when he attempts to redirect Alicia. ("Cuenta todas las de la orilla." – Count all the edge ones.) But at this point Alicia appears to be too caught up in events to evaluate or take up Bernardo's suggestion. Ms. Davis, for her part, chooses not to introduce any mathematical language or notation of her own (avoiding, for example, using "like terms" or "units"), thereby keeping the entire conversation grounded in the students' own present understanding of the mathematics.

It's worth noting that the students are working hard at understanding the connections between the representation using 1's and x's and the physical objects, and then explaining those connections clearly. Those are two central mathematical practices.

2. Much of the "action" in episode 3 centers around the classroom norms. What evidence is there about the norms we've been focusing on and how they're being reinforced?

Contrasting episodes 2 and 3 gives us back-to-back examples of our three focal norms. Here we see different kinds of pressures and challenges brought to bear on the students and their understanding of the mathematical task. Ms. Davis's actions, although relaxed, soft-spoken, and intended to keep the students at ease and productive (it's hard to convey tone, but we see her statement "I'm confused... maybe I'm confused" as an attempt to ease tensions and take some pressure off Alicia), reinforce norms 1, 2, and 3.

> **Norm 1:** When a group calls the teacher over to explain, this certifies that the group representative is ready to present the group's explanations on their own and without help. The teacher will stop attempts from other students to scaffold a group representative's efforts during the explanation.

This norm is clearly in place. When she returns to the group at the beginning of this episode, Ms. Davis picks up where she had left off, addressing the "Where's the ten?" question to Alicia; Alicia is the only student who initially responds to the question. Carla also signals her expectation that Alicia needs to be the one who explains "where the ten" is, by saying explicitly, "she wants you to explain it." The responsibility of explaining remains squarely on Alicia's shoulders.

The strength of the norm is tested over the course of the episode. Initially, Alicia appears confident that she can explain and is eager to take on the role of explainer – she "Shhhhhh"'s her group-mates when they intrude on her explanation. But things change when it's clear that Alicia is still confused. Alicia pushes back on norm 1 by first seeking help from her peers, asking "And what if I want you to help me explain it?" and then by suggesting that Bernardo should be the one to talk since "he's the one who can explain it." But the final ten lines of the episode make it clear that explaining the group's thinking is Alicia's responsibility.

Norm 2: All group members are responsible both for understanding the mathematics themselves and for making sure the group representative understands it and is ready to explain it.

This particular episode focuses on Alicia, so we have to infer that things would be the same with other group representatives. But the norm itself is clear. When Carla says that Alicia is confused and Alicia, frustrated, says "Well, why don't you all explain it then?", Carla's response is, "Because she wants you to explain it." And when Alicia says, "And what if I want you to help me explain it?" Ms. Davis jumps back into the conversation, saying "That's right. You help her and I'll be back." Even as Alicia continues pushing back, "Why are you asking me? Why don't you ask them? He's the one who can explain it," Ms. Davis counters with, "Yeah. So listen to him."

Norm 3: The mathematical standards of the original task are high, and they are maintained. Students are expected to address the question as posed, and to provide clear, coherent, and detailed explanations.

As we've noted, the very question "where is the 10?" sets a high bar; it would have been easy to accept the answer "$10x + 10$" for the perimeter of the object and move on. Ms. Davis keeps the bar high by pointing to the fact that Alicia was counting the x's, not the 1's, and that Alicia has to be able, by herself, to show where the 10 comes from. It's clear that Alicia would rather have one of her group-mates take on the burdens of understanding and explaining, but Ms. Davis is firm: "You need to be able to…" indicates that the burden of explanation is hers. That's reinforced in the parting exchange, where Alicia says that Bernardo can provide the explanation and Ms. Davis responds, "So listen to him."

3. **How are students positioned (either tacitly or explicitly) by each other when they interact with Ms. Davis? Is there anything to note about how Ms. Davis deals with the tensions that arise in interaction?**

Alicia began this episode with confidence, backed by that of her group-mates. All of the students in the group seemed to agree they were ready to call Ms. Davis over and Alicia was eager to share her thinking as soon as she joined them. Alicia's smiling and shushing of her group-mates indicated comfort with the situation and confidence in herself.

That didn't last. But even as some of Alicia's comfort faded and her frustration increased, she was determined to play the role of group explainer, saying to her teammates as they corrected her: "I know that, but let me explain it, then argue with me." Carla's response, "Ok, well, you're just confused," was, even though it was tossed off casually, rather harsh, and positioned her as needing help and being unable to explain.

By the end of the episode, Alicia's frustration outweighed her confidence. She gave in and accepted the positioning given to her by her peers, expressing a desire to relinquish the explainer role: "Why are you asking me? He's the one who can explain it" (referring to Bernardo). Ms. Davis, on the one hand, made it clear that she felt that Alicia could live up to the classroom norms and be able to explain where the ten is. On the other hand, her final comment, "So listen to him" did acknowledge that Alicia wasn't ready and had work to do.

Carla positions herself as an authority, both with regard to the mathematics ("You're not going to get a simple, 'ah, yeah. The 10 is right here.'") and as compared to Alicia ("you're just confused"). There isn't much evidence one way or another about Carla's understanding in this episode, so what we see for the most part is a reflection of her own self-perception.

Although Bernardo spoke very little during this episode, he was positioned throughout this episode as a reliable source of knowledge and help. The only time he spoke was to offer Alicia authoritative support when Ms. Davis said she was confused: "Cuenta todas las de la orilla. (Count all the edge ones"). This was tacitly accepted as correct by everyone, though it wasn't followed up on because it was Alicia's responsibility to provide the group's explanation.

At the end of the episode, Alicia positioned Bernardo as an expert explainer when she tried to shift the responsibility for explaining to him ("Why are you asking me? He's the one who can explain it."). Ms. Davis affirmed this positioning when she responded, "So listen to him."

While a lot of this goes by fast, one has to wonder about the impact of such positionings if they're consistent over time.

3.3.4 Episode 4 (from 5:27 to 9:10 in the Lesson Segment)

Overview

Alicia, Bernardo, and Carla dive back into their work after Ms. Davis leaves their table. They are all engaged, with Bernardo and Alicia working actively to provide Alicia with a solid explanation for Ms. Davis. When Ms. Davis returns Alicia is able to explain where the 10 comes from and, when Ms. Davis follows up to make sure, to explain where the $10x$ comes from. Ms. Davis confirms that the students "get a stamp" and they move on to the next task.

Narrative

Bernardo picks up the pencil and points to the figure. "Cuenta todas estas. Todas las de la orilla. (Count all of these. All the ones on the outside.)" Alicia has her hands on her face but pulls herself together and focuses on the diagram as Bernardo continues.

"Asi van a aser 18 todas (That gives a total of 18)."

"Aja" agrees Carla.

"Menos 8 (Minus 8)" interjects Alicia.

Bernardo continues, "Luego menos 8 van a aser 10 (After you take away 8 there are 10 left)."

"Menos 8 que son estos? (Minus these 8?)" asks Alicia. She appears to be pointing at the 1's highlighted in Figure 3.17, though it is hard to be sure.

Bernardo says, "No. No. Tu nomas dile menos 8. (No, No. you don't have to tell her more than minus 8.)"

Gesturing emphatically, Alicia responds, "Me va a decir de donde agarraste el 8 (She is going to ask me where I got the 8)."

Carla, yawning, turns to Bernardo and asks, "where did you get the 8?"

Bernardo rotates the paper so that it is now facing him and says, "Son 18 (There are 18)"

Alicia repeats her concern. "Aja, son 18, no. Son 18. Le quitamos 8. De donde agarraste el 8 (Yeah, there are 18, no. There are 18. We take away 8. Where did you get the 8 from?)"

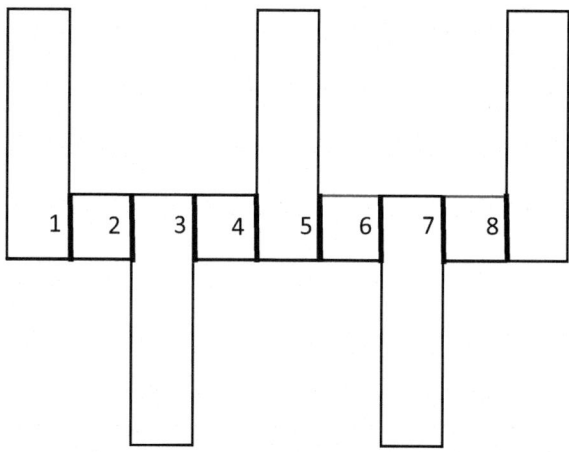

Figure 3.17 The points of overlap

Bernardo asks, "cuales 8 (which 8)?"

This throws Alicia. Frustrated, she covers her face with her hands, laughs nervously and utters, "ayyyyy"

Bernardo continues, speaking rapidly: "Le quite dieciocho no agarre 8. Le quite dieciocho porque aqui hay 8 menos. (I took away from 18. I did not come up with 8. I took away from 18 because there are 8 less.)"

As Bernardo talks he points to a place on the paper where "18" or "18-8" may be written, we can't see. "Y de las dieciocho le quite 8 para ponerselas aqui. (From the 18 I took away 8 to put them here)."

Carla, half following and half working by herself, lets out a frustrated "ayyy."

Alicia, Picking up on Bernardo's comment about 18-8, says "Para agarrar 10 (To get 10)."

Bernardo confirms, "Y luego ya se va a convertir este en 10. So ya va a aser 18-8 es diez (And then this will result in 10. This will be 18 minus 8, that's 10)."

Carla, who has been following working in parallel, looks up and notes, "OK, the 8. Pero eso no es lo mismo (Isn't that the same thing)?"

Alicia takes this as confirmation that she and Carla are on the same page about wanting an explanation for the 8, as Bernardo continues on:

"ya? Ella te esta preguntando con el 10... (OK? She is asking you about the 10)," tapping his pencil on the 10 he had written as part of the answer.

Alicia brushes her hair away from her eyes and pursues her request,

> Pues yo quiero algo facil. Voy a decir 18 le quito 8. Pero dime de donde y como le quito el 8 (Well, I want something easy. I am going to tell her, 18 minus 8. But tell me where and how I take away the 8).

Talking over Carla as Carla starts to interject, Alicia continues, "No porque no quiero darle tantas vueltas (No because I do want to go around in circles)."

And, as Carla starts to clarify "You have to – explain her..." Bernardo continues,

"Mira tu le dices. Tu le dices que son 18 de estas y le quitas 8 y se convierte en 10." (Look, you tell her. You tell her that there are 18 of these and you take away 8 of these and that converts to 10.)

"Y luego ella te va a preguntar porque le quitas el 8. ("And then she's going to ask you why you're taking away 8.") He says "Porque hay 8 de estas ("Because there are 8 of these"), pointing to the 8 sections that are $(x - 1)$ in length) "y necesitan una cada una (and each one needs one of the 1's.)". He says, "Necesita uno de estos (It needs one of *these*)", pointing to the "18" on the sheet of paper – the 18 1's. See Figure 3.18.

Alicia gets what Bernardo was explaining. Smiling, she responds, "ohhhhhh okay."

The tension is broken. Carla has also followed Bernardo's explanation. She smiles, looks at Alicia, and says, "Duh doofy."

Alicia responds in kind, "Be quiet," and Bernardo leans back laughing.

"Doofy," Carla says once again. She and Alicia laugh.

Bernardo tells Carla to raise her hand, "Levantala (Raise it)."

Alicia, still laughing, physically helps Carla raise her hand and says, "Stop laughing."

Carla, still laughing, continues, "doofy." She calls out for Ms. Davis.

Ms. Davis from afar asks the group if they are ready.

Bernardo leans backward a bit and quickly responds, "Yeah."

Carla, chuckling, says "She said she's ready. I mean, I'm not sure."

Alicia, responding, says "I'm not going to do it."

Bernardo dives in while they wait for Ms. Davis, "Ahora tu explicalo a nosotros (Now explain it to us)."

"There are 8 of *these* (the 8 dark vertical lines, each of which is x-1 units long)
And each one needs one of *these* (the 18 dotted segments, each of which is 1 unit long)"

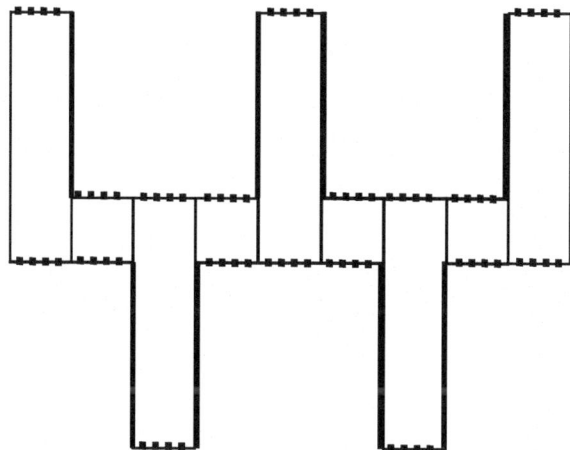

Figure 3.18 Bernardo's argument: Each of the 8 (*x* – 1)s "needs" (takes away) one of the unit lengths from the original 18

Carla, taking the role of the instructor with a somewhat businesslike tone, says "Okay, explain it to us. Where did you get 10?"

Without hesitation Alicia says, "Son 18 en total verdad. (There are 18 in total, right.) So... hmmm..." But before Alicia gets a chance to practice her explanation, Ms. Davis arrives and kneels between Alicia and Bernardo. Ms. Davis begins, "Okay."

Alicia responds quickly. "Alright. I'll tell you how I got 10." She rolls up her sleeves and continues.

> I'll tell you how I got 10. All of these little things (pointing to the 1's on the outside of the figure) equal to 18. So, like, we donate all the parts and (pointing to an (*x* – 1) as she continues speaking) we need to take one away to use it in the other and (pointing to another (*x* – 1) one away to use in the other. So (with her gestures indicating that all of the (*x* – 1)s would be included) we take away 8 and it equals 10.

Ms. Davis, with a big smile, says "Very good, Alicia." Realigning the algebra tiles, she asks, "Now, can you show me..." she pauses for a second ... "10*x*?"

Alicia (we think, referring to her first interaction with Ms. Davis, where she did show her 10*x*) says "I did it already..."

Ms. Davis, apparently referring to the "ten" Alicia has just shown her, says, "I thought that was..." Both she and Alicia laugh and Alicia puts her head down, perhaps in mock despair.

Carla jumps in. "No, no, no. It's not the same, It's not the same. Don't get confused. Don't get confused."

Alicia sits straight again, still with her hands over her mouth and face. She is laughing.

Ms. Davis continues, "It's Ok. Breathe. Breathe. Now show me the 10*x*. Breathe."

It appears to us that Ms. Davis is trying to have the students show her the 10*x* as well as the 10 so that she can wrap up the problem neatly. Alicia has just shown her the 10; earlier, she had shown her the 10*x*. But that was a while ago, and now was the time to clean things up. In what follows Ms. Davis is no longer strict about requiring only Alicia to explain for the group.

Carla takes the lead this time and starts to explain, "Oh, yes, yes, yes, yes. So ... Let me show you." She leans in and with her pencil is ready to begin to explain.

Alicia says, "I'm gonna get some tissues." She points to the direction of the tissues and stands up.

"Wait. No, Alicia." She wants the whole group together for the explanation. Alicia sits back down and looks at the diagram in the center.

Carla begins explaining, "18. Ok, we have 10. Everybody knows…"

Ms. Davis clarifies what she is looking for. "I've got the 10. The regular 10. So where is the $10x$ part?"

Carla shaking her head responds, "We have $8x$."

"$8x$, $8x$, where is the $8x$?" Ms. Davis asks Carla.

As Carla says "No, no, no. That's the other point," Alicia jumps in and takes over.

She rolls up her sleeves, leans in and using her pencil begins to count. Pointing to the eight green $(x - 1)$'s in Figure 3.19, she counts "One, two, three, four, five, six, seven, eight…"

As Alicia is counting, Carla tries to clarify the task, asking Ms. Davis, "Do we have to get 10 or $10x$?" But Alicia knows what the task is. She continues counting, pointing to the two full x's on the outside of the figure, and says "9, 10. x." She looks confidently at Ms. Davis as she does so. See Figure 3.19.

Ms. Davis begins to wrap up. She asks,

> Okay, and, and you … Do you remember what you've already told me why these make full x's? Because from here I only see 2 full x's. But, like, if I was going around the perimeter here, this side, this piece could be-

"Took away 1." says Alicia, finishing Ms. Davis's thought. Bernardo is also leaning in with his hand pointing at the model.

"Right." Ms. Davis agrees. Carla rolls on with the explanation:

"But we took [it away] already though, the number 1. That's the reason because we have 10. So we have 10x. Complete all the thing." She ends her explanation pointing at the diagram, "This (pointing to a 1 that was covered), it's not any more, it's in the past." See Figure 3.20.

Alicia asks, "So we get a stamp?" Carla continues, "This (pointing to something else we can't make out) is in the future." Ms. Davis laughs.

They're clearly done. Alicia asks again, "So we get a stamp?"

Smiling, Ms. Davis says, "Ok, yes, you get a stamp. Very good." This is greeted by a "Whoo" from Carla.

As Ms. Davis rises to move to another group, her parting words are "You need to make sure you keep everyone together and everyone understands."

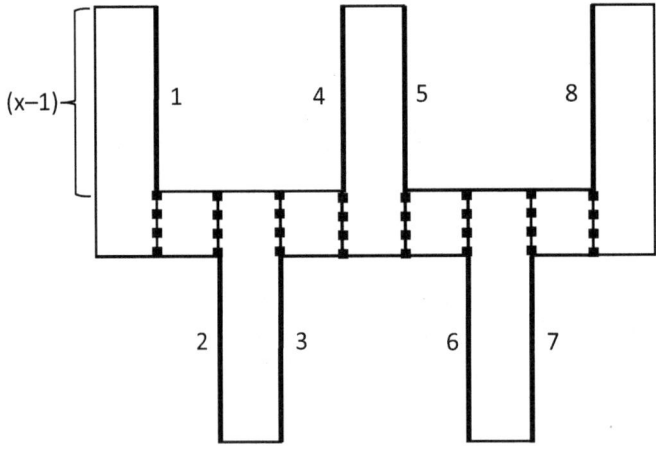

Figure 3.19 Alicia's explanation

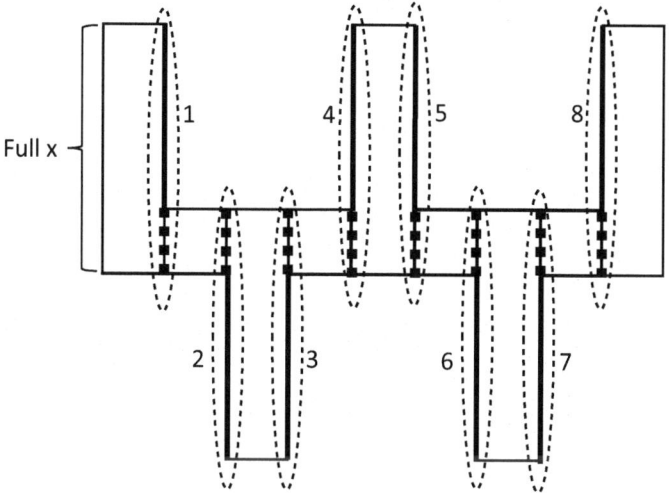

Figure 3.20 The 8 missing 1's, which are now in the past

Focus Questions for Episode 4

 In episode 4, Alicia, Bernardo, and Carla dig into the representation once again. This time, things seem to click. Our goal for these focus questions is to develop our collective understandings of what happened so that we're in a position to discuss their implications.

Focus Question 1
Again, let's start with the mathematics. What was the group's final explanation for why there are 10 1's in the perimeter, and 10 x's? Is it fully correct? Might there be more to say?

Focus Question 2
We'd like to revisit the classroom norms and their impact. Here are the norms we've discussed:

Norm 1: When a group calls the teacher over to explain, this certifies that the group representative is ready to present the group's explanations on their own and without help. The teacher will stop attempts from other students to scaffold a group representative's efforts during the explanation.

Norm 2: All group members are responsible both for understanding the mathematics themselves and for making sure the group representative understands it and is ready to explain it.

Norm 3: The mathematical standards of the original task are high, and they are maintained. Students are expected to address the question as posed, and to provide clear, coherent, and detailed explanations.

Can you describe how the students' actions did or didn't live up to those norms? Similarly, how did Ms. Davis's actions support or undermine those norms? Can we say anything at this point about the impact of those norms on student understanding?

Focus Question 3
Things were tense at times in the first three episodes. The three students were obviously comfortable with each other, but at times Bernardo and Carla were somewhat dismissive of Alicia… to the point where one might be tempted to intervene as a teacher. How did things work out? What can we say about each student's participation, and possibly their mathematical senses of self?

Our Commentary

1. Again, let's start with the mathematics. What was the group's final explanation for why there are 10 1's in the perimeter, and 10 x's? Is it fully correct? Might there be more to say?

Alicia's answer to the question "Where is the ten?" is

I'll tell you how I got 10. All of these little things (the dotted lines of length 1 on the outside of Figure 3.21) equal to 18. So, like, we donate all the parts and (pointing to one of the interior (x – 1)s indicated by a dark vertical line) we need to take one away to use it in the other and (pointing to another (x – 1) one away to use in the other. So (with her gestures indicating that all of the (x – 1)s would be included) we take away 8 and it equals 10.

This is completely correct, including specific references to the 18 1's that are visible and the 8 1's that have to be subtracted. Her answer to the question "Now show me the 10x" is straightforward. She had already shown that the eight pieces in red, with a 1 added to each, became 8 x's. There were two whole x's on the side of the figure. That made 10x.

The mathematics is fully correct and, after no small amount of collective work, Alicia's explanation is lucid and complete. The connections between the symbolic representation (the x's and 1's) and the algebra tiles have been solidly made. This is a nice outcome after work that was sometimes challenging and frustrating.

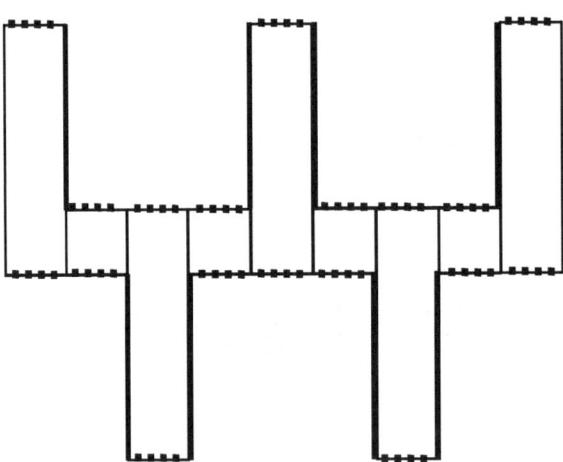

Figure 3.21 Alicia's explanation, unpacked

Might there be more to say? Yes, there almost always is! As we saw in Section 5.1, there are different ways to show why the perimeter is $10x+10$. Perhaps these would emerge in later whole-class conversations; we've only seen one small-group session. But Alicia, Bernardo, and Carla's collaboration has resulted in a coherent and complete explanation of the "$10x$" and "10" parts of the answer.

2. We'd like to revisit the classroom norms and their impact. Here are the norms we've discussed:

> Norm 1: When a group calls the teacher over to explain, this certifies that the group representative is ready to present the group's explanations on their own and without help. The teacher will stop attempts from other students to scaffold a group representative's efforts during the explanation.

> Norm 2: All group members are responsible both for understanding the mathematics themselves and for making sure the group representative understands it and is ready to explain it.

> Norm 3: The mathematical standards of the original task are high, and they are maintained. Students are expected to address the question as posed, and to provide clear, coherent, and detailed explanations.

> Can you describe how the students' actions did or didn't live up to those norms? Similarly, how did Ms. Davis's actions support or undermine those norms? Can we say anything at this point about the impact of those norms on student understanding?

With the single exception of Alicia's frustrated plea at the end of episode 3 ("Why are you asking me? Why don't you ask them?"), the three students embraced the norms. For the most part, Alicia took on her role as explainer with enthusiasm. She not only welcomed but at times demanded Bernardo and Carla's help – she was responsible for presenting the group's work, and their role was to help her understand the math so that she could do it correctly (norms 1 and 2).

Beyond that, Alicia understood that a superficial answer wouldn't do. Here is a summary of what happened at the beginning of this episode:

> Bernardo starts with a quick explanation of where the 10 comes from, saying Alicia should "count all the 1's on the outside ... [and then] after you take away 8 there are 10 left"
> Alicia pointed to the 1's in the middle of the diagram, asking "These 8?"
> Bernardo implied that she didn't have to worry about that detail: "No. You don't have to tell her more than minus 8."
> But Alicia responded emphatically, "She is going to ask me where I got the 8."
> Alicia persevered. When Bernardo didn't understand what she was asking for, she asked him again, saying "Well, I want something easy. I am going to tell her, 18 minus 8. But tell me where and how I take away the 8."

In sum, Alicia didn't rest easy until she was confident in her explanation of *why* you subtract the 8, and *how* she could show it. She is living up to the high mathematical standard of norm 3.

At the same time, Bernardo and Carla took their responsibilities for sharing their understandings seriously. They provided consistent support for Alicia as she worked to make sense of the mathematics (norm 2). And, they were invested in her being able to provide a good explanation. After Carla called for Ms. Davis, the students had some time before Ms. Davis made it to their table. Bernardo turned to Alicia, saying "Now explain it to us"; Carla, possibly role-playing a teacher, said "Okay, explain it to us. Where did you get 10?"

In these ways, Bernardo and Carla worked to support Alicia in understanding and explaining the mathematics.

Ms. Davis was a calm and positive presence. At the same time, she was clear in enforcing the norms. During the bulk of their exchanges, she focused her questions on Alicia, turning aside possible contributions from the others, because it was the group's responsibility to make sure that Alicia understood and could explain by herself (norm 1). We can't know whether Ms. Davis was tempted at times to make things easier for Alicia. What we do know is that, despite the students' occasional discomfort, Ms. Davis kept to the original question "where is the ten?", maintaining high expectations for the explanation she expected Alicia to produce (norm 3). Her discussion at the end covered both the "10" and "10x" parts, and in those exchanges she was able to see that all three students were capable of producing solid explanations. Her final comment, "You need to make sure you keep everyone together and everyone understands," was direct reinforcement of norm 2.

Although it was clearly challenging for the students at times, adhering to the norms appeared to pay off. By the end of episode 4 Alicia confidently produced a coherent and complete explanation, Bernardo's explanation was richer and more complete than that at the beginning of the case study, and Carla's understanding had deepened.

3. **Things were tense at times in the first three episodes. The three students were obviously comfortable with each other, but at times Bernardo and Carla were somewhat dismissive of Alicia... to the point where one might be tempted to intervene as a teacher. How did things work out? What can we say about each student's participation, and possibly their mathematical senses of self?**

We want to start by saying that teaching constantly involves making judgment calls based on what you know about the students and their history as well as what's happening in the moment. We don't know what Ms. Davis knows about the students and their relationships, and we're in no position to second-guess her. Also, we saw and heard a lot more than Ms. Davis did. A lot of what we see (as recorded on camera) took place while Ms. Davis was working with other groups. So what we're doing here is making observations and raising issues to think about.

Some of the exchanges that took place in these episodes could be seen as problematic. For example, Carla's comments about Alicia, "She's confused" and "She *thinks* she's ready" could contribute to Alicia's seeing herself as mathematically weak. Some of this appears to be good-natured snark; but, such comments could have a long-term impact.

At the same time, it is clear that the three students get along well. They share frustrations and make wisecracks, sometimes at each other's expense; but they laugh together and, more importantly, the three students clearly work together as a team, as discussed in our response to Focus Question 2. The students really do support each other in their work. We see that in the explanations they provide, in the fact that they're willing to push each other, and the fact that Bernardo and Carla offer Alicia a chance to rehearse her explanation before Ms. Davis reaches their table.

Deciding whether to intervene in student-to-student exchanges is a judgment call; there's no simple right answer. The same is the case with regard to whether and when a teacher might offer a struggling student hints. It's clear that Alicia was struggling at times; when students are struggling, we have a natural inclination to offer support by way of hints. At the same time, providing hints might deprive Alicia, or other students in her position, of opportunities for productive struggle. This too is a judgment call. As we've seen here, Alicia's struggles turned out to be productive – she grappled successfully with the task and produced a solid explanation. As you'll see in the epilogue, the short-term impact of this success was a clear boost in Alicia's mathematical sense of self and her willingness to engage. The challenge for us as teachers is to build and maintain learning environments that offer students meaningful and attainable challenges, and a social context in which they can work productively on them.

3.3.5 Epilogue

Overview

When Ms. Davis joins Alicia, Bernardo, and Carla to talk about the next problem there is a noticeable change in Alicia. Having successfully completed the first task in episode 4, she appears happier and more confident. The three students continue to be fully engaged as a team.

Narrative

The successful completion of the "Where is the ten?" task in episode 4 meant that Alicia no longer bore the responsibility of being sole explainer for the group. The three students start afresh on the next problem.

In the next problem, the students are supposed to use particular algebra tiles (an x-by-y rectangle called an "xy" and two 1-by-y's, called "y's") to put together a figure whose perimeter is $2x+2y+4$. The tiles are shown in Figure 3.22.

As Ms. Davis joins the group, there is a figure made of x pieces and y pieces in front of Carla, and the students are writing on their worksheets. The figure doesn't use the assigned tiles (an xy and two y's), suggesting that they haven't understood the task. Ms. Davis explains that the problem is different from the previous problem: instead of finding the perimeter of a figure, "you have this perimeter [pointing to the expression $2x + 2y + 4$ on the paper] and you want to find the shape." You can hear Carla say "Oh!", and Bernardo and Carla start erasing what they had previously written on their worksheets.

Alicia puts down two y-pieces, saying "OK, we've got 2 y already," and begins to look for two x's in the box of algebra tiles. Ms. Davis stops her.

"Be careful," she says, picking up the two y's. "Alicia, you're right, this is 2y. But what we're doing is the perimeter." Putting the two y's down on the paper next to each other, Ms. Davis asks, "So what would the perimeter of this (Figure 3.23) be?"

Without hesitation, Alicia says "y, y, 2, and 2"

Ms. Davis asks, "and how much would that be all together?"

Alicia says "2y+4."

Ms. Davis then points to the assigned task. "But here – pointing to the expression $(2x + 2y + 4)$ – we're missing the $2x$." Alicia points to the two sides of length x in the xy piece, saying "$2x$".

Ms. Davis then asks, "How can we put these things (the xy tile and two y tiles on the table) together, so that we get $2x + 2y + 4$?"

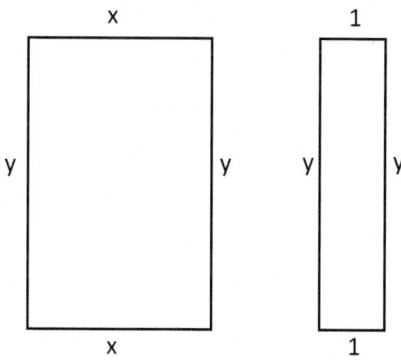

Figure 3.22 *"xy"* and *"y"* tiles

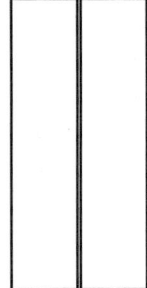

Figure 3.23 What is the perimeter of this object?

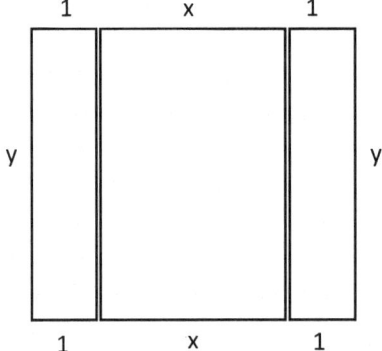

Figure 3.24 The perimeter of the new object

As she is asking the question, Carla puts the three pieces together to make a rectangle. (Figure 3.24). Alicia reaches over and confidently counts the perimeter: "2y" (pointing to the left and right sides), "2x" (pointing to the "x" parts of the top and bottom), "and 1, 2, 3, 4" (enumerating the four units adjacent to the x's).

Ms. Davis asks Alicia to show her again. Alicia takes a pencil and marks the units on the side of the figure as she counts: "y, y, x, x, 1, 2, 3, 4, which would account for $2y + 2x + 4$," which she writes with complete confidence.

Ms. Davis leaves the group, saying "so that's what you want to do for the other two (assigned problems)."

If you had walked into the classroom as Ms. Davis knelt down to work with Alicia, Bernardo, and Carla on this task, you'd have seen little to distinguish the three students. They had all misinterpreted the task, as evidenced by Alicia reaching for x tiles at first and Bernardo and Carla erasing parts of their worksheets when Ms. Davis explained what was intended. When they understand the task, Alicia confidently finds the perimeter of the figure that Carla has put together. In this exchange, Alicia appears completely comfortable with the content and her capacity to do the assignment. She doesn't need to rely on Bernardo and Carla as support for her interactions with Ms. Davis.

All told it appears that, despite her having struggled on the previous problem, Alicia's success in rising to the occasion and meeting the standards maintained by Ms. Davis has given her a boost in confidence. We don't want to read too much into this; it's just one 12-minute episode. But, the bottom line appears to be that all three students learned from their interactions in episodes 1–4, and that Alicia gained a boost in agency from having met the stringent standards the group was held to. (It's hard to say anything about any changes in Bernardo or Carla's agency, since Alicia was the one who faced the most pressure in this lesson.)

3.4 Reflecting on "Where is the ten?"

A lot of things happened during this lesson. If we were meeting in person, we'd have watched the video a few times, stopping the tape at places we thought were important or confusing to make sure we understood what was happening. That's what we tried to convey in Section 3.4.

Here we'd like to take a more distanced view, examining the whole case through the five lenses of the TRU Framework.

Dimension 1, The Mathematics

| The Mathematics | • What are the big ideas in this lesson? |
| | • How do they connect to what I already know? |

 As we saw in Section 2.1, there's the potential for a lot of rich mathematics to emerge from discussions of the algebra tiles task at the heart of this case study. What do you think about the mathematics that emerged in the conversations between **Alicia, Bernardo, Carla, and Ms. Davis? In what ways was it challenging and rich? How might the choice of mathematical topic and task design have shaped what the students experienced? What kinds of mathematical expectations were established for and by the students, and how did they play out?**

Please think about and discuss these issues before you read our commentary.

Our Commentary

One of the interesting things about the mathematics task that provides the backdrop for this vignette is how easy it would be to miss most of the substance that *can* emerge from a discussion of the task. This was the first task:

As we saw in Section 3.1, you can label all the pieces in the original figure by their lengths. With 2 x's, 8 (x – 1)'s, and 18 1's, the perimeter is seen to be $[2x + 8(x – 1) + 18] = 10x + 10$. If Ms. Davis had simply checked that the students had arrived at the right answer and had them Ion to the next task, much of the potential richness of the task would have gone to waste.

Her question "where is the ten?," asked with real intent – *"show me where the ten is"* – is what opens up the potential of the task. Explaining where the 10 comes from calls for making the connections between the algebraic symbols and the objects they represent. When students learn to make such connections, they begin to do the kinds of algebraic sense-making that really matters – e.g., a length of x in the figure can be visualized as a length of (x – 1) plus a length of 1. Without making such linkages, students can get in the habit of "symbol pushing" or "empty symbol manipulation," which puts them at risk of getting answers that don't make sense at all. (How many of us have seen students solve a word problem and write down that someone is 160 years old, that someone runs at 100 miles per hour, or that the height of a building is a negative number?)

Making connections is one key mathematical practice; providing clear and coherent explanations is another. The question is, what are the standards for such explanations, and how are they maintained, if they are? In short, what matters is how mathematics comes to life in classroom discussions.

We believe all three students in this lesson engaged meaningfully with mathematics and that they gained a lot from that engagement. Alicia is the student at the center of the case, and it's worth remarking on what she learned. At the beginning of the lesson segment she was confused about the difference between 10s and x's, thinking she had shown where the 10 is when in fact she was counting 10 x's. When that was pointed out, she struggled to explain the source of the 10. But in the interactions with her group-mates she developed a robust understanding of where the ten came from, and made it her own. The most important evidence for that is when Alicia talks about "donating" some of the 18 1's to compensate for the covered parts of the x's. That's her own language, an indication that Alicia "owns" the explanation. The same is the case when Carla points to the covered 1's and says they're "in the past." The students can and do provide explanations using the formal academic language, but the fact that they provide explanations for why 8 of the 1's get subtracted *in their own words* indicates both sense-making and ownership. This is notable for Carla too. Her grasp of the reason for subtracting 8 seemed shaky at times, but it seemed more solid at the end of the lesson. And we think Bernardo may have gotten better at explaining. At minimum, his explanations grew more focused in response to Alicia's and Carla's specific requests for help.

We think that the classroom norms played an important role in helping the mathematical richness of the task come to life. Specifically, consider norm 3:

> **Norm 3.** The mathematical standards of the original task are high, and they are maintained. Students are expected to address the question as posed, and to provide clear, coherent, and detailed explanations.

From the beginning, Ms. Davis made her expectations for student explanations very clear. This was frustrating for Alicia, Bernardo, and Carla at times. It would have been much easier for everybody to say the answer is $10x+10$ and move on to the next problem. Yet, Ms. Davis's decision to maintain high standards paid off. Yes, it was a challenge for the students. It took significant time and effort for Alicia, Bernardo, and Carla to get to the point where they could all produce cogent explanations, in which they related the x's and 1's to parts of the figure and combined them appropriately. Maintaining the standards (norm 3), while also making sure that Alicia had the resources that would enable her to meet them (norms 1 and 2; see the discussions below) established the context for the progress that all three students made.

But there's more – there's also the question of the ways in which the students have internalized the mathematical standards. Consider this exchange at the beginning of episode 4:

> Bernardo says, "No. No. Tu nomas dile menos 8. (No, No. you don't have to tell her more than minus 8.)"
>
> Gesturing emphatically, Alicia responds, "Me va a decir de donde agarraste el 8 (She is going to ask me where I got the 8)."
>
> Carla, yawning, turns to Bernardo and asks, "where did you get the 8?"
>
> Bernardo rotates the paper so that it is now facing him and says, "Son 18 (There are 18)"
>
> Alicia repeats her concern. "Aja, son 18, no. Son 18. Le quitamos 8. De donde agarraste el 8 (Yeah, there are 18, no. There are 18. We take away 8. Where did you get the 8 from?)"

"Where did you get the 8 from?" is the mathematical question that links the symbols to the algebra tiles – and here we see Alicia insisting on knowing the answer and being able to explain it. She herself insists on making the connection (and Carla asks as well). Part of mathematics learning is internalizing high mathematical standards – expecting to understand, to make connections, to provide clear explanations. Alicia's comment indicates she's serious about understanding the mathematics. Alicia, Bernardo, and Carla are engaged in meaningful mathematics learning.

Dimension 2, Cognitive Demand

Cognitive Demand	• How long am I given to think, and to make sense of things? • What happens when I get stuck? • Am I invited to explain things, or just give answers?

 The idea behind cognitive demand is that students learn best from "productive struggle" – that it's good for them to be stretched, but not to the breaking point. What individual and/or collective challenges did Alicia, Carla, and Bernardo encounter? What resources were available to each of them to meet those challenges?

Please think about and discuss these issues before you read our commentary.

Our Commentary

"Productive struggle" may sound simple in theory, but it's anything but simple in practice. Just what *is* within reach for any student or group of students? If we over-scaffold we risk taking away the challenge and the potential for growth. If we leave students to struggle for too long we risk their getting bored and disaffected. Worse, if they continuously struggle unsuccessfully they may develop the sense that they're just not good at mathematics. Finding the right level of cognitive demand for your students means knowing them individually and collectively: one student may be up for a significant challenge, for example, while another may need to take smaller steps at present. It also means keeping an eye on the progress they're making, so you can make adjustments as they seem necessary.

The first thing we have to say is that Ms. Davis knew her students and we don't, so it's not at all appropriate for us to make judgments about what she should or should not have done. The second is that things appeared to turn out fine in terms of the students' learning, at least this time. But our job isn't to say "that's great" and to move on. It's to understand what seemed to work, and why; *and* to understand the challenges and risks involved. In this case study, Ms. Davis adhered very strictly to the three norms. Consider norms 1 and 2:

Norm 1: When a group calls the teacher over to explain, this certifies that the group representative is ready to present the group's explanations on their own and without help. The teacher will stop attempts from other students to scaffold a group representative's efforts during the explanation.

Norm 2: All group members are responsible both for understanding the mathematics themselves and for making sure the group representative understands it and is ready to explain it.

Norm 1 put a lot of pressure on Alicia. She was definitely on the hot seat. At the same time, norm 2 meant that Alicia was not on her own; she had the support of her tablemates. From their interactions, it was reasonable for Ms. Davis to believe that Bernardo and Carla had things to offer Alicia, meaning that Alicia should be able to make some progress with their help. In group work, a student's tablemates can be a tremendous resource. They *can* be – whether they turn out to be depends on the group dynamics and the norms in place. Here, norm 2 really has taken hold and is part of the group's "social contract." Alicia knows that she does have to provide the explanation to Ms. Davis (norm 1), but she also has the right to full support from Bernardo and Carla. And as we've seen, she calls on them for help until she truly understands. Norm 2 definitely helped Alicia's work become a productive struggle.

Even so, it's still a judgment call for Ms. Davis regarding whether or when to provide some scaffolding for Alicia, who demonstrates obvious frustration at times. In this case, Ms. Davis doesn't. She adheres to norm 3,

Norm 3. The mathematical standards of the original task are high, and they are maintained. Students are expected to address the question as posed, and to provide clear, coherent, and detailed explanations.

It's worth noting how she does so. The norms are in place, so they serve as a backdrop; it seems as if Ms. Davis is there to remind the students of the rules rather than to impose them. If Alicia were playing chess and moved a knight the wrong way, Ms. Davis would be saying, "Alicia, the knight has to move like this." There's no affect involved on her part, just a gentle reminder. The result is that enforcing the norms doesn't seem punitive; it's just playing by the rules of the game. That's important in terms of Alicia's personal reaction to her struggles (see our discussion of Dimension 4). Adhering to the high standards results in growth for all three students.

Let's start with Alicia. Alicia found the task of understanding and explaining the mathematics challenging – and, at the end of the lesson, manageable. She struggled at times, and there were hardly any guarantees that she would be successful at the end. But, Alicia was clearly working "in the zone" of what could make sense to her. She was strongly motivated to make sense of the mathematics in deep ways. Her insistence on understanding *why* she was subtracting 8 and where the 8 units being subtracted could be seen in the figure ultimately paid off. At the end of the vignette she "owned" the explanation of where the 10 is, despite having been frustrated at times.

When a student is working on things that might well be in reach, the situation is an example of *potential* productive struggle. In this case Alicia's struggles did turn out to be productive, thanks to the collective resources of the group.

Carla may have assumed more about her own understanding than was justified at some points. Yet she did remain engaged, following the conversation at all times and commenting when she felt like it. Some of Alicia's questions resulted in Carla realizing that she didn't fully understand where the 10 came from, as when (in episode 4) she turned to Bernardo and asked "where did you get the 8?" Her explanation to Ms. Davis at the end of episode 4, where she explained that 8 of the 1's were "in the past," demonstrates that she too had developed deeper understandings. So, although the spotlight in this case study was on Alicia, we see that Carla, too was positioned to engage in productive sense-making.

The issues related to Bernardo are more subtle. At the beginning of the lesson segment he appeared to have a good grasp of where the ten came from, so it's impossible to tell how much of a challenge the mathematical content was. But, it's important to remember that explaining is a core part of doing mathematics. Bernardo's explanations get more nuanced as the lesson proceeds. His early explanations don't have a sense of linkage to the figure; they're largely about the 8 "minuses" in the 8 $(x - 1)$s. In episode 4, he makes the linkage explicit *and* makes it clear to both Alicia and Carla. He too is getting better at a core mathematical practice, because he's being nudged to get better at explaining to his group-mates.

All in all, the three students were each moved forward. The environment provided room for growth and some pressure to grow. And this time, at least, the pressure wasn't too great; Alicia's struggle was productive, and Bernardo and Carla made progress as well.

In concluding this discussion of cognitive demand we should note that it would be unreasonable to expect such growth all the time! There are times when we practice skills, times when we hone or solidify our understandings, times when we find ourselves in over our heads. That's life, perfect learning is impossible. The goal is to arrange things so that, a large percentage of the time, students are grappling with ideas that may stretch them. Doing so is ongoing work. We'll return to that issue in our discussion of formative assessment, Dimension 5.

Dimension, 3 Equitable Access

Equitable Access to Content	• Do I get to participate in meaningful math learning? • Can I hide or be ignored? In what ways am I kept engaged?

Equitable Access is "the extent to which classroom activity structures invite and support the active engagement of all of the students in the classroom with the core mathematical content and practices." To what degree were each of the three focal students in this case study invited to engage in the core content and practices described in Section 2.5.1, what opportunities did they have to participate, and in what ways were they each supported in them?

Please think about and discuss these questions before you read our thoughts.

Our Commentary

An important point about equitable access, as indicated by the figure below, is that it does not mean that everybody gets the same things – it's that everyone gets the support they need to engage with important content and practices. The task the students grappled with in this case study, to carefully explain why there are only 10 1's in the perimeter, is challenging. It was never made easier, and the difficulty wasn't "scaffolded away" by hints from the teacher.

So the questions for us here are: what opportunities did Alicia, Bernardo, and Carla have, and what kind of support did they get? We'll focus here on opportunities to engage with the content and practices relevant for the algebra tiles task. In the next subsection (Dimension 4), we'll discuss the impact of the discussions on the students' agency, ownership, and identity.

Let's start with Alicia. Once she was chosen as the group representative, Alicia had significant responsibilities for understanding and explaining the important mathematical ideas in the perimeter problem. But were they *opportunities?* That depends on the context. If the "Where is the ten?" task turned out to be far beyond her reach and she wasn't supported in addressing it, then serving as group representative wasn't really an opportunity to learn. What turned her role into an opportunity was the set of support structures that were in place – her group's collective responsibilities and Bernardo and Carla's willingness to pitch in and support Alicia in understanding and explaining where the ten came from. And, we shouldn't neglect Alicia's drive to learn. She really wanted to understand why they should subtract 8 from 18, and kept at the issue until she could paraphrase what was happening in her own words.

In sum, Alicia had opportunity *and* support. While she might have been lost if the challenge was to figure out by herself where the 10 came from, the combination of classroom norms, her group-mates' pitching in (even to the point of giving Alicia a chance to rehearse her explanation before Ms. Davis came over to check their work), and her own drive to learn resulted in Alicia's being able to engage productively with the task. She grappled meaningfully with both the mathematical content and with developing a solid explanation of it.

Let's turn to Bernardo. As far as we can tell, he had figured out an answer to "Where is the ten?" by the time Ms. Davis joined the group. And, Bernardo was positioned by the others as someone who "gets it." But that didn't mean that he understood fully, or that he got to sit back and wait while the others worked. Bernardo's responsibilities as a team member – making sure that Alicia understood and could explain where the ten came from – provided him with opportunities to learn. As the lesson progresses, we see him understanding more of what is challenging for Alicia, and we watch his explanations getting more thorough. At the beginning he says

> "No. No. Tu nomas dile menos 8" – in essence, "Just tell her you're taking 8 away. No need to worry about more than that."

In episode 4, however, we see him providing a thorough explanation that includes pointing to the 8 1's that get subtracted; and, we see him checking that Alicia "gets it" and can explain. Thus the environment provided Bernardo with opportunities to grow, due in part to the existence of the classroom norms (particularly norms 2 and 3), with Alicia and Carla calling upon Bernardo for more clear and thorough explanations.

Finally, Carla. Because Carla wasn't serving as the group representative in this lesson segment, she wasn't on the hot seat in the same ways that Alicia was. Yet, the norms supported her too. As a group member, Carla was responsible for understanding and being able to explain the same things as Alicia. At first, Carla acted as though she understood things fully, and some of her comments to and about Alicia were dismissive (see our discussion of AOI below). And yes, Carla deferred to Bernardo for the most part, letting him provide most of the explanation to Alicia. At the same time, Carla was monitoring her own understanding as well as Alicia's. At the beginning of episode 4, when Alicia told Bernardo she wanted to be able to explain which 8 1's get subtracted, Carla stopped her own work, looked up, and asked "where did you get the 8?" She followed Bernardo's evolving explanation actively alongside Alicia, and had the same moment of enlightenment when the explanation came together. (That's when the tension breaks and Carla says "Duh Doofy.") And, when Ms. Davis joins the three students, Carla is happy to volunteer an explanation. Thus we see that Carla too had significant opportunities to learn and was supported in learning.

In sum, equitable access plays out in very interesting and different ways for these three students. Alicia, Bernardo, and Carla begin engaging with the "Where is the ten?" challenge with different understandings. But, the ways things play out, each has opportunities to grow, and collective support in doing so, while the high mathematical and explanatory standards of the task are maintained. All three students contribute meaningfully, albeit differently, to the group's collective understanding and explanation. That exemplifies what's meant by equitable access to engage with (and learn) central mathematical content and practices.

We make one concluding comment. As we've noted, Ms. Davis's decision to let the students work things out themselves, while offering little guidance of her own, was hardly risk-free. Things worked out well this time; there is no guarantee they always will. Again, it's important to recognize that she knows her students and we don't, so we're not about to second-guess her decisions. What's critically important in this case study is that Ms. Davis has shaped the context in which the students work, developed a task that challenged the students but also provided handholds for them to make progress, and built the norms and support structures that enabled Alicia, Bernardo, and Carla to learn collectively and individually.

Dimension 4, Agency, Ownership, and Identity

Agency, Ownership, and Identity	• What opportunities do I have to explain my ideas? In what ways are they built on? • How am I recognized as being capable and able to contribute?

 Issues of agency concern students' willingness to dig into mathematical challenges. Ownership relates to students making the math their own, rather than implementing procedures they're given. Identity concerns how the students see themselves as doers of mathematics. These are all shaped by the classroom interactions – how students are positioned, what they're responsible for, which resources (including linguistic resources) they feel comfortable bringing to bear, and the kinds of progress they make. What evidence do we have about these issues, for each of the three students?

Please think about and discuss these questions before you read our thoughts.

Our Commentary

As teachers we face challenging situations and decisions related to AOI every day – students can say hurtful things to each other; they can get disheartened if they feel the work is too challenging; and more. Mindful of the fact that things can work out differently and that we don't really know Ms. Davis's students, we discuss some of the tensions faced by Ms. Davis, the decisions she made and the risks and outcomes.

One observation we want to emphasize is that Ms. Davis displayed a tremendous amount of confidence in her students' abilities to work things out by themselves – in fact, more than they themselves display at times! Consider, for example, the exchange that closed out episode 3:

> Carla [commenting on a flawed explanation from Alicia] interjects matter-of-factly: "Ok, well, you're just confused."
>
> Alicia, clearly annoyed, continues. "Well, why don't you all explain it then?"
>
> Carla responds, "Because she wants you to explain it."
>
> Alicia, now a bit more relaxed, says, "And what if I want you to help me explain it?"
>
> Ms. Davis jumps back into the conversation and says, "That's right. You help her and I'll be back."
>
> Alicia puts her head down on the desk and protests, "No. You stay here so I can explain it to you."
>
> Ms. Davis doesn't give in to Alicia's plea. "Nooooo. You need to be able to-"
>
> Alicia tries again, "Why are you asking me? Why don't you ask them?" She points to Bernardo and says, "He's the one who can explain it."
>
> Ms. Davis builds on Alicia's observation, saying, "Yeah. So listen to him," as she walks away from the group.

There was a great deal of tension in this exchange. As teachers, we might be tempted to ease that tension by letting Bernardo or Carla pitch in with parts of the explanation. Or, we might be tempted to give a hint or two. We might also be tempted to intervene when we hear Carla making disparaging comments about Alicia. Ms. Davis's decisions to not yield to either temptation are clearly based on her knowledge of the students. We can tell, even from our brief observations of their interactions, that Alicia and Carla are very comfortable wisecracking with each other. Given that, Ms. Davis may not want to step in to referee their interactions – unless their give-and-take seems to get out of hand.

It's equally important that the students, despite their occasional frustrations, have embraced the classroom norms. Alicia wants to understand why you should subtract 8 1's from the 18 that are visible, and to show that she can provide a clear explanation of why. Bernardo and Carla see themselves as part of a team and take their responsibilities to the collective effort seriously. So, the conditions have been established for productive struggle: Bernardo has the mathematical understanding (even if he hasn't yet come up with a way of explaining it to Alicia's satisfaction), and Carla has a partial understanding and is ready to pitch in as a coach. The resources for productive collaboration are there. As we discussed above, there's always a risk to leaving the students on their own to sort things out. But based on what she knows about Alicia, Bernardo and Carla, Ms. Davis has confidence that they can make progress (and she can always intervene later on if they don't).

Because she happened to be the group representative this time, Alicia played the central role in this vignette. We saw her struggle – but we also saw in the epilogue that she displayed a significant amount of agency in working on the next task, undoubtedly a result of her success in dealing with the "where is the ten?" question. Again, it's important to note that in the vignette she made the explanation very much her own, coming up with the idea that the covered 1's are "donated" to the perimeter. Her ownership of this explanation contributed to her confidence and at least for the moment, to her identity as someone who can (even with hard work) understand and explain mathematics.

The issues are more subtle for Bernardo and Carla. Both appeared to have a sense of agency, willingly digging into the tasks and explanations. Bernardo certainly seemed to own the mathematics, and presumably had a good sense of himself as an explainer. Indeed, he got better at explaining as the lesson progressed – and repeated experiences of this type contribute to a strong mathematical identity. There isn't much solid evidence about Carla, but it's worth noting that she actively monitored her own understanding, learned during the exchange, and happily jumped in with explanations when she had a chance.

Reviewing this case as a whole, we see Alicia, Bernardo, and Carla as engaged learners and teammates. They take their responsibilities to each other seriously. Not only is that a very positive and productive aspect of who they are mathematically and academically, it means that the three students serve as meaningful resources for each other (see also Dimension 5). Their knowledge and understandings are available to share.

Further, the fact that the classroom environment supports their conversing in Spanish as well as in English makes additional resources available to them. One thing for sure is that Alicia, Bernardo, and Carla were working hard during this lesson. They took the task of understanding and explaining where the ten comes from seriously, and that's not an easy task! If they were restricted to communicating in English, the task would have been more difficult – for no good reason, since Alicia's ultimate explanation was in English, and met the high standards of being linguistically and mathematically correct. That they could think and discuss the mathematics in whatever way is most comfortable, and then produce an answer in English that meets the relevant standards, makes a lot of sense. But there's more to it than that. Alicia, Bernardo, and Carla are a team. They're more relaxed bantering in Spanish, and feeling more at ease almost certainly contributes to their ability to dig in and do a good job of figuring out the mathematics. Ultimately, being able to dig in that way helped them make the mathematics their own. That feeling of ownership is essential.

Finally, it's worth thinking about identity in very broad terms. Identity is always a "work in process." Twelve minutes in a math classroom is a tiny part of a student's total mathematical experience. We can't expect the brief experiences in this lesson segment to have significant impacts on Alicia, Bernardo, and Carla's senses of mathematical agency, ownership, and identity. But, each student's mathematical history is made up of thousands of such experiences, and those experiences do have a cumulative impact. The success Alicia,

Bernardo, and Carla had in confronting the challenges of the "where is the ten?" problem could be seen in the ways in which they dug into the next problem, as described in the epilogue. Such experiences add up!

Dimension 5, Formative Assessment

Formative Assessment	• How is my thinking included in classroom discussions? • Does instruction respond to my ideas and help me think more deeply?

 Formative Assessment involves making students' thinking public and, if necessary, adjusting classroom activities to support students in engaging meaningfully and productively with the content. The resources available are not only the curriculum and the teacher, but other students as well. In what ways did formative assessment play out in this case study, with regard to rich mathematics and the opportunities each of the three students had to engage meaningfully with it?

Please think about and discuss these questions before you read our thoughts.

Our Commentary

We start by emphasizing some of the big issues related to TRU and to formative assessment.

TRU's fundamental focus is on how each student experiences the learning environment – which includes the teacher, the curriculum, fellow students, and the social environment as orchestrated by the teacher. All of these are potential resources. So, when we think about formative assessment, the key question is *not*, "how does the teacher respond to student thinking?" It *is*, "How does the teacher orchestrate classroom materials, activities, and all available resources to bring forth student thinking and respond. productively to it?"

From that perspective, we see formative assessment at work throughout the lesson segment. Ms. Davis has arranged the environment so that the mathematics is important, student thinking is revealed, and resources are arrayed to support each student's progress.

All through the lesson, Ms. Davis's questions are aimed at key mathematical content and practices. Her opening prompt "show me where the ten is" is aimed at making linkages between the algebraic symbols and the algebra tiles, and her call for verbal explanations (supplemented by gestures that highlight particular parts of the figure) results in the students' sense-making being "out in the open." Moreover, Ms. Davis is looking at both the big picture and at specific details. We saw that at the beginning when Alicia was counting x's instead of 1's, and at the end, when Alicia produced a clear and complete explanation of why she subtracted eight 1's. Moreover, it wasn't just Alicia whose thinking was revealed; Bernardo's understanding was clear and Carla's explanation at the end provided evidence that she, too, understood why eight units are subtracted from the 18 visible units. Throughout the lesson, the students are focused on important mathematics and their thinking is out in the open and accessible.

What makes formative assessment so powerful in this case study is how Ms. Davis organizes the students to serve as resources for each other. In a traditional classroom, she might have provided explanations when the students got stuck. The students would undoubtedly have learned something, but a good case can be made that when the students serve as resources for each other, as they did here, their learning is deeper and broader.

Alicia had to explain the mathematics to her peers *and* to Ms. Davis. Developing an explanation with Bernardo and Carla resulted in an explanation that made sense to all three of the students, in ways that were personal and meaningful. The feedback that Ms. Davis provided when Alicia's explanations were incomplete pointed to places where they needed to work further, and the explanations Alicia, Bernardo, and Carla ultimately produced in response to her feedback met high mathematical and expository standards. All three students contributed to and learned from their exchanges.

Part of formative assessment involves the teacher judging and if necessary adjusting the level of difficulty so that students can engage in productive struggle. Ms. Davis's judgment was that the group resources were up to the task. When she provided feedback, she used the students' language and, while she did focus their attention on what needed work, she did not inject new information into the conversation. As we've seen, the students rose to the task. Mathematics-wise, things worked out fine, but there's more than just mathematics learning taking place here. In addition to crafting an environment (including the classroom norms) in which students served as resources for each other, Ms. Davis shifted some of the responsibility for developing the mathematical connections to the students themselves. That each of them succeeded in their own ways resulted in their individual and collective growth in agency, ownership, and identity. In this way, formative assessment plays more than a purely mathematical role. It helps students continue to develop as mathematical thinkers, learners, and community members.

Note

1 These will be described in some detail later in the chapter.

Chapter 4

Graphing Quadratic Functions

A Focus on Formative Assessment

4.0 Introduction

This chapter focuses on the first 30 minutes of a "Formative Assessment Lesson" (FAL) in an Algebra II classroom. The FAL is a review/synthesis lesson that addresses what students understand about the algebraic representations of quadratic equations and their graphs. In previous classes, the students have graphed a range of functions including linear functions and their absolute values (e.g., $y = |x{-}3|$), parabolas, hyperbolas (e.g., $y = 1/x$), exponentials, logarithms, and square roots.

The entire 90-minute FAL can be downloaded at no cost from https://www.map.mathshell. org/lessons.php?unit=9245&collection=8. It is aimed at helping students review and synthesize what they know about parabolas and the different forms in which they can be represented algebraically. In the first 30 minutes of the lesson that provide the substance of this case study, the class discusses three questions designed to reveal and reinforce their understandings of central ideas related to the graphs of quadratic functions. The mathematical goals for the lesson are summarized for the teacher in the FAL's extended lesson plan as follows:

> This lesson unit is intended to help you assess how well students are able to understand what the different algebraic forms of a quadratic function reveal about the properties of its graphical representation. In particular, the lesson will help you identify and help students who have the following difficulties:

- Understanding how the factored form of the function can identify a graph's roots.
- Understanding how the completed square form of the function can identify a graph's maximum or minimum point.
- Understanding how the standard form of the function can identify a graph's intercept.

This case differs from the cases discussed in Chapters 2 and 3 in at least one important way. The "Where is the Ten?" and "Car Value Problem" lessons focused on students making sense of new material. In contrast, the students in this classroom begin the quadratic equations FAL knowing a significant amount about quadratic functions and their graphs. However, it would be wrong to think of this FAL as a "review" lesson that simply provides opportunities for practice. In the lesson the students are challenged to deepen their understandings by comparing and contrasting

DOI: 10.4324/9781003375197-6

graphs, synthesizing what they know, and exploring the kinds of information that different representations of parabolas can offer. The teacher, alert to what the students say as they discuss the tasks, asks the whole class to consider some of the issues they raise. What we see is formative assessment in action, as students refine and build on their mathematical understandings.

Our focus, then, includes the mathematical possibilities inherent in algebraic descriptions of parabolas; we zoom in on the use of formative assessment as students explore those ideas. Here we expand somewhat on the discussion of formative assessment in Chapters 1 and 2. The short description of formative assessment that we gave in Figure 1.1 is

> The extent to which classroom activities elicit student thinking and subsequent interactions respond to those ideas, building on productive beginnings and addressing emerging misunderstandings. Powerful instruction "meets students where they are" and gives them opportunities to deepen their understandings.

We've noted that the five dimensions of TRU are somewhat independent, and that each has its own integrity. Yet, formative assessment is unique in the way it works: attending to what your students say and do, and adjusting accordingly enhances each of the other four dimensions. In broadest terms, the uses of formative assessment cut across all the dimensions of the TRU Framework. Figure 4.1 captures these multiple relationships. Our discussions will reflect the way FA plays out in the lesson.

To presage the discussion that follows, we suggest the nature of each arrow in Figure 4.1 with a question:

- (Dimension 1, The Mathematics): In what ways are the things that students reveal about their understandings used to lead the class into rich mathematical explorations?
- (Dimension 2, Cognitive demand): In what ways is classroom discourse leveraged to help students engage in sense-making that is grounded in what they know?
- (Dimension 3, Equitable Access): In what ways are opportunities to participate arranged so that every student has opportunities to engage meaningfully with core content?
- (Dimension 4, Agency, Ownership, and Identity): In what ways is classroom discourse supported so that students have a chance to contribute to collective understandings and see themselves as powerful mathematical thinkers?

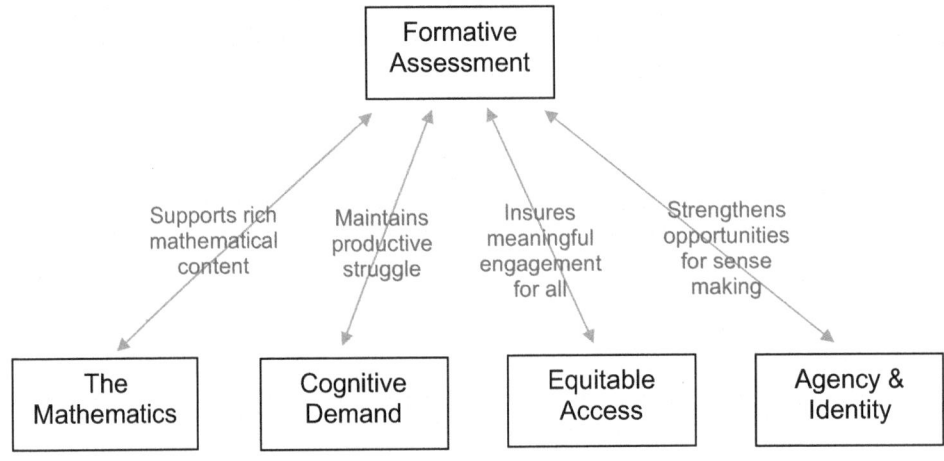

Figure 4.1 The key roles of formative assessment in powerful mathematics classrooms. Reproduced with permission from Burkhardt and Schoenfeld (2019)

A general feature of this case is our consistent attention to the following type of question: what does something a student says reveal about the student's understanding, and how might that issue be pursued?

This question underlies the design of the FALs. To begin, the tasks themselves are designed in ways that engage students in thinking and generating ideas, not simply "answer-getting." This design feature means that there are significant opportunities for students to reveal what they do and don't understand. As you'll see, suggested class structures include group work, student presentations, and whole-class conversations. In setting up classroom dynamics that way, the lesson plan provides multiple opportunities for students to engage with each other over the content. Part of the FAL development process includes research into the aspects of the content that students find challenging, and ways to address them. Pilot testing of the materials revealed places where students ran into difficulties – some of which are perfectly natural! (The idea is not to avoid errors such as incorrect generalizations, but to be prepared for them, so that martial understandings can be refined.) The lesson plan for this and every other FAL includes descriptions of common student errors and suggestions to the teacher about ways to address them. But as we'll see, listening to student thinking can lead to discussions that extend beyond a lesson plan – even a lesson plan that is 17 pages long!

4.1 Thinking about the Mathematics in the Quadratic Functions Formative Assessment Lesson

We're going to approach the mathematics at the core of this lesson somewhat differently than we did in the first two case studies. In "Car Value Problem" and "Where is the Ten?" the students were introduced to new mathematical concepts or applications through the use of carefully chosen introductory problems. The natural question to ask under those circumstances was: "What important mathematical ideas do we want to emerge from the students' work on this problem?"

In contrast, the FAL the students work on in this case study is a review/synthesis lesson that comes after the students have engaged with quadratic functions. The teacher can assume that the students have some, although not always solid, knowledge of the basics, including definitions of function, domain, and range; and that they have some understanding of the standard forms for graphing parabolic functions. (In fact, the graphs on the classroom wall indicate that the class has also studied functions such as $y = 1/x$ and its translations.)

In Section 4.1.1, we begin by grappling with the big question: just what do we want students to know and be able to do when it comes to graphing quadratics? In Section 4.1.2, we explore the specific problems that the students are assigned and the ways in which those problems can be used to reveal what the students know. In Section 4.1.3, we briefly describe the subsequent part of the FAL, in which the students work a task that provides practice on the understandings discussed in Section 4.1.2.

Before we delve into the mathematics, it's important to note that there's a *lot* here – you might want to take your time going through it. We will cover a month or more worth of math in Section 4.1.1, so it may be slow going in places. But, we've been told by readers and by participants in our workshops that it was really worth getting their heads around the mathematics in Section 4.1.1. In particular, we are convinced that the more deeply we understand the mathematical potential of the lesson, the better we will be positioned to (a) anticipate what our students might say and help them build on it, and (b) capitalize on the potential of the task to help the students develop deeper understandings and engage in productive mathematical practices.

4.1.1 General Issues Regarding the Graphs of Quadratic Functions

Please think about the following question about the graphs of parabolic functions before reading some of our thoughts. If you're working with a group, please discuss your responses to the question together. It's a big question – take your time to consider it! Our response follows.

 Imagine that you've finished teaching a unit on graphing parabolas. What would you like students to know and be able to do? What skills, understandings, and fluencies are desirable?

Here are some of our thoughts about the mathematics that students should know.

At this point in an Algebra II course the students should be familiar with basic definitions related to function, domain, range, and quadratic expressions. They should understand the basics related to graphing, e.g., that the point (a, b) is on the graph of $y = f(x)$ if and only if $b = f(a)$.[1]

In what follows, we'll assume for the sake of simplicity that we're graphing expressions of the form $y = f(x)$, which means that parabolas are quadratic expressions in x.[2] That is, we're interested in graphs of functions that can be expressed in the form $y = ax^2 + bx + c$, where a, b, and c are constants (and $a \neq 0$).

In this discussion we'll start with the graphical properties of parabolas, addressing questions of what the graphs of quadratic functions look like. After that we'll focus on the algebraic properties of quadratics, including the different forms in which the equation of a parabola is usually written. For reasons of space, we'll just describe what's important. It goes without saying that in instruction, students should come to understand how and why all this fits together.

Graphical properties of quadratics

The general shapes and properties of parabolas.

If we're thinking in terms of $y = f(x)$, then every quadratic function $y = ax^2 + bx + c$ is defined for all values of x. (Formally, the domain of the function is all real numbers.) That means that the graph "spans" the entire x-axis – even if it doesn't look like it! – and that it has a y-intercept at the point $(0,c)$. Generally speaking, the graph of a parabola is U-shaped, with the "U" opening upward if the lead coefficient a is positive and opening downward if a is negative. The vertex of the parabola is the lowest y-value of the parabola when the parabola opens upward, or the highest value when it opens downward. If we know the vertex of the parabola, then we know the range of the function. The range consists of all values that are greater than or equal to the y-value of the vertex if a is positive, and all values less than or equal to the y-value of the vertex if a is negative. The value of a also indicates how wide or narrow the parabola appears to be. The larger the value of $|a|$ is, the narrower it appears. The closer the value of $|a|$ is to zero, the wider it seems. (Explaining why this is the case is a good exercise for students.)

A few illustrative graphs are sketched in Figure 4.2.

Here are the equations of the four functions sketched in Figure 4.2:

1. $y = x^2 + 4x - 3$ 2. $y = 5x^2 - 7x + 1$ 3. $y = -x^2 - 4x - 3$ 4. $y = -.1x^2 + x - 4$

Can you identify which sketch corresponds to which equation and justify your answer briefly? This is a good exercise for our students. We'll address issues like this in the next part of this discussion.

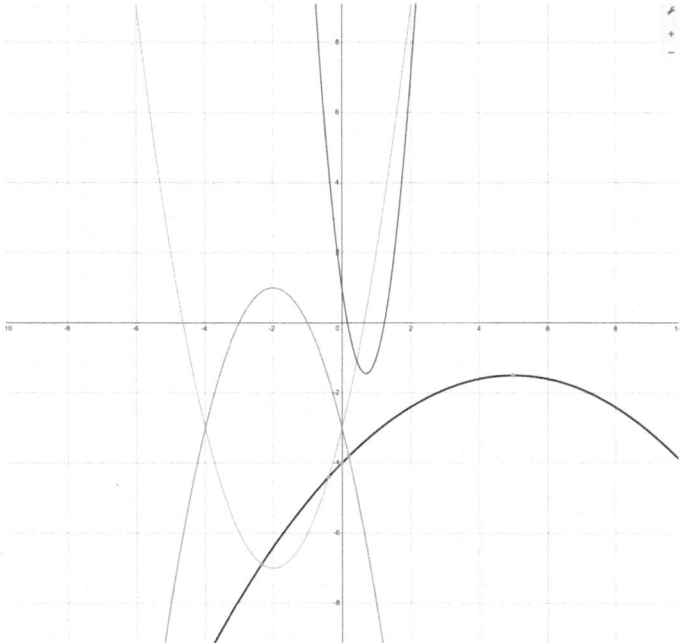

Figure 4.2 (Partial) graphs of four quadratic functions

Key properties of the equations and graphs

It's useful to remember that our perception is limited by what we can draw or see on a screen. Every quadratic function is defined for all values of x (the domain of the function is all real values of x), so if you "zoom out" to see more and more of the parabola, the x-values of the parabola will eventually cover the whole x-axis. That means in particular that every quadratic curve has a y-intercept, whether or not you see it in the graphing window. Finding the y-intercept is straightforward – it's the value of $f(x)$ when $x = 0$. That means in particular that if the parabola is written in the standard form $y = f(x) = ax^2 + bx + c$, the coordinates of the y-intercept are $(0,c)$.

If you want to sketch a parabola (or graph it to any desired degree of accuracy), then a few features are critical. The first is the vertex. If you know the coordinates of the vertex and you know whether the lead coefficient is positive or negative, you've got a good start.

What else? The next things to look for are roots of the quadratic. The solutions to the equation $y = f(x) = ax^2 + bx + c = 0$ may be real or imaginary. If they are imaginary, then the parabola doesn't cross the x-axis: either the vertex is above the x-axis and the parabola opens upward, or the vertex is below the x-axis and the parabola opens downward. If there's just one (double) root, then the parabola is tangent to the x-axis (and faces up or down depending on the lead coefficient). If there are two distinct roots, then either the vertex is below the x-axis and the parabola faces upward, crossing the x-axis twice, or the vertex is above the x-axis and the parabola faces downward (again crossing the x-axis twice).

Knowing the vertex and the coordinates of the y-intercept gives you a pretty good sense of how "wide" the parabola is. In fact, knowing these two pieces of information can help you draw a pretty good sketch, especially if you take advantage of symmetry. That's because of the following.

The vertical line drawn through every parabola's vertex is that parabola's line of symmetry. If you know what the "left hand side" of the parabola looks like, then the right hand side of the parabola is its mirror image. (A brief proof of this is given below.) That information also ties together the roots and vertex of a parabola: the x-coordinate of the vertex is half-way between the x-values of the roots. If, for example, you know that the roots of a parabola are 3 and 7, then you know that the x-coordinate of the vertex of the parabola is half-way between (that is, the average of) 3 and 7, or 5. So, the vertex is somewhere on the line $x = 5$ and the parabola is symmetric with respect to that line.

Let's pursue this algebraically for a moment, in anticipation of the next section and to suggest how everything fits together. If you know that $x = 3$ and $x = 7$ are the roots of the quadratic function $f(x)$, then $(x-3)$ and $(x-7)$ are factors of $f(x)$, so $f(x) = a(x-3)(x-7)$, where a is the lead coefficient. Since the vertex of the quadratic is the point on the graph with x-coordinate 5, the vertex is $(5,f(5)) = (5,a(5-3)(5-7)) = (5, -4a)$.

Those are the graphical basics. We'll look at exercises that involve sketching after we turn to what's revealed by the algebra.

Exploring the algebraic properties of quadratics

The basic question regarding algebraic representations of parabolas is,

> Any quadratic expression can be rewritten in a number of different ways. What information about the graph can we get *easily* from any of the different forms in which quadratics are usually written?

The standard form for the parabola is $y = f(x) = ax^2 + bx + c$. (If you're given the equation of a parabola in any other form, you can always multiply out and collect terms to write the standard form.) When you calculate the value of y for $x = 0$, the result is $y = 0a + 0b + c = c$. So, we know that $(0,c)$ are the coordinates of the y-intercept of the parabola. (That fact alone takes us a long way toward identifying the graphs in Figure 4.2. But, generating the graphs takes some more work – see below.)

Suppose the parabola is written in the form $y = a(x-h)^2 + k$. If $a > 0$, then the minimum value of the parabola occurs when $x = h$, and the vertex of the parabola is the point (h, k). If $a < 0$, then the maximum value of the parabola occurs when $x = h$. Again, the vertex of the parabola is the point (h, k).

Also, writing the function in this form – and every quadratic expression can be written this way, by completing the square – shows why the parabola is symmetric around the line $x = h$. Consider the two values of x that are distance d to the left and right of h. Calculating $f(h-d)$ gives us

$$f(h-d) = a\left[(h-d)-h\right]^2 + k = a(-d)^2 + k = ad^2 + k;$$

calculating $f(h+d)$ gives us

$$f(h+d) = a\left[(h+d)-h\right]^2 + k = a(+d)^2 + k = ad^2 + k.$$

Figure 4.3 indicates the symmetry. Admittedly, we don't want to dwell on algebraic operations like these, but when we draw parabolas, we often use their symmetry (rather than point-by-point plotting) in order to sketch them quickly. It's good to know why we can depend on that property, and the other properties discussed in this section.

Finally, if the parabola is written in the form $y = f(x) = a(x-p)(x-q)$, then $f(p) = f(q) = 0$, so the roots of f are $x = p$ and $x = q$.

So how does all this information help us? If we know the information discussed above and we have some fluency manipulating algebraic expressions, we can sketch the graphs of algebraic expressions very rapidly. And, we can "read" the algebraic properties of the functions from their graphs.

A worked example.

Here is one example, which we chose so that the algebra works out nicely:

> Sketch the graph of the function $y = f(x) = x^2 - 6x + 8$,
>
> making use of the information in the various algebraic re-expressions of $f(x)$.

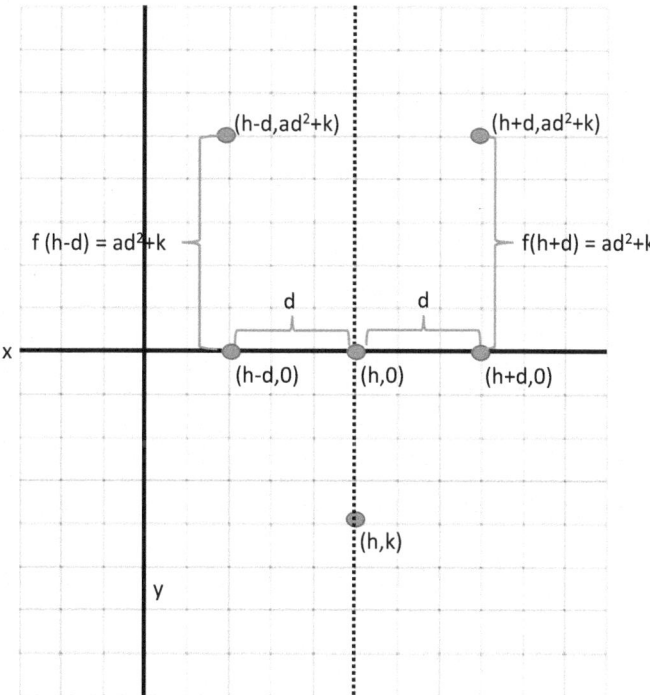

Figure 4.3 If (h, k) is the vertex of a parabola $y = f(x)$, then $f(h-d) = f(h+d)$

Since $f(x)$ is in standard form, we can set $x = 0$ and see that the graph's y-intercept is the point $(0,8)$.

The function factors nicely: $f(x) = x^2-6x + 8 = (x-2)(x-4)$. That says that the roots are 2 and 4; in graphical terms, the points $(2,0)$ and $(4,0)$ are on the graph. Thanks to the symmetry of quadratic functions, we know the x-coordinate of the parabola's vertex lies half-way between the two roots, so the x-coordinate of the vertex is 3. We can use either the standard form or the factored form to compute

$$f(3) = 3^2 - 6(3) + 8 = (3-2)(3-4) = -1,$$

so the vertex of the parabola is $(3, -1)$.

The understandings we developed above allowed us to determine the vertex, the roots, and the y-intercept of $f(x)$ very efficiently. Although there's no need for it in this case, another way to derive the vertex is by have to complete the square:

$$f(x) = x^2 - 6x + 8 = \left(x^2 - 6x + 9\right) + (8 - 9) = (x - 3)^2 - 1$$

This calculation confirms that the vertex of the parabola is indeed $(3, -1)$. Finally, since the parabola is symmetric around the line $x = 3$, and we know $f(0) = 8$, then $f(6) = 8$.

With all this graphic information, we have a very good sense of what the parabola looks like. If we wanted greater precision we could plot more points – but these five points give us a pretty accurate idea of what the graph would look like. See Figure 4.4.

That's the kind of fluency we'd like students to develop.

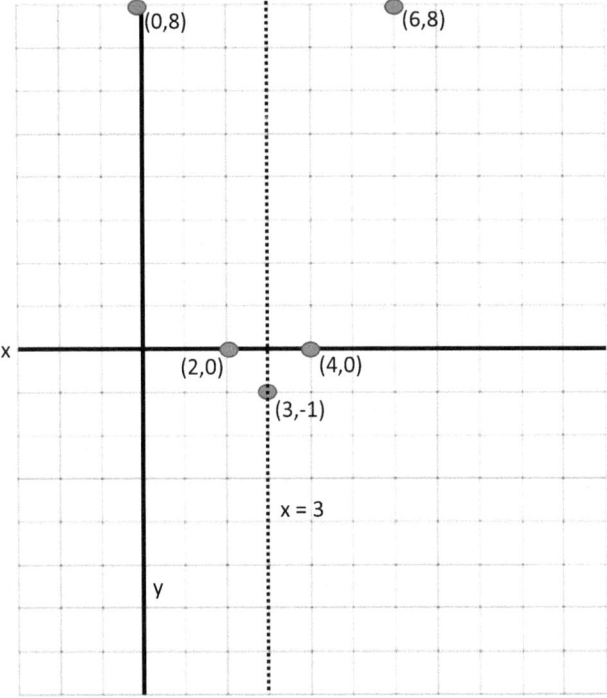

Figure 4.4 The information obtained from the different forms of
$y = f(x) = x^2 - 6x + 8$ provide quick and accurate information
about the graph of the parabola

First Extension: Some "Dynamic" Thoughts

We'd like to think about some of the parameters that appear in the different forms of quadratic expressions.

Let's begin with the constant, c, because it's the most straightforward one to visualize. Here's a "starter" question students should be able to answer:

> Consider the function $y = f(x) = x^2 + c$. What happens to the graph of $f(x)$ as the value of c varies?

We'd like students to recognize that for any value of c, the shape of the parabola remains the same – every graph is a vertical translation of the graph of the function $y = f(x) = x^2$, but with vertex at $(0,c)$. That is, the graph of $g(x) = x^2 + 7$ is precisely what you'd get if you took the graph of $f(x) = x^2$, picked it up, and "lifted" it 7 units higher – for each x, the y-value for $g(x) = x^2 + 7$ is 7 more than the y-value for $f(x) = x^2$. The same is the case for any two functions that only differ by a constant. For example, the graph of $g(x) = 4x^2 - 2x - 6$ has precisely the same shape as the graph of $f(x) = 4x^2 - 2x - 2$, but the graph of $y = g(x)$ is 4 units below the graph of $y = f(x)$. So, students should be comfortable with the idea that changing the constant in a quadratic function "moves the graph of the function up or down," and they should be able to say what point "anchors" the graph.

Similarly, the parameter a tells us how fast the parabola "grows" (alternatively, how narrow or wide the parabola appears to be). Students should be able to describe what happens to the graph of $y = f(x) = ax^2$ as the value of a changes from, say, -10 to 10. (We can imagine them gesturing with their hands, making a narrow downward-facing U-shape, widening it until it flattens out (for $a = 0$) and then turning upward, narrowing again as the value of a gets larger.)

All this helps students "see" parabolas as objects, which is important – they often get mired in plotting point after point, a process that is both tedious and error-prone.

The impact of the parameter b is harder to visualize in the standard form than the impact of a and c. We won't pursue that here, because it's not very useful for purposes of graphing. But, there is a form for which all three parameters provide useful information. If you're given the algebraic vertex form

$$y = a(x - h)^2 + k$$

(or if you start with the standard form and complete the square so that the equation is written as above), then all three parameters a, h, and k in that form have clear visual impact. The role of a is precisely as above: as the value of a changes, the parabola gets thinner or fatter; the parabola flips up or down when a passes through 0. The point (h, k) is the vertex of the parabola, and the line $x = h$ is its axis of symmetry. As the value of h changes, the shape of the parabola remains the same, but the parabola "moves" left or right; as the value of k changes, the parabola "moves" up and down.

All this means that we can have a very good intuition regarding the shape of a quadratic written in the form $y = a(x-h)^2 + k$, *without plugging in any values to calculate.* If we are asked

What does the graph of $y = f(x) = -4(x + 5)^2 + 2$ look like?

we can answer immediately, "it's a rather thin downward parabola whose vertex is at the point $(-5, 2)$." If we want to say more, we can plug in $x = 0$. This allows us to continue,

When $x = 0$, $y = -98$, so the y-intercept is $(0, -98)$. Both roots are negative, since one root lies between the vertex and 0, and the curve is symmetric around the vertical line $x = -5$ (the vertical line through the vertex).

In fact, since $f(-5) = 2$ and $f(-4) = -2$, we know one root of the function lies between $x = -5$ and $x = -4$; by symmetry, the other root lies between $x = -5$ and $x = -6$.

That's the kind of fluency, and the kinds of intuitions, we'd hope students would develop.

Second Extension: How Much Information Do You Need to Determine the Equation of a Parabola?

What follows may feel like a stretch for most high school classrooms, but it's worth mentioning just to round out the picture regarding quadratics and their graphs. The reason that we mention it is that it could be relevant when students discuss the first problem assigned in the case study, which asks them to draw "different" quadratics.

When students first study linear equations, they often develop the impression that the different forms of the equations for a line, such as:

- the general form $ax + by = c$,
- the slope-intercept form $y = mx + b$,
- the two-intercept form $x/a + y/b = 1$,
- the two-point form, which is often written as

$$y - y_1 = \left[\frac{y_2 - y_1}{x_2 - x_1} \right] (x - x_1)$$

but can also be written as

$$y = \left[\frac{x - x_2}{x_1 - x_2} \right] y_1 + \left[\frac{x - x_2}{x_2 - x_1} \right] y_2$$

are all completely different, have little to do with each other, and need to be memorized and used separately. What students need to learn, over time, is that the equation of every line is completely determined by any two pieces of information about that line. (Formally, we say that there are "two degrees of freedom" in determining the equation of a line.) For example, knowing any two points on the line (one or both of which may happen to be intercepts) allows you to determine the equation of the line; so does knowing the slope of the line and any other point on the line (which may happen to be an intercept). Once you know the equation of the line, in any form, you can determine the coordinates of all the points on it, and you can write the equation in all of the other forms. Which form you would want to use is a matter of what's easiest for you to use at the time.

In parallel, there are three degrees of freedom in determining the equation (and therefore the graph) of a quadratic. Once you've specified three pieces of information about a parabola, the parabola is determined. Students can see this by looking at the three forms used to express parabolas,

$$y = f(x) = ax^2 + bx + c, \ y = f(x) = a(x-h)^2 + k, \text{ and } y = f(x) = a(x-p)(x-q)$$

Although the forms look different, they share the property that they each have precisely three parameters. That is: Once you've specified a, b, and c in the standard form; or a, h, and k in the vertex form; or a, p, and q in the root form; you've determined the parabola. No more variation is possible. This explains why, when students try to sketch quadratics that are "quite different" from each other, they can create no more than three differences.

Just for the record, we note there is a "3-point formula" for parabolas that is analogous to the 2-point formula for lines given above. The form can be written as follows:

$$y = \left[\frac{x-x_2}{x_1-x_2}\right]\left[\frac{x-x_3}{x_1-x_3}\right]y_1 + \left[\frac{x-x_1}{x_2-x_1}\right]\left[\frac{x-x_3}{x_2-x_3}\right]y_2 + \left[\frac{x-x_1}{x_3-x_1}\right]\left[\frac{x-x_2}{x_3-x_2}\right]y_2.$$

This form lets you see directly that:

- when you substitute x_1 for x, the result is $(1)y_1 + (0)y_2 + (0)y_3$;
- when you substitute x_2 for x, the result is $(0)y_1 + (1)y_2 + (0)y_3$; and
- when you substitute x_3 for x, the result is $(0)y_1 + (0)y_2 + (1)y_3$.

Finding this formula would be a challenging extra credit assignment for a high school class!

4.1.2 The Specific Mathematical Foci of this Case Study

Turning from this general discussion toward the specifics of the current lesson, we ask: what kinds of exercises will provide students with the opportunities to develop the kinds of understandings discussed in Section 4.1.1?

That's where the lesson in this case study begins. For extended detail on the lesson, see the formative assessment lesson at https://www.map.mathshell.org/lessons.php?unit=9245& collection=8. (We hasten to add that our discussion has been very broad, so the FAL doesn't address everything that we've discussed here.)

Here, in essence, are the four main issues the students in the lesson segment were asked to address:

- Sketch two quadratic curves that look quite different from each other, and explain the fundamental similarities and differences between them
- Can the graph of a quadratic stay entirely in one quadrant?
- Here are the equations of three quadratic functions:

$$y = x^2 - 10x + 24 \quad y = (x-4)(x-6) \quad y = (x-5)^2 - 1$$

- Without performing any algebraic manipulations, write the coordinates of a key feature of each of their graphs. For each equation, select a different key feature.

- Compare and contrast the following functions:

$$f(x) = -(x+4)(x-5) \quad g(x) = -2(x+4)(x-5)$$

These questions set up the main activity in the FAL, which is called "dominos." The full lesson takes an hour beyond the 30 minutes discussed in this chapter. We don't have the space to work through the whole lesson, but we encourage you to look at the FAL to see the way in which the task design provides students with engaging opportunities to use and refine the understandings highlighted above. (Because the task involves a matching game, students discover when their answers don't "line up"; they don't have to be given an answer key or be told by the teacher that they're mistaken. Thus the activity itself provides them with feedback but not answers, supporting formative assessment.)

In what follows we discuss the mathematical potential of the four issues bulleted above. Those issues are the mathematical focus of the case study.

The following question from the FAL is used to start the lesson. (The class does not have whiteboards, so the students are asked to draw the parabolas on a pair of index cards).

Today, we are going to look at the key features of a quadratic curve. On your mini-whiteboards [in this class, on index cards], draw the x- and y-axis and sketch two quadratic curves that look quite different from each other.

What makes your two graphs different?
What are the common features of your graphs?

Imagine asking your students to work on this question in groups. What mathematical points might you want to emerge from the discussion of this task, and how might you head in that direction? What would you want to emphasize in a class discussion? Please think about this before reading our discussion.

Here are some of our thoughts about the task.

The FAL itself provides useful guidance for digging into the problem. The lesson plan suggests calling on students whose curves have these properties:

- one of which has a maximum point, the other a minimum,
- one of which one has two roots, the other one or none,
- that are not parabolas,

prior to asking the two questions italicized above. This approach focuses on some key features (*maxima and roots*) and provides an opportunity to catch misunderstandings (*graphs that are not parabolas*) early in the lesson.

Note the way the question is framed. The fact that students are asked to generate examples on their own means that (as we'll see) there's lots of room for students to reveal what they consider important, as well as what they understand and don't understand. "Closed" questions such as "find the vertex of this parabola" are narrowly focused and may not reveal nearly as much about student thinking. In the FAL lesson plan the teacher is encouraged to "elicit responses from the class and try to keep your own interventions to a minimum." Giving students a chance to make their understanding public is a key aspect of formative assessment.

Reflecting on Images of Practice (The Case Studies)

The lesson plan calls for asking questions that provide students with opportunities to address some of the larger questions about the shapes of graphs. The expectation is that students will explain their thinking. Not only is explanation an important mathematical practice, but giving explanations (in the appropriately supportive climate) provides students with opportunities for the development of mathematical agency and identity. The FAL suggests that students can be induced to consider issues such as the following.

Ask about turning points:

How many turning points does each of your graphs have?

Is this turning point a maximum or minimum?

Can the curve of a quadratic function have more than one turning point/no turning points? Show me.

If all students have drawn graphs with minimums, ask students to draw one with a maximum. Ask about roots:

How many roots does each of your graphs have?

Where are these roots on your curve?

Does anyone have a graph with a different number of roots?

How many roots can a quadratic have?

If all students have drawn graphs with two roots, ask a student to draw one with one or no roots. Ask about y-intercepts:

Has anyone drawn a graph with different y-intercepts?

Do all quadratic curves have a y-intercept?

Can a quadratic have more than one y-intercept?

These explorations into student thinking highlight key mathematical ideas. They help uncover and address incorrect or incomplete understandings, and they provide opportunities for students to contribute to group discussions[3]. In sum, there's a *lot* of mathematics that can be explored given this kind of open-ended question.

The second task emerged in conversation with the students when the class discussed a student's sketch showing a parabola entirely in the first quadrant.

A student suggests that the graph of a quadratic can stay entirely in one quadrant. What are some of the mathematical ideas that can emerge or be reinforced by a discussion of this possibility?

Please consider this issue before reading our discussion.

Here are some of our thoughts about this issue.

We like this question a lot, because it opens up some interesting issues. A conversation stimulated by this question can go in various and sometimes unexpected directions, some of which you'll see as the lesson evolves. One issue considered in our opening discussion is that anything we draw or sketch is only a small part of the "whole thing" – our graph paper or computer screen only shows a small part of the actual parabola generated by the function. Just what does the "rest" of the graph look like?

The question raises questions of what the "U" shape of a parabola really represents, and questions about domain and range. We might ask if the student could imagine what the equation of a parabola with that graph might be. Whatever it is, it'll be of the form

$$y = f(x) = ax^2 + bx + c$$

at which point we can ask if the graph has a y-intercept. Or, we can ask what $f(-1)$ is and what that means. There are many directions in which that conversation can go.

In this classroom, two things emerge. The first is the concept of asymptotes. One student suggests that an "upwards" parabola whose vertex is in the first quadrant might be asymptotic to the y-axis, and thus remain in the first quadrant. That's an interesting comment! It gives rise to the question of what generates asymptotes; it also raises issues of symmetry, because the parabola would also have to have a vertical asymptote to the right of its vertex. As discussed above, a quadratic function is defined for all values of x, so it can't have asymptotes. The student's question provides an opportunity to explore why, and for reviewing which algebraic forms do give rise to asymptotes, and why. (The students have studied functions with vertical asymptotes, so this question is answerable.) So, we might pose a question like this: is it possible to write the equation of a function part of whose graph is an infinitely tall "U" shape with two vertical asymptotes? That pushes the content into a new area, but that kind of graphing is done in algebra. The result would be real sense-making that builds on student thinking, and is anchored in the curriculum. (Moreover, with graphing technology the students can try various conjectures, and see what the graphs of the functions they create actually look like.)

Moreover, if time permits, there are natural extensions of this problem. A more comprehensive version might be posed as,

a. Is it possible for the graph of the function $y = f(x) = ax^2 + bx + c$ to be entirely contained in one quadrant? If your answer is yes, give the equation of such a function and describe its graph. If your answer is no, explain why.

b. Is it possible for the graph of the function $y = f(x) = ax^2 + bx + c$ to be entirely contained in two quadrants? If your answer is yes, give the equation of such a function and describe its graph. If your answer is no, explain why.

c. Same question for three quadrants.

d. Same question for four quadrants.

The second concept to emerge, which gets mentioned in passing in a brief conversation between the teacher and a student, is the notion of real-world applications. If the function $y = f(x) = ax^2 + bx + c$ is being used to model some real-world phenomenon (e.g., factory production, the volume of boxes, or some other physical objects that only have non-negative values), then the domain of the function may only be values of $x > 0$. If that's the case, what does the graph look like? Is it a "complete" U-shape? Pursuing such issues can lead to a deeper understanding of graphs and their properties, and of the assumptions we make when building mathematical models.

We stress here that we're thinking in the abstract about the space of possibilities. There's far too much in what we've been discussing for one lesson. The teacher has to make decisions about which issues to pursue in real-time! And, those decisions are based on the teacher's knowledge of the students, which we don't have.

Here's the third problem to consider. What do you take the point of the problem to be? Please consider your response before reading ours.

 Here are the equations of three quadratic functions:

$$y = x^2 - 10x + 24 \quad y = (x-4)(x-6) \quad y = (x-5)^2 - 1$$

Without performing any algebraic manipulations, write the coordinates of a key feature of each of their graphs. For each equation, select a different key feature.

Here are some of our thoughts about this problem.

As we see it, this task is designed to highlight one of the key points we began our discussion with – that some properties of graphs can be seen directly when quadratics are written in specific forms. Specifically,

- The y-intercept of any function $y = f(x)$ is the value of $f(x)$ when $x = 0$. In the first expression, $y = f(x) = x^2 - 10x + 24$, every term except the constant, 24, is 0 when $x = 0$. So, you can "read" the y-intercept of 24 right from the equation.
- Similarly, since $(x - 4) = 0$ when $x = 4$ and $(x - 6) = 0$ when $x = 6$, the expression $y = f(x) = (x - 4)(x - 6)$ will equal 0 when $x = 4$ or $x = 6$. The factored form of the quadratic shows you the roots of the parabola, with hardly any work at all.
- The third equation represents a parabola that opens upward, with its minimum value when $(x - 5) = 0$, or $x = 5$. When $x = 5$, $y = -1$. That says the vertex of the parabola $y = f(x) = (x - 5)^2 - 1$ is $(5, -1)$.

There's some nice TRU-related pedagogy embedded in this problem, in which everyone works on the problem and then students come to the board to write and explain their answers. In our opening discussion, we highlighted the mathematics that is central in this task; we won't repeat that discussion. What matters here is that students are doing the explaining. Choosing which students are selected to write and explain their answers is a mechanism for equitable access – if such choices are made carefully and with the right classroom norms in place. Under the right conditions, being asked to explain their ideas or comment on someone else's ideas provides students with the opportunity for the development of agency and identity. In addition – a core function of formative assessment – student thinking is made public this way, so the teacher gets a chance to see whether any of the issues that emerge in the discussion need to be clarified, reinforced, or discussed further.

Although the task says not to perform any algebraic manipulations, some students won't be able to stop themselves. At least one student is bound to notice that $(x - 4)(x - 6) = x^2 - 10x + 24$, which means that the first two expressions represent the same function; that will lead to (or the teacher can suggest) multiplying out the third expression, revealing that all three expressions represent the same function.

So, one important insight students can gain from working this task is that the different forms in which a function can be written make it easy to identify key points on the graph of the function. With hardly any work at all, you have the vertex, roots, and y-intercept of the function. If you want a very accurate graph you might calculate a few more values, but you know a lot already. That sure beats unsystematically plugging in tons of values for x! (Or even systematically – many students will calculate the values for $x = 1$, $x = 2$, $x = 3$, etc., to plot graphs.)

The fourth question

The fourth question, discussed (briefly) in the third part of the case study, asks the students to compare and contrast the functions $f(x) = -(x + 4)(x - 5)$ and $g(x) = -2(x + 4)(x - 5)$. This question and its discussion are much more "closed" than those of the questions highlighted above – the point of the task is to highlight the fact that the roots of the parabola (and thus the axis of symmetry) remain the same, but that the parabola itself is "thinner." This *could* lead to a general discussion about how, in general, the graph of the function $g(x) = 2f(x)$ compares to the graph of the function $f(x)$ – but that's going pretty far afield.

It's one thing to have the understandings described in this section; it's another thing to develop the habit of looking for graphical information in the algebraic forms of equations, and to develop the intuitions about graphical properties that come from experience working with them. That's what the next activity in the FAL is designed to provide. It's important as an aspect of mathematical fluency, a part of "symbol sense." The class session devoted to that part of the FAL is long, so we don't have the space for a transcript of the discussion – but we do want to show you what the activity is like.

4.1.3 Putting these Insights to Work: The "Dominos" Part of the Lesson (Not in the Case Discussion)

Here's the introduction to Dominos. The lesson plan allocates about 10 minutes for the introduction, and 45 minutes for the domino and other related activities.

> Organize the class into pairs. Give each pair of students cut-up 'dominos' A, E, and H from Domino Cards 1 and Domino Cards 2. (See Figure 4.5)
>
> Explain to the class that they are about to match graphs of quadratics with their equations, in the same way that two dominoes are matched. If students are unsure how to play dominos, spend a couple of minutes explaining the game.

The graph on one 'domino' is linked to its equations, which is on another 'domino'. Encourage students to explain why each form of the equation matches the curve:

> *Dwaine, explain to me how you matched the cards.*
> *Alex, please repeat Dwaine's explanation in your own words.*
> *Which form of the function makes it easy to determine the coordinates of the roots/y-intercept/turning point of the parabola?*
> *Are the three different forms of the function equivalent? How can you tell?*

The parabola on Domino A is missing the coordinates of its minimum. The parabola on Domino H is missing the coordinates of its y-intercept. Ask students to use the information in the equations to add these coordinates.

> *What are the coordinates of the minimum of the parabola on Card A? What equation did you use to work it out? [(4, –1).]*
>
> *What are the coordinates of the y-intercept of the parabola on Card H? What equation did you use to work it out? [(0,16).]*

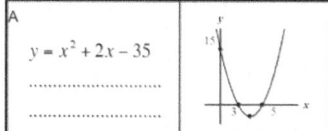 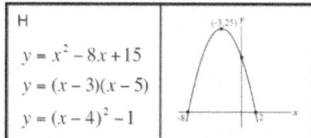 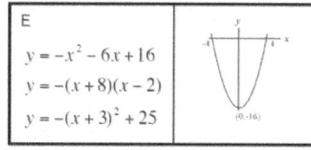

A
$y = x^2 + 2x - 35$
....
....

H
$y = x^2 - 8x + 15$
$y = (x - 3)(x - 5)$
$y = (x - 4)^2 - 1$

E
$y = -x^2 - 6x + 16$
$y = -(x + 8)(x - 2)$
$y = -(x + 3)^2 + 25$

Figure 4.5 Three sample "dominos"

At this stage, students may find it helpful to write what each form of the function reveals about the key features of its graph.

If you think students need further work on understanding the relationship between a graph and its equations, then ask students to make up three different algebraic functions, the first in standard form, the second in factored form, and the third in completed square form. Students are then to take these equations to a neighboring pair and ask them to explain to each other what each equation reveals about its curve.

The domino activity uses 10 cards formatted like the cards A, H, and E above. Like card A, each of the left-hand sides of the cards has at least some information missing. The students work on the task in small groups, following these instructions:

1. Take turns at matching pairs of Dominos that you think belong together.
2. Each time you do this, explain your thinking clearly and carefully to your partner.
3. It is important that you both understand the matches. If you don't agree or understand, ask your partner to explain their reasoning. You are both responsible for each other's learning.
4. On some cards an equation or part of an equation is missing. Do not worry about this, as you can carry out this task without this information.

The students then compare their results with other groups. In a final discussion, the teacher orchestrates a whole-class conversation that highlights some of the issues that arose during the student work.

4.1.4 Concluding Discussion

In concluding this discussion we want to point to some of the design features of the lesson, not only with regard to the mathematics involved, but also with regard to the way that the lesson plan supports all five dimensions of TRU.

We've already highlighted the central mathematical idea involved – that you can "read" features of certain graphs directly from the forms of the equations, a fact that makes it much easier to sketch the graphs of parabolas than trudging through point-by-point plotting. A nice feature of the lesson design is that this information emerges from the tasks themselves, as the students discuss their work on them. And the "drill and practice" part of the lesson comes by way of an entertaining, somewhat self-correcting activity. In the video of the lesson the students are fully engaged with the graphical "dominos," and they often realize something is wrong when the Dominos don't line up in the ways they expect them to.

And, as they say in the late night commercials, there's more. The vast majority of the lesson consists of small-group work and student presentations, so student thinking is made public – the base condition for formative assessment. Working in small groups helps assure that students serve as resources for each other, lessening the demand on the teacher. This is an important and often overlooked aspect of formative assessment.

The lesson plan itself points to places where students may find the mathematics challenging, and ways the teacher can address those issues. For example, see the paragraph that begins "If you think students need further work on understanding" in the lesson segment quoted above. That kind of activity scaffolds learning without taking the initiative away from students and helps to prepare them for the Dominos task – all making sure that the students can profit from productive struggle, the heart of cognitive demand. If opportunities to present are carefully distributed, there are opportunities for equitable access and for the development of agency and positive mathematical identities. In these ways, the lesson supports the use of formative assessment as illustrated in Figure 4.1. Our concluding discussion, in Section 4.4, will highlight the ways in which formative assessment contributes to all aspects of the lesson.

4.2 Some Background about the Classroom We're Visiting

The lesson featured in this case takes place in an Algebra II class in an urban, public, specialized high school (a school for which there is a competitive admissions test). More than 95% of the students are students of color, and 75% of the school population is eligible for free and reduced lunch. About 15% of the students are identified as having special needs. Less than 5% of the student body are English language learners. In the videotape of this class, all whole-class and small-group conversations take place in English. The teacher is a member of the *Math for America* professional development community (see https://www.mathforamerica.org/).

The walls of this classroom are covered with various kinds of information – rules for group work, a description of the problem-solving process, "the seven habits of good readers," laminated sheets with key pieces of mathematics (e.g., graphs of trig functions, information about imaginary numbers and logarithms). There are some inspirational statements including a large boldfaced "Failure is not an option," and a poster saying "check your attitude at the door." Hanging from the ceiling are a collection of posters describing functions the students have graphed recently: $f(x) = x^2$, $f(x) = x^3$, $f(x) = |x|$, $f(x) = 1/x$, and more.

The students' desks are pushed together so that they can collaborate in groups of 3 and 4. The atmosphere is casual, and in both group work and whole-class discussions the students are actively engaged. The conversations at each table are lively, and numerous students volunteer to present their work to the class. It's clear that the students are accustomed to sharing their thinking both in small groups and with the whole class, for example by holding up their work so everybody can see it or presenting at the front of the room. The students who volunteer appear to feel comfortable venturing ideas, even when those ideas turn out not to be correct. In that sense, the classroom appears to be a safe space for mathematical conjectures and conversations.

4.2.1 Preview of the case study

As discussed in Section 4.1, this case study presents the first 30 minutes of a total of 90 minutes devoted to the Formative Assessment Lesson "Representing Quadratic Functions Graphically" (https://www.map.mathshell.org/lessons.php?unit=9245&collection=8), a review/synthesis lesson that addresses the relationships between the algebraic and graphical forms of quadratic equations. The case study splits naturally into three episodes, aligning with the problems discussed in Section 4.1.2.

In episode 1 (Section 4.3.1) the teacher introduces the first activity, in which students are asked to sketch two quadratic curves that look quite different from each other, and explain the fundamental similarities and differences between them. The bulk of the episode consists of small group work, with students working on the problem as the teacher circulates through the room. This episode provides access to classroom norms, a range of student thinking and the ways the teacher and students work on ideas collaboratively.

In episode 2 (Section 4.3.2) two students present their pairs of "quite different" graphs to the whole class. In the first presentation a student highlights specific differences between the parabolas drawn on her two index cards (one opens up, one down; one has a positive y-intercept, the other a negative y-intercept. Both are symmetric about the y-axis). The claim by the second student that the graph of a parabola can sit entirely in the first quadrant is highlighted by the teacher. This leads to some very interesting conversations.

In episode 3 (Section 4.3.3) the teacher introduces the problem that asks students to identify points that are on the graphs of specific algebraic expressions, without doing any algebraic manipulations to the expressions. The students recognize that all three expressions refer to the same function, so the discussion elaborates on the utility of the different expressions in providing information about key points on the graph. Here too, what's interesting is what students say – their discussions reveal comfort with the use of the parameters and what they imply about the graphs.

As always, what matters is the classroom experience from the student's point of view.

4.3 What Happened, in Detail (With Some Reflection Questions and Comments along the Way)

4.3.1 Episode 1 (from 00:00 to 09:15 in the Lesson)

Overview

This episode begins with the teacher explaining the first activity of the lesson. She asks the students to sketch two quadratic curves that look "quite different from each other" and to discuss the fundamental similarities and differences between their two sketches with their groupmates. The bulk of the episode consists of small-group work, with the students discussing the problem as the teacher circulates through the room. The teacher spends most of her time listening, answering clarifying questions, and noting parts of the student conversations that might merit whole-class discussion.

Figure 4.6 shows the classroom layout. There is space in both the front and in the rear of the room for someone to present to the entire class. We'll call the teacher Ms. Smith.

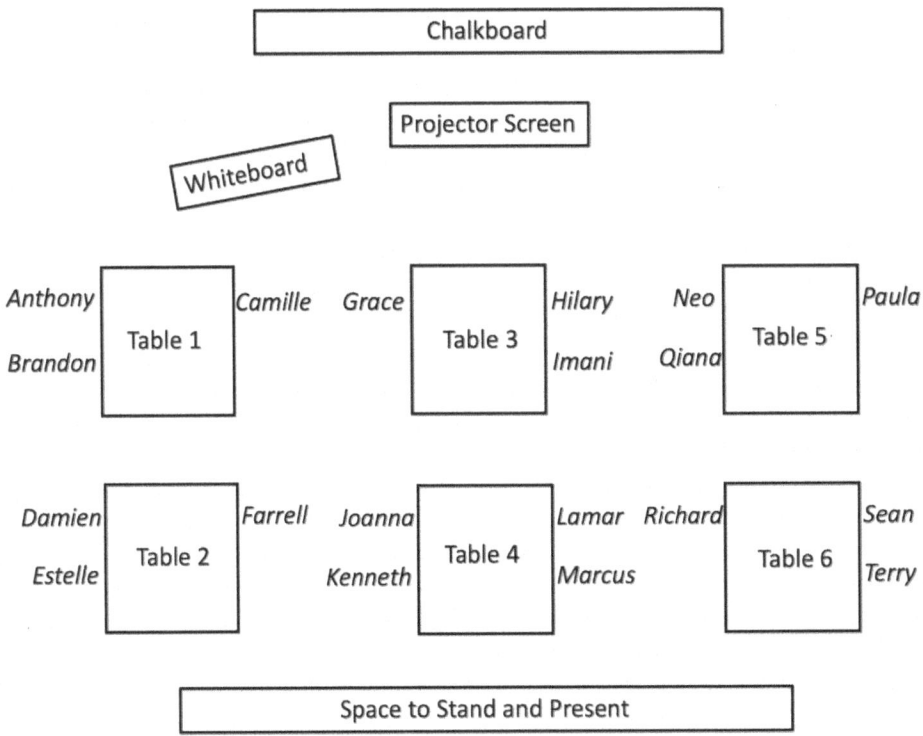

Figure 4.6 The classroom layout

Narrative

Work on the FAL begins with Ms. Smith pointing to the "Do Now" assignment projected in front of the class (see Figure 4.7) and saying,

> Alright guys, I'm going to modify this a little bit. You're going to take 2 index cards. OK? So each person is going to take 2 index cards. And each index card, on the blank side, not the lined side. On the blank side, you're gonna do that.

She continues, "OK? I'm gonna give you a few minutes. If there are any questions, let me know." Two students at Table 2 have their hands raised. When Ms. Smith reaches their table, Estelle asks for a clarification about what gets written on each index card. Ms. Smith picks up the two index cards in front of Estelle and addresses the whole class:

> You are going to have 2 separate... you're going to have 2 index cards... On each index card, you're going to draw one quadratic, and they have to be different in some way. And don't forget you're also drawing your axises on each index card. Is that clear?

The class doesn't respond, so she says it again, in a singsong voice: "Is that cleeeear?"[4] This time there are a number of "yeses" in response.

As Ms. Smith circulates through the class she notes student work, hands out sharpies (markers) to tables that don't have them, addresses minor issues ("Yes, use your sharpie on the blank side"), and monitors progress. When a student re-enters the classroom from the hall, she tells that student about the assignment. She then clarifies the "Do Now" with the students at Table 1. Their exchange (2:50) is typical:

Brandon:	"So we're just coming up with our own parabola..."
Ms. Smith echoes:	"Your own parabola,"
Brandon:	"Ohhh, so we want to make them... so they're different."
Ms. Smith:	They're *quite* different. Yeah." Turning to Brandon's tablemate: "You got that, Anthony? OK?"

Table 2 conversation 1 *(Starting at 3:20)*

Ms. Smith moves past Table 2, glancing at their sketches. As she moves on, she says "make sure your curves are quite different."

Referring to his sketches Damien says "One touches the x-axis, and one doesn't." Estelle responds, "Yeah. It has different y-intercepts."

Estelle then takes her turn. Pointing to the first index card, she starts to explain, "So, what I did is I made one with a... I forget what it's called." As she pauses to search for the right word, Damien quickly suggests "a double root?"

Estelle continues, "Yeah, a double root! And this one is narrower, and it's concave upwards" (see Figure 4.8). Using her pencil to show where the second curve crosses the x-axis, she says "this one... it's concave downwards, and the roots are further apart."

DO NOW

On your index card:

1. Draw the x- and y-axes.

2. Sketch two quadratic curves that look quite different
   From each other.

Figure 4.7 The "Do Now" exercise that starts the lesson

When Estelle has finished, Damien clears his throat, picks up his cards, and begins. "Alright, for mine, the *a* coefficient will be positive for one, and this one would have a negative *a* coefficient because it's more narrow" (see Figure 4.9).

Estelle quickly interjects, "It's not gonna be narrow because the *a* value… It's concave downwards, and others are the same." Damien acknowledges this with a quick "yeah" and continues as he points to his two sketches: "basically, I gave 'em both double roots."

Table 6 conversation 1 *(Starting at 4:40)*

As the camera moves to Table 6, the three tablemates are examining each other's index cards, contrasting their sketches with each other's. The sketches are shown in Figure 4.10.[5]

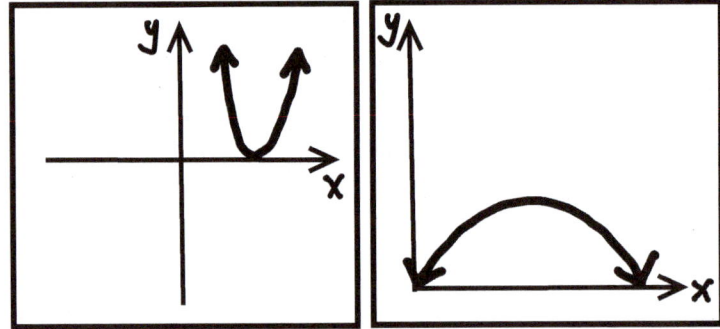

Figure 4.8 Estelle's sketches

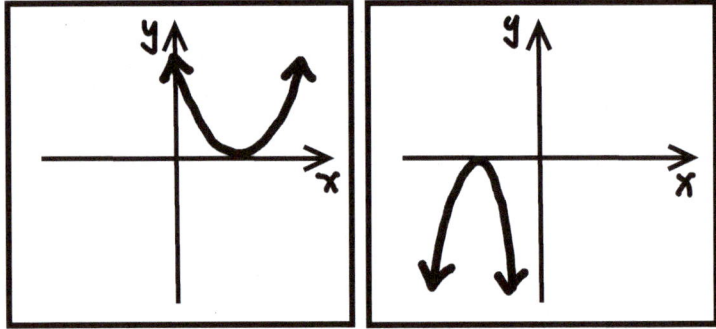

Figure 4.9 Damien's sketches

Richard's sketches   Sean's sketches   Terry's sketches

Figure 4.10 Table 6's sketches

They note that each of them has drawn one graph concave up and the other concave down. Sean says, "If you want to look at similarities... I don't have any roots. The only thing that separates mine is concave up and concave down." Richard notes that both of his sketches cross the *x*-axis (but on opposite sides of the *y*-axis), and Terry says "one of my graphs doesn't have any negative roots." They then veer off into a conversation about the previous day's test.

Table 3 conversation 1 (Starting at 6:10)

The camera moves to Table 3, where the teacher is in conversation with Hilary about one of her sketches. It appears that Hilary has said that her first curve does not cross the *x*-axis (see Figure 4.11).

Grace says, "But it's touching the *x*-axis" and Hilary responds, "Oh, I meant to draw it lower." The teacher asks "But will it have *x*-intercepts if you draw it lower?" Pointing to Hilary's sketch, Imani says, "Yes. If it extends, it will have one." Hilary says "Ohhh," understanding what Imani has just pointed out. She is dismayed that her two graphs are similar. Picking up on this, the teacher says, "They are not as different as you want them to be? You have the time to replace one of your cards if you want to do that. That's cool."

As Ms. Smith moves to another table, Hilary starts working on another sketch. As she does, she turns to Imani and says "You can discuss yours if you want to." Imani then shows her tablemates the two sketches in Figure 4.12.

As she begins to talk about her sketches, Ms. Smith's voice commands the attention of the class:

> Some of you want to do...Listen up! If your parabolas don't look different enough, you can do another ...

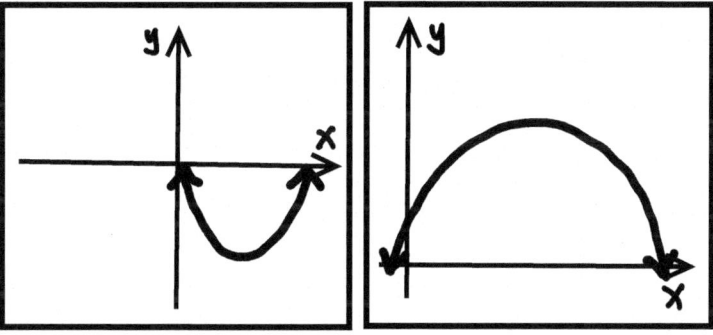

Figure 4.11 Hilary's sketches

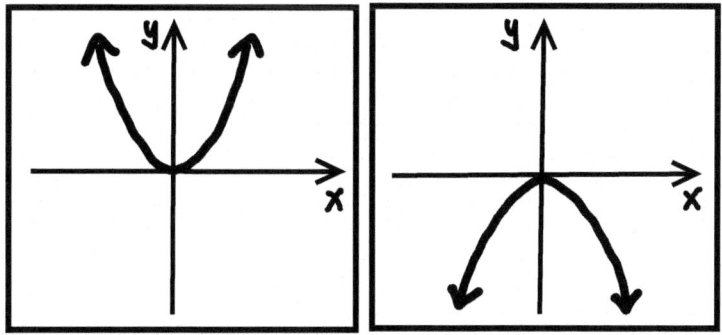

Figure 4.12 Imani's sketches

The students return to their work, and Ms. Smith joins Table 4.

Table 4 conversation 1 (Starting at 7:00)

Seeing that Joanna is creating a new sketch alongside her two finished sketches, Ms. Smith asks, "what happened?" Joanna says, "I'm making another one." Somewhat mischievously, Marcus says "She wants to be special." The teacher responds, "she is special" as she moves toward another group. Smiling at Joanna, Marcus says "Yeah, special girl."

As the camera moves around the table for a better view of Joanna's sketches, she holds her three index cards up so her tablemates can see them (see Figure 4.13). An animated conversation follows, with the students talking over one another and looking at each other as they speak.

Kenneth and Marcus start discussing whether the parabolas in Joanna's sketches have roots, with Kenneth saying "it has roots. It's just the coordinates…" Pointing to the two cards on the right hand side of Figure 4.13, Lamar says, "this has roots, it does have roots, it's just they're zero." Kenneth and Marcus look at each other and then the others, and nod in agreement. Lamar goes on, "and it looks like the *a* for both of them are the same." Lamar looks and says, "it looks close to me."

Looking at the board for the precise wording, Lamar says, "No, for them to be 'quite different,' it has to be the right numbers." Marcus adds, "that's what I'm talking about, see?" Lamar says "so I'm gonna draw another one." As this is taking place, Marcus takes another index card as well.

Table 6 conversation 2 (Starting at 8:07)

The camera moves to Table 6. Ms. Smith is talking to Sean, who has 3 index cards in front of him (see Figure 4.14).

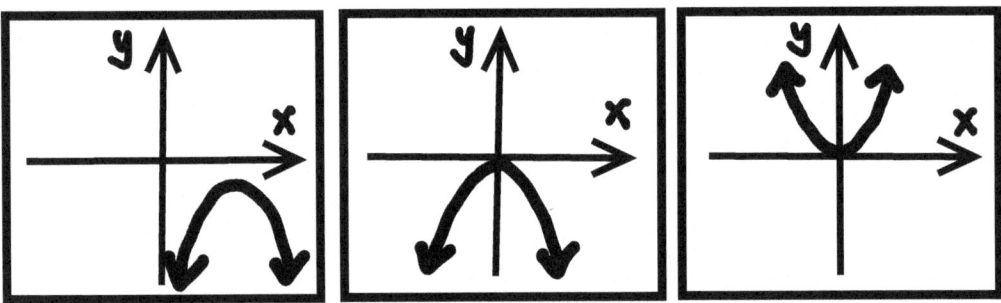

Figure 4.13 Joanna's sketches

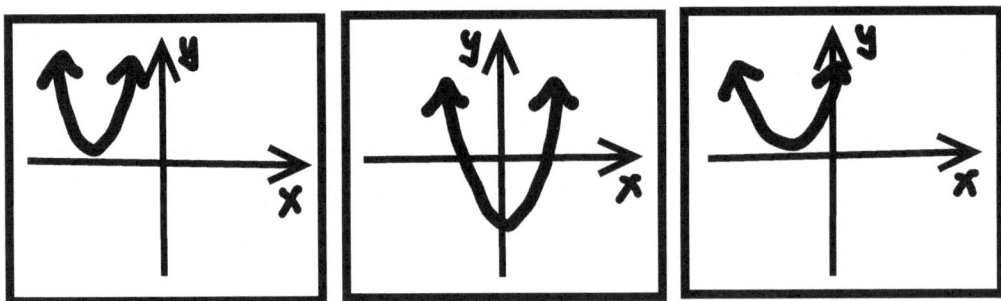

Figure 4.14 Sean's sketches

Sean appears to have created a third sketch because his first two sketches seemed too similar. Ms. Smith says "How so?" and he explains, "Apparently these two graphs, they were only... they were small enough to fit entirely in one quadrant."

Ms. Smith says, "They have to stay in one quadrant, that's such a good question to ask..." Since Sean has three sketches in front of him, she asks, "which ones stay in one quadrant?" Sean says "these two," pointing to the two sketches on the left of Figure 4.14. Ms. Smith then asks the group, which has been following along, "do these two stay in one quadrant?"

Both Richard and Terry respond immediately (see Figure 4.15).

At the left of Figure 4.15 you can see Richard extending the parabola across the y-axis in one of Sean's sketches as he says, "this does not, it keeps going on and on and on and on." At the same time, Terry responds to Ms. Smith's question saying

> They don't, they don't, because this one [pointing to the parabola on the right of Figure 4.14] is gonna go this way [moving his finger across the y-axis] and this one [retracing what Richard had done on the left] is gonna go that way.

An "Ohh..." from Sean indicates he has clearly registered what they have said. As the students continue talking the teacher says "Can a parabola live in one quadrant? That's the question I'm gonna ask the whole class." She moves toward the front of the room.

The discussion continues, with all three students speaking simultaneously – Richard and Terry arguing, with gestures, that the curves will cross the y-axis, but Sean not looking convinced. Terry explains again, saying, "It doesn't matter which way you look at it, it's still gonna have the y." At this point Richard concedes, putting his palms in front of himself, saying... "OK OK OK" and then to salvage some of his position, "So, in this case, it would only be in 2 quadrants then."

Table 6's discussion ends as Ms. Smith brings the class together, looking for someone to share their sketches with the class.

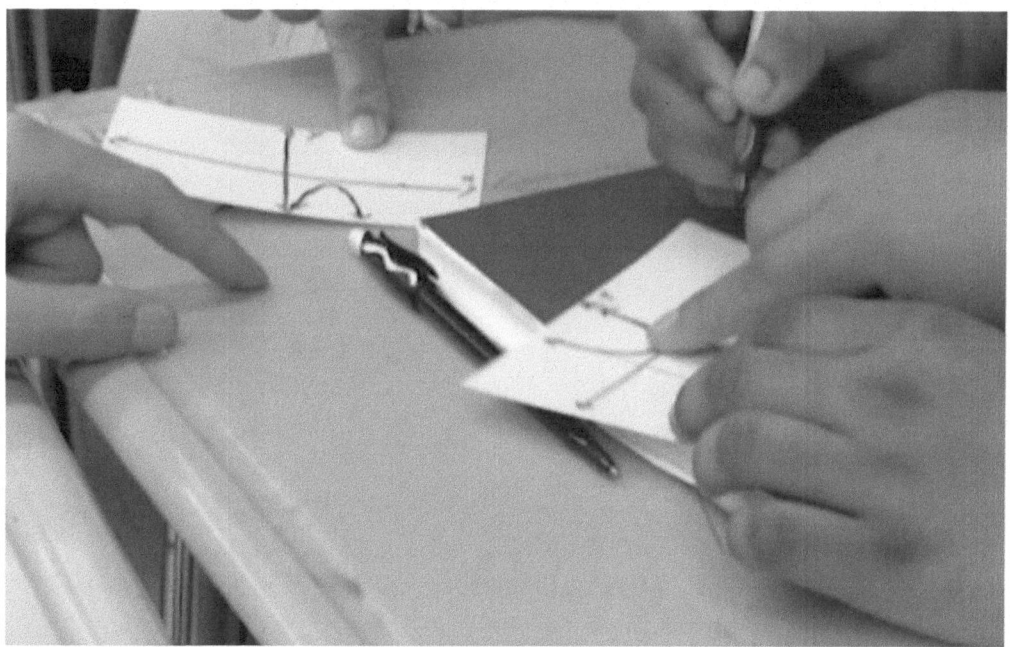

Figure 4.15 Richard and Terry extending the parabolas in Sean's sketches

Focus Questions for Episode 1

 In this first set of questions we look at the task the students were asked to address, the conversations that took place in response, and the potential for formative assessment to build on what is revealed about student thinking in order to dig more deeply into that thinking and help the students develop deeper understandings.

As always, please think about your responses (and discuss them with your colleagues) before reading our commentary.

Focus Question 1
The "Do Now" that began the lesson was:
On your index card:

1. Draw the x- and y-axes.
2. Sketch two quadratic curves that look quite different from each other.

The students were then asked to explain how the curves were different.

A. This is a very open task. What do you think the advantages of posing the task in this way might be? What might some of the challenges of using a task like this be?
B. Imagine you were discussing this task with your students. What are some of the features of quadratic functions you would want to differentiate, and how might you talk about them?

Focus Question 2
One of the goals of instruction is for students to become fluent with the use of algebraic and geometric descriptors of functions. Can you point to specific instances where the students' mathematical talk indicated some degree of comfort and fluency? If there were challenges, how were they dealt with? (Looking for such evidence is a way to hone our skills at formative assessment.)

Focus Question 3
As you read through the description of episode 1, were there mathematical ideas and/or questions you noted that might be worth following up on, either with individual students or in whole-class discussion? The idea is to identify potential errors, incomplete thinking, and/or connections that might be deepened.

Focus Question 4
Errors and incomplete understandings occur all the time. Can you identify what happened when students made them? What might the implications of such interactions be for student learning and AOI?

Our Commentary

1. The "Do Now" that began the lesson was:

On your index card:

1. Draw the *x*- and *y*-axes.
2. Sketch two quadratic curves that look quite different from each other.

The students were then asked to explain how the curves were different.

> A. This is a very open task. What do you think the advantages of posing the task in this way might be? What might some of the challenges of using a task like this be?
>
> B. Imagine you were discussing this task with your students. What are some of the features of quadratic functions you would want to differentiate, and how might you talk about them?

A. The effectiveness of this "Do Now" depends very much on when the task is assigned and how much the students are expected to know.

The task might be useful at the very beginning of a unit on quadratics, just to see what the students know. Assigning it then could reveal prior knowledge and give a sense of the base that you have to build on as you start the unit. But, that is not the case here.

The Do Now task is very open, leaving a lot to the discretion of the students. That means it's not very useful for determining specifics – e.g., if the students know about the role of some particular parameter. Typically, more focused questions are used to determine the particulars of student learning. (We'll see examples of specific questions in episode 3.)

We think this is a terrific question for the beginning of a review/synthesis lesson, precisely because of its open nature. The students have studied the material, and they have some grasp of the details. So, an important question is, how much of the big picture do they have? The open character of the task allows the teacher to see which features of quadratic functions the students consider to be important, and how they justify their choices. This is useful diagnostic information. Student discussions of the task can provide a window into the ways that students connect different representations and invoke these connections. The task is explicitly framed in terms of the graphical representations of quadratics, but it provides opportunities to see when and in what ways they invoke their algebraic representations. (This is an aspect of the fluency discussed in Focus Question 2.)

In addition, as the dialogue in the small groups reveals, the students' discussions point both to strengths and weaknesses of their understanding of details – more so, in fact, than very pointed questions could. (You can't anticipate all the things students will say, and you can't possibly test them all.) At this point, an open question – especially one answered by small groups, so that numerous different issues can emerge – provides both the opportunity to identify details of understanding that might have gone unnoticed and larger themes that would benefit from whole-class discussion. This is a general point.

It's also important that this opening question is posed to students for work in small groups. That gives the students much more of a chance to express their ideas and to get feedback from their fellow students. How effectively this works depends on the norms for student discussions. If students interact productively, there are opportunities for explanation and for students to both critique and build on each other's ideas. This has implications related to dimension 4 of TRU, agency/ownership/identity (AOI). But that's an "if."

That last "if" leads us to think about the challenges of using tasks like this. One set of challenges concerns classroom management. It's hard to monitor and support half a dozen independent conversations at the same time! That's why building productive norms for classroom conversations is so critical. In this classroom we see students taking turns, correcting each other gently, and building on what each other says. The conversations and their outcomes might have been very different if the students hadn't learned to do that.

Finally, this kind of task calls for being very nimble as a teacher. As the teacher joins conversations midstream she needs to pick up what's important in them and decide (a) how much to say to nudge the conversations along and (b) which ideas to have the whole class discuss. Those are skills that develop with time, and they can seem almost overwhelming at first. (But they're worth it!)

B. Not surprisingly, we'd focus on the features of quadratics that were discussed in Section 4.1.

Before we proceed, we note the phrasing of the task. It could have said, "Sketch two parabolas…" Instead, it said "Sketch two quadratic curves…" That choice of words serves as a reminder that the algebraic form of the curve, as well as its geometric shape, is important. In very subtle ways, it's a reminder that the parameters of the quadratic functions are important.

In visual terms, the most distinctive features of a parabola are whether it opens upward or downward, and how narrow or wide it seems to be. We'd consider having one of the parabolas open upwards and one downwards, one being comparatively narrow, and one wide. Since the parameter a [in all of the forms $y = f(x) = ax^2 + bx + c$, $y = f(x) = a(x-h)^2 + k$, or $y = f(x) = a(x-p)(x-q)$] indicates the direction in which a parabola opens and how rapidly it grows, we might choose one quadratic with a large positive value of a, the other with a small negative value (or large negative and small positive).

The issue of roots is more complex (no pun intended), since there is a choice between parabolas that have no real roots, a double root, or two real roots. There's also the issue of placement – is a double root "unique" if it's centered at the origin? Should two distinct roots be placed on the same side of the y-axis, or on opposite sides? Should they be placed symmetrically (in which case the vertex is on the y-axis)? In algebraic terms, this relates to the choices of p and q in the form $y = f(x) = a(x-p)(x-q)$.

Placing the vertex and y-intercept add two more considerations – or, do they? If, for example, we've settled on the roots, then the vertex determines the y-intercept, and vice versa. Although the lesson plan doesn't address the issue, there's a lovely mathematical question lurking in the background: just how much freedom do you have when you're characterizing the properties of a parabola? (This was discussed in the final part of the mathematical discussion in Section 4.1.1, "How much information do you need to determine the equation of a parabola?")

One way to think about this task, then, is that its deliberate under-specification opens the doors for multiple conversations. It is not an exercise in that it does not call for the application of a procedure or calculation. Nor is it a typical "problem." The way the task is posed, students have the freedom to choose which "differences" between quadratics they want to emphasize.

As we've discussed, there is no "one answer" to the problem, because no two quadratics can demonstrate all of the differences discussed above. As a result, different students can present different approaches to the problem. Attempts to make the parabolas "even more different" may lead to questions of just how many differences are possible. This opens the door to a very different kind of conversation than one aimed at "answer-getting." Moreover, the freedom to respond in a wide variety of ways provides opportunities for students to reveal incomplete understandings that might not appear if they were responding to questions that are narrower in focus.

2. One of the goals of instruction is for students to become fluent with the use of algebraic and geometric descriptors of functions. Can you point to specific instances where the students' mathematical talk indicated some degree of comfort and fluency? If there were challenges, how were they dealt with? (Looking for such evidence is a way to hone our skills at formative assessment.)

Consider the first conversation at Table 2. At the beginning of this exchange Damien refers to his two sketches, saying "One touches the x-axis, and one doesn't." Estelle's response, "Yeah. It has different y-intercepts" shows her natural use of the appropriate technical term. Then Estelle points to one of her sketches, a parabola whose vertex is on the x-axis. She says she forgets what the term is called. Damien quickly interjects "a double root" and Estelle continues, "yeah, a double root, and this one is narrow, and it's concave upwards, and this one (takes 2nd index card)… it's concave downwards, and the roots are further apart." Estelle's saying that she forgets the technical term suggests that she expects to be held accountable for it, and she picks up the phrase immediately when Damien provides the term; her use of the terms "concave upwards" and "concave downwards" is natural and conversational. Similarly, Damien's first words are "Alright, for mine the a coefficient will be positive for one and this one will have a negative a coefficient…", an indication that he is accustomed to describing the graphs in terms of their algebraic parameters. These exchanges and others suggest some degree of algebraic fluency and the understanding that is useful to describe functions with reference to their algebraic parameters.

Likewise, at Table 6 (T = 4:40), Sean refers to his two curves as being concave up and concave down; pointing at Sean's sketches, Terry says "you have the vertex…"

At Table 4 (T = 7:22), as he points to a graph where the vertex is on the x-axis, Lamar says "it does have roots, because they're zero." The fact that he refers to roots in the plural means she knows that the vertex being on the x-axis implies the presence of a double root.

It appears, then, that the students have some command of descriptive algebraic and geometric language. At the same time (see Focus Question 3), there may be other issues as we peer beneath the surface of their conversations.

Before we proceed to Focus Question 3, it's worth noting that the corrections and adjustments that we noted in the examples discussed above all took place in student-to-student conversations. They didn't require the teacher to be present! This is important for two reasons. First, teachers often worry when they begin doing formative assessment that unearthing and building on or challenging incorrect or incomplete things students say is an impossible task, because there are so many students and so many things to look for! Here we see that students play a major role in formative assessment. When there are open (and respectful) conversations, many issues get ironed out without needing intervention by the teacher. Second, when the entire classroom functions as a learning community, students profit from the give-and-take in the conversations. Contributing to the discourse, and learning from it, are productive mechanisms for the development of agency, ownership, and identity.

A number of exchanges in episode 1 caught our attention. It's important to say that we can't expect a teacher to follow up on all of them – there's just not enough time. In fact, being able to read and reflect on the description, which includes small-group conversations the teacher wasn't present for, puts us in a very different position than the teacher. This extended look, without time pressure, gives us a chance to practice thinking about the possible implications of what we see and hear.

3. As you read through the description of Episode 1, were there mathematical ideas and/or questions you noted that might be worth following up on, either with individual students or in whole class discussion? The idea is to identify potential errors, incomplete thinking, and/or connections that might be deepened.

The first conversation at Table 2 makes us wish that the camera had lingered for a while longer. When Damien started speaking he said, "Alright, for mine the *a* coefficient will be positive for one and this one will have a negative *a* coefficient cause it's more narrow." Estelle responded, "it's not gonna be narrow because the *a* value…" as Damien continued, "I made 'em both double roots."

Damien's statement, "this one will have a negative *a* coefficient cause it's more narrow" is incorrect; it's the magnitude of the coefficient *a*, not its sign, that determines the width of the parabola. Estelle caught this, but Damien continued speaking about the roots. Was this a slip, or did his statement represent a misunderstanding?

Similarly, there are unanswered questions during the teacher's stopover at Table 3. Hilary's upwards parabola stops at the *x*-axis, indicating its roots. When we join the conversation she is saying, "Oh, I meant to draw it lower." Picking up on this, the teacher asks, "But will it have *x*-intercepts if you draw it lower?" Hilary's tablemates say yes, but it's not at all clear what Hilary actually thinks.

Finally, we have the discussions at Table 6, in which Sean implies that the parabola he has drawn could lie entirely within the first quadrant. What lovely fodder for a whole-class discussion!

In fact, we can abstract a bit from these last two examples. The comments by Hilary and Sean both indicate a significant reliance on their sketches – in particular, on the parts of the graphs that we can see, as drawn on their index cards. This raises the larger question of what we're seeing. The index card (or a computer screen) only captures a small window, while the function is defined for all *x*. It's worth thinking about what our sketches or graphs capture, and what they don't. And it's worth having that conversation with our students!

4. **Errors and incomplete understandings occur all the time. Can you identify what happened when students made them? What might the implications of such interactions be for student learning and AOI?**

We found two patterns to be very interesting. First, many errors (or lapses of memory) occurred in small-group conversations without the teacher present, and most of them were corrected by the student's tablemates. Thanks to the classroom norms, these errors were addressed in a low key and collaborative way, with positive implications for individual and collective agency, ownership, and identity. As we noted earlier, students are an essential resource in formative assessment!

The second pattern occurred when the teacher joined a group and found something problematic. We'll discuss two examples.

Consider the first discussion at Table 3:

> The camera moves to Table 3, where the teacher is in conversation with Hilary about one of her sketches. It appears that Hilary has said that her first curve does not cross the *x*-axis. (see Figure 4.11. once more)
>
> Grace says, "But it's touching the *x*-axis." Hilary responds, "Oh, I meant to draw it lower." The teacher asks "But will it have *x*-intercepts if you draw it lower?" Pointing to Hilary's sketch, Imani says, "Yes. If it extends, it will have one." Hilary says "Ohhh," understanding what Imani has just pointed out. She is dismayed that her two graphs are similar. Picking up on this, the teacher says, "They are not as different as you want them to be? You have the time to replace one of your cards if you want to do that. That's cool."

We find it very interesting that Ms. Smith didn't correct Grace; she asked the question "But will it have *x*-intercepts if you draw it lower?" and Imani answered it.

Similarly, consider this exchange during Conversation 2 at Table 6:

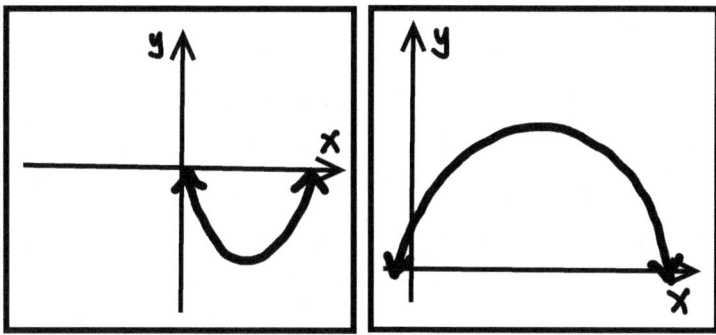

Figure 4.11 Hilary's sketches

As with some of the other groups, Sean appears to have created a third sketch because his first two sketches seemed too similar. The teacher says "How so?" and he explains, "Apparently these two graphs, they were only... they were small enough to fit entirely in one quadrant."

The teacher says, "They have to stay in one quadrant, that's such a good question to ask..." Since Sean has three sketches in front of him, she asks, "which ones stay in one quadrant?" Sean says "these two," pointing to the two sketches on the left of Figure 4.14. Ms. Smith then asks the group, which has been following along, "do these two stay in one quadrant?"

Here Ms. Smith raised questions about whether two of Sean's parabolas stayed in one quadrant – and Sean's tablemates answered the question. She went on to ask the question to the whole class, highlighting (but not answering) a central question raised by a student.

In both cases, Ms. Smith chose not to impose her mathematical authority to correct a student error. She posed a question, and students answered it, providing the necessary information. This subtle move conveys a very powerful message: the students are capable of addressing this issue, so it doesn't call for the teacher's using her authority to render a judgment. Shifting authority to the students in this way has very strong positive implications for student AOI.

4.3.2 Episode 2 (from 09:15 to 19:30 in the lesson)

Overview
This episode consists of the whole-class follow-up to the "Do Now" exercise in episode 1, in which each student sketched two quadratics that were "quite different" and explained the differences to their tablemates. Two volunteers to discuss their pairs of graphs. As the first student shows their graphs to the class Ms. Smith notes their vocabulary on the whiteboard, and the class clarifies the role of the parameter a in the graphs of quadratic functions. The second student's presentation raises issues of the roots of quadratic functions and whether a quadratic can reside entirely in one quadrant. This leads to a discussion of the domain and range of quadratic functions, and whether quadratic functions can have asymptotes.

Narrative
First student presentation (Starting at 9:15)
After getting the class's attention, Ms. Smith asks, "Who would like to share – somebody with two really really different graphs?" Grace raises her hand and the teacher says "Go ahead

Grace. You have to hold them up so that everybody can see. You probably should stand up actually."

Grace moves to the front of the room and starts speaking (Figure 4.16).

"Alright, well, my first graph is a parabola that goes downward, it's concaving down, it's vertex is the *y*-intercept, and it has a positive *x*-intercept and a negative *x*-intercept..."

[While Grace speaks, Ms. Smith writes the technical vocabulary terms that she mentions on the whiteboard in the front of the room]

... and it has a wide... [Grace looks somewhat unsure of herself and glances at the teacher]... a wide curve. A student says "a what?" and Grace repeats "a wide curve."

Ms. Smith asks the class, "a wide curve, what does Grace mean when she says 'a wide curve'?"

There is some talking at the tables, after which Ms. Smith calls on Brandon who says, "that means that the lead coefficient, *a*, is closer to zero." In response the teacher asks, "Does everyone understand?" There is a chorus of yeses.

Grace continues: "And my other graph is a parabola concaving upwards and it's narrow, also the vertex is the *y*-intercept."

Ms. Smith says "thank you" as she continues writing "narrow" and "wide" on the whiteboard. Grace takes her seat to the applause of her classmates.

Second student presentation *(Starting at 10:50)*

Ms. Smith then asks, "All right, can I have another volunteer?" Marcus signals his readiness and stands, saying "should I go up to the front or..." and the teacher says, "why don't you go to the back of the room?"

Marcus does. Holding his index cards so the class can see them (See Figure 4.17), Marcus says, "The one to my right, it is concave down, and it has a *y*-intercept and two roots, and it passes through all four quadrants, but the one on my left, it's only in the second quadrant. It has no *x*-intercept and no *y*-intercept" (Figure 4.17).

Ms. Smith responds, "Hmm. What do you guys think about that?"

There are about 30 seconds of give-and-take between Marcus and various groups of students, as he clarifies that he meant for the upward-facing parabola to not cross the *y*-axis. He starts to return to his seat, but Ms. Smith stops him, saying,

"Can you go back up there, because you said something interesting. You said, for the one on your, left, I guess, for the concave up one, you said there are no *y*-intercepts, I think you said. Or *x*-intercepts." She addresses the class. "Do you agree or not?" Turning back to Marcus, she says "You can call on somebody."

An off-camera student says, "I have a question. Is it possible to have an *x*-intercept but no roots?" Some hands shoot up in response.

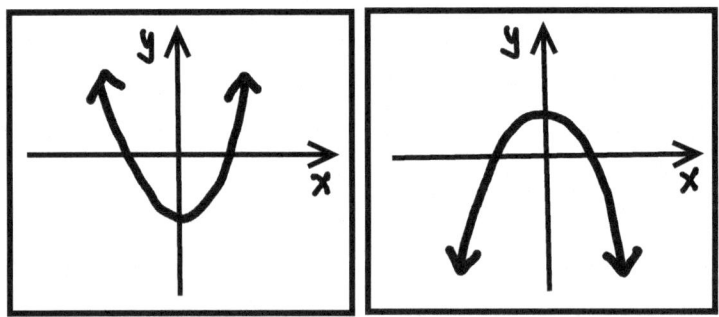

Figure 4.16 The sketches Grace shows the class

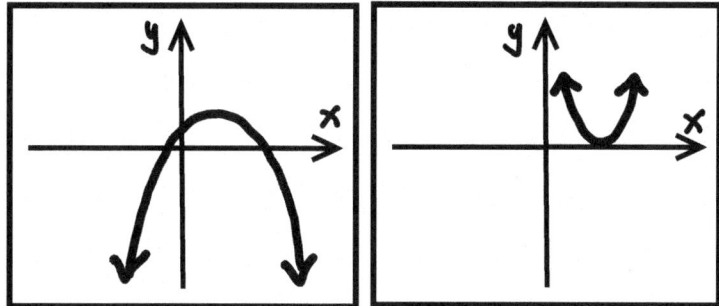

Figure 4.17 The sketches Marcus shows the class

With emphasis, Ms. Smith says, "Oooh. Is it possible to have roots and no x-intercepts? That is your colleague's question. Oh my goodness. Pick on somebody to [address] that." Marcus points to Damien.

Damien says,

Ok, I think we went over that, like, it has roots, and then, [if] it doesn't have an x-intercept, that means the roots are imaginary… And the other thing I wanted to speak on was, you said it doesn't have a y-intercept. It does have a y-intercept but your graph just doesn't show it in its window.

Ms. Smith follows up: "Hmm. What kind of y-intercept would it have? First of all, do you agree with that? Are you OK with that? [pause] Would you have a positive or a negative y-intercept for that graph?"

A student responds, "positive," to which the teacher says, "Positive. Do we see that?" There is a chorus of yeses.

Ms. Smith starts to say "I also asked" but, seeing that Marcus is still standing, says "Thank you Marcus" and he returns to his seat.

Discussion: In how many quadrants can a parabola reside? (Starting at 13:40)
Ms. Smith continues, with pauses and repetitions for emphasis:

I asked a question to one of the groups, I think it was that group, and I said, can a parabola be in one quadrant?… Only one quadrant?… Can a parabola be in only one quadrant? [3 sec pause] Think about that. Can a parabola…

… Maybe on the other side of your card, you can try to draw a parabola that's only in one quadrant. Can a parabola only reside in one quadrant? Can it, can it, can it… That's a good question…

The students discuss the question in groups.

The students at Table 4 are talking over each other as they look at Joanna's index card, which has a very narrow parabola drawn on it. See Figure 4.18.

Marcus says "Why does it have to be so thin?" as Kenneth says "Will it matter if…" and Joanna, pointing at her index card, says, "… [if it's] wider it will probably go across the line."

Again overlapping, Lamar says "But eventually it is gonna keep going," Kenneth says "If it goes straight up it's only in one quadrant" and Joanna says, "Yeah, it's going straight up."

Continuing what he had been saying, Lamar says, "Because, think about it, so, all right, eventually, it's gonna, it starts to move slowly to this side [pointing at the index card, suggesting the LHS will cross the *y*-axis]."

Marcus adds, "Yeah, because mine is a little bit curved, but hers is going straight up."

Kenneth, with the group watching closely as he points to the index card, says "If you think about it, it's thinner, but as they go up, they do get slightly wider. So, eventually, probably gonna take a really long time, but, eventually it's gonna pass the *y*-intercept."

Joanna builds on this, with emphasis: "All right, so, if you had an equation for this [pointing at the index card], and if you plugged in zero for the… [inaudible]…" Marcus, asks, "Zero for what?" but the group's deliberations are cut short when Ms. Smith re-convenes the class.

She asks, "Can a parabola only be … only reside in one quadrant? What a good question! What do you think?"

Joanna says "I say no" and Ms. Smith asks, "Why not?" Joanna says, "um, because, eventually the parabola, even if it's going straight up like this…

The teacher interrupts, "We can't see." Joanna shows her card to the whole class and continues, "… going straight up."

Ms. Smith invites Joanna to the front of the room: "Do we all see that? You want to come up? Can you show it on the board?" Saying "sure," Joanna: goes to the board and draws a very narrow parabola in the first quadrant. See Figure 4.19.

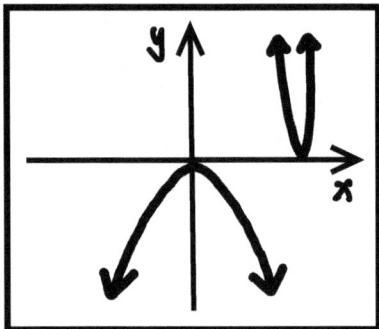

Figure 4.18 The very narrow parabola

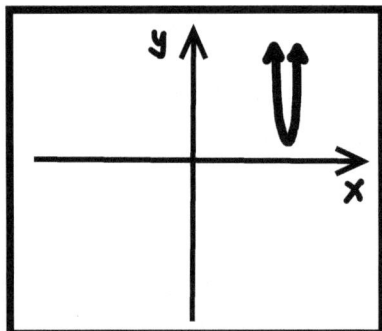

Figure 4.19 Joanna's parabola

She says, "If it was going straight up like this, eventually it would open up because ... because, if it had an equation and you plugged in zero for x's, you'd probably find the y-intercept. Alright?" [She looks around, uncertain, and glances at the teacher for possible confirmation.]

The teacher turns to the class. "Think about that, what are your reactions to it? Do you agree? Disagree? [To Joanna:] Call on somebody." Joanna Calls on Kenneth, who says, "So, feeding off of what she was saying, although it's thinner, eventually it's gonna get wider, it'll probably take a very long time, but eventually it's gonna cross the y-intercept and it's going to go to the second quadrant."

Ms. Smith affirms, "So, no, not just one quadrant," and Kenneth repeats, "No, it can't be in just one quadrant."

Turning to the class, the teacher says, "Do you agree with that?" Responding to a hand in the air, she says, "Damien, you're itching to say something, what is it?"

Damien says,

> I, um, I agree with Kenneth's point that a parabola isn't a, like a perfect U, they expand outward in both directions, so eventually, it's going to um, cross over to another quadrant, but i feel like the only way that it will be done, restricted to one quadrant is like, there's like circumstances for like a situation like, where the values can't be negative.

Ms. Smith asks for clarification: "Ah, like what?," to which Damien says, "Like, if you're talking about in a, um, in um, circumstances, if you're using data with people, there can't be a negative amount of people."

Turning to the class, Ms. Smith says "So, when we restrict the what? We talked about this the other day, when we restrict the what, guys?"

Damien, overlapping with other students, says "The domain and range."

The teacher, emphasizing the vocabulary, says "The what? I heard words."

A student responds, "The window"

Ms. Smith continues, "The window, which is a reflection of the what? I don't hear it." Various voices say "domain and range" at which point she exclaims "Awesome! Thank you Joanna. Yes, Sean, and then we gotta move on."

The question of Asymptotes (Starting at 16:50)

Sean ventures, "I might be wrong but, um, what about asymptotes?"

In response to "What?" from Ms. Smith, Sean responds "Asymptotes. They are like um, they are like the places where you can't have a solution, so, wouldn't a graph like stop before it gets to the asymptotes?"

The teacher says, "An asymptote, sure. Do you guys recall what an asymptote is?"

Various voices say "No."

The teacher follows up with Sean: "Can you elaborate?" He responds, "Um, from my knowledge, uh, I guess, I am thinking that an asymptote on a graph is a specific place where you can't go any higher or lower, that's like a boundary?"

["Uh-hm" from the teacher].

Sean: "So, if this graph [pointing at G(4,2)'s sketch on the board] had an asymptote, it would stop at a point, so, we could make it stop at the first quadrant if we wanted to."

Ms. Smith sums up: "So the question is, do quadratic equations have asymptotes?"

As Sean says "Or is it like the ones with a..." the teacher continues, "Well, let's look for asymptotes." She points to the posters describing functions that hang from the ceiling (see Figure 4.20). "We look at our parent functions up there (Figure 4.20). Do any of those functions have asymptotes? Damien, give me one."

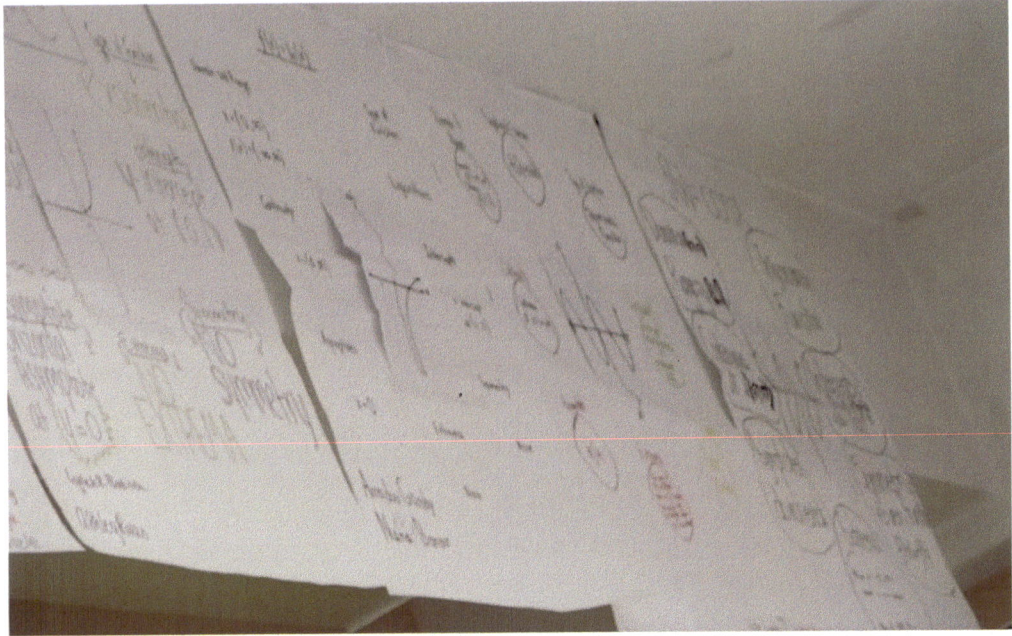

Figure 4.20 A part of the "parent functions" list

Damien says "The logarithm," and the teacher points at the poster: "The logarithmic function ln(x), do you see that? You guys see that?... Yes, I know that one is really small... The third one from the right, there is an asymptote there, what is the asymptote there?"

A student calls out "$x = 0$," and Ms. Smith pushes a bit: "$x = 0$, which is also known as…"

Students call out "the y-axis, the y-axis," which the teacher confirms: "The y-axis, right." She goes on to ask, "Any other function up there have asymptotes? We have some prior knowledge… or are you just looking up there?" In response to some indistinct utterances, she says lightheartedly: "I can't hear you."

A student says "$f(x) = e^x$," which Ms. Smith follows with "$y = e^x$, the exponential one, the one right before it, what's the asymptote there?"

A dialogue similar to the previous one follows. A student says "$y = 0$"; she says "$y = 0$, also known as…", a student says "the x-axis." Ms. Smith confirms this, "the x-axis," and moves on to her next question. "And do we have a quadratic function up there?"

One student says yes and the teacher asks, "Where is the quadratic function, which one is it?"

When the same student starts to say "$f(x) =$…" the teacher, with some emphasis, says "Only [S] sees a quadratic function up there? Can you see it?" More students call out as the teacher says (19:02) "Which one is it?." Through the chorus of voices she echoes, "The second one. Are there any asymptotes for quadratics[6]?" See Figure 4.21.

The students chorus "no," and the teacher continues, "No. If they did have asymptotes – Damien raises a great point – perhaps the domain is restricted. For quadratics, we don't have them. And as we study the other functions later this year, we'll get into that.

All right… let's move on. Are there any questions? [None] Good, that was a good refresher…"

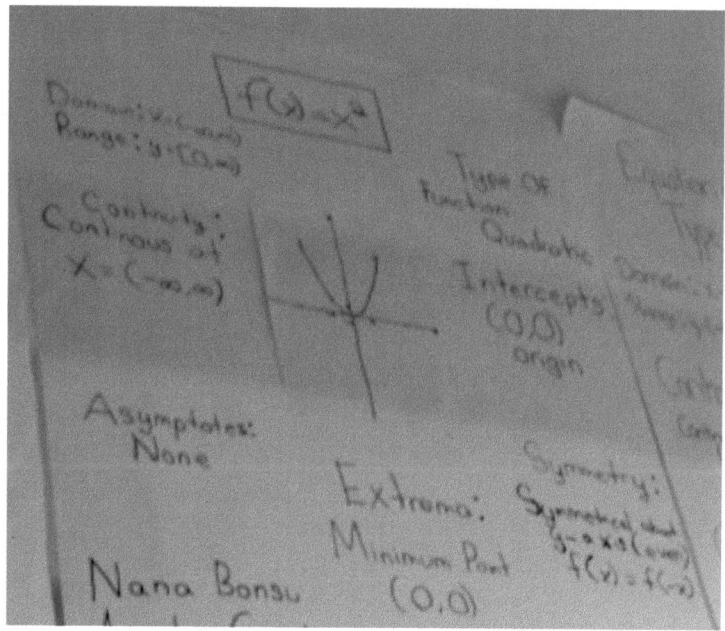

Figure 4.21 The parent function x^2

Focus Questions for Episode 2

In this second set of questions we explore the ways in which the student thinking revealed in episode 1 played a role in whole-class discussions of the "Do Now" exercise. These four questions focus on the specifics of the four sub-episodes of Episode 2, emphasizing the ways in which formative assessment feeds into instruction. As always, however, there's a lot more to the classroom dialogue than the mathematics. We focus on opportunities for the development of students' mathematical agency, ownership, and identity as well.

As always, please think about your responses (and discuss them with your colleagues) before reading our commentary.

Focus Question 1
Please discuss how you see student thinking being revealed and utilized during the first student presentation. Consider the intentions for the task discussed in episode 1, and impromptu opportunities provided by the student's presentation. Please note as well interactions that are relevant to TRU dimensions 2 through 4.

Focus Question 2
Please discuss how you see student thinking being revealed and utilized during the second student presentation. Consider the intentions for the task discussed in episode 1, and impromptu opportunities provided by the student's presentation. Please note as well interactions that are relevant to TRU dimensions 2 through 4.

Focus Question 3
Please discuss how you see student thinking being revealed and utilized during the "one quadrant" discussion. Consider the intentions for the task discussed in episode 1, and impromptu opportunities provided by the student's presentation. Please note as well interactions that are relevant to TRU dimensions 2 through 4.

Focus Question 4
Please discuss how you see student thinking being revealed and utilized during the "asymptotes" discussion. Consider the intentions for the task discussed in episode 1, and impromptu opportunities provided by the student's presentation. Please note as well interactions that are relevant to TRU dimensions 2 through 4.

Our Commentary

1. Please discuss how you see student thinking being revealed and utilized during the first student presentation. Consider the intentions for the task discussed in Episode 1, and impromptu opportunities provided by the student's presentation. Please note as well interactions that are relevant to TRU dimensions 2 through 4.

The first student presentation lasts about a minute and a quarter. On the surface, there's little to note. But a close look indicates some of the ways in which opening things up for students to say for themselves what they think is important about mathematical content can lead to interesting and useful discussions and support the development of AOI. Because the conversations are grounded in what students say, there's a good chance the level of cognitive demand is appropriate for productive struggle.

In the brief time that Hilary speaks, the teacher notes six vocabulary terms – concave up, concave down, vertex, intercept, "narrow" and "wide." The last two are in quotes, as an indication that they're not technical terms. In fact, Hilary's mention of wide and narrow curves and the way the discussion built on that mention are worth discussing.

Hilary was talking about the first of her two sketches when, looking somewhat unsure of herself, she paused and said "a wide curve." Another student asked "a what?" At that point Ms. Smith broke into the conversation, saying "a wide curve, what does Hilary mean when she says 'a wide curve'?" The teacher gave the students some time to chat in their groups. She then called on Brandon who said, "that means that the lead coefficient, a, is closer to zero."

Let's examine this exchange. It began when the teacher noticed a potential confusion. The teacher chose not to clear it up herself. Instead she gave the class a chance to work through it, in groups. The exchange ended when another student gave the mathematically rigorous characterization of "wideness" in terms of the parameter a.

This error identification gave the class a chance to make the link between the term "wide" and the nature of the parameter. We can't know how many students beyond Hilary needed this reminder, but the discussions in the small groups were lively. The fact that the topic was discussed gave the issue some prominence, so (this is just a guess, we can't say for sure) the role of the parameter may have been more memorable to the students than if the teacher simply summarized it. In addition, the teacher's choice not to clarify "wide curve" but to make its meaning a topic of classroom discussion signaled that the topic was worthy of discussion and likely to be a point of confusion for many of the students. This took some pressure off Hilary, in essence saying "Don't feel concerned about not having this at your fingertips. It's an issue for a lot of people."

In sum: Numerous relevant terms were mentioned and highlighted. A confusion was surfaced, discussed, and resolved, all in a very low-key way, with the student appearing to be comfortable during the process and being applauded as she returned to her seat. This advanced the mathematical review objectives of the lesson (Dimension 1), avoided potential AOI risks to the student who presented (Dimension 4), doing so by building on what the students revealed (Dimension 5).

2. **Please discuss how you see student thinking being revealed and utilized during the second student presentation. Consider the intentions for the task discussed in Episode 1, and impromptu opportunities provided by the student's presentation. Please note as well interactions that are relevant to TRU dimensions 2 through 4.**

The second student presentation also provides an opportunity to clarify student thinking. In this exchange, Marcus describes his two sketches and starts to return to his seat. As he does, however, the teacher says, "Can you go back up there, because you said something interesting. You said, for the one on your left, I guess, for the concave up one, you said there are no y-intercepts, I think you said. Or x-intercepts." She then turns to the class and asks, "Do you agree or not?" and suggests that Marcus call on somebody.

In some ways this conversation parallels the conversation discussed in Focus Question 1. In calling Marcus's comment "interesting" rather than indicating that it is incorrect, the teacher avoids labeling Marcus's contribution as wrong. In calling for the class to discuss the question, she signals its importance *and* her judgment that the class has the knowledge to resolve it – a deflection of her own authority. Further, in giving Marcus the option to "call on somebody," the teacher positions Marcus as someone with some level of power and decision-making rights. All these moves are protective of Marcus, and good for his AOI.

The next student's question – "I have a question. Is it possible to have an x-intercept but no roots?" – gets to a key issue. The teacher's "Oooh" and "Oh my goodness" when she repeats the question underscore its importance and highlight the importance that the question came from a student. This too is an AOI issue. The teacher could have answered the question, but once again she deflected her authority back to the class. Damien's response to the question posed, "Ok, I think we went over that, like, it has roots, and then, [if] it doesn't have an x-intercept, that means the roots are imaginary," provides the relevant mathematical response.

Damien then went on to address the question of the y-intercept: "And the other thing I wanted to speak on was, you said it doesn't have a y-intercept. It does have a y-intercept but your graph just doesn't show it in its window."

The teacher follows up on this by asking if the class is comfortable with that explanation, and confirms that the y-intercept in the sketch would be positive. It appears that the discussion of intercepts has come to an end, as she thanks Marcus.

The issues that emerge in this short conversation, including some partial understandings about graphs touching the x-axis, go beyond the list of "common issues" (common student misconceptions) highlighted in the FAL. The open nature of the task, and the teacher's speed in identifying topics for discussion, resulted in useful clarifications. That's formative assessment in action.

At the same time, Damien's comment about the y-intercept went by pretty fast. We might wonder if that comment was enough ... but then we see the teacher's follow-through, in the "one quadrant" discussion. As we'll see, the unexpected issue raised by Marcus is given extended attention.

In sum, the discussion addresses some key content issues (TRU Dimension 1); it does so based on student thinking (TRU Dimension 2); it places the responsibility on all of the students for thinking about key issues (TRU Dimension 3) and relies heavily on student contributions (TRU Dimension 4).

The "one quadrant" discussion is related to the question of whether every parabola has a y-intercept. The teacher brings it up afresh, saying that one of the groups (the group at Table 6) had raised the question of whether a parabola could only be in one quadrant.

3. Please discuss the ways in which you see student thinking being revealed and utilized during the "one quadrant" discussion. Consider the intentions for the task discussed in Episode 1, and impromptu opportunities provided by the student's presentation. Please note as well interactions that are relevant to TRU dimensions 2 through 4.

In response, the teacher asked the students to try to draw such a parabola. This is a concrete task that serves to ground the conversation and gives the students a basis from which to visualize and extrapolate. This enhances the likelihood of productive struggle. As we saw, there was real sense-making taking place in response. The dialogue at Table 4 shows the question of whether a parabola can stay in one quadrant is a very "live" issue for students: they ask each other if you can make the parabola really thin, so that it goes "straight up," because "if it's wider it'll probably cross" the y-axis. Group 4 is grappling with the issues, including finding $f(0)$, and they have not resolved the issue when the class re-convenes.

The whole-class discussion led by Marcus ultimately addresses all these issues – and it cycles back to the earlier comments by Damien, that the quadratic function is defined for all values of x, and thus must have a y-intercept. Moreover, the teacher solicits and reinforces the relevant information from the class: they are only seeing a "window," and what matters when you consider the graph in its totality are the domain and range of the function. A lot of important mathematics gets discussed and resolved.

We note three things here by way of recap. They will be pursued in greater detail in the final discussion.

One of the teacher's frequent rhetorical moves is to resist making authoritative statements herself, but to toss questions back to the class. When Joanna looked uncertain while saying "If it was going straight up like this, eventually it would open up because, because, if it had an equation and you plugged in zero for x's, you'd probably find the y-intercept. Alright?", the teacher's response was, "Think about that, what are your reactions to it? Do you agree? Disagree? [To Joanna:] Call on somebody." This move places the resolution of the question in the hands of the class – a move with very significant implications for agency, ownership, and authority. The teacher is very clearly an authority – but not in a top-down authoritative way.

We also note that the teacher often asks the class such questions when a student has indicated confusion or said something incorrect. The way the questions are handled ("This is an interesting issue, what do you think?") has the property of making the classroom a safe space for inquiry. It eases the pressure on the students who raised the question, and it lowers the threshold for bringing up further issues students may be confused about. We see that in the next sub-episode, concerning asymptotes.

Last, this sub-episode and the next make very clear just how valuable it was for the stimulus task (to sketch two very different quadratic functions) to be somewhat under-specified. Who would have imagined, in thinking about a lesson focused on the graphical implications of the parameters of quadratic functions, that issues related to quadratics remaining in one quadrant, and of asymptotes, would surface? It's a good bet they wouldn't have if they are given specific, focused questions to work on.

4. **Please discuss the ways in which you see student thinking being revealed and utilized during the "asymptotes" discussion. Consider the intentions for the task discussed in Episode 1, and impromptu opportunities provided by the student's presentation. Please note as well interactions that are relevant to TRU dimensions 2 through 4.**

Sean's opening comment, "I might be wrong but, um, what about asymptotes?", shows how important it is to give students safe space to express their thoughts, and to follow their lead if what they reveal is problematic (or just plain thought-provoking). The purpose of formative assessment as a whole is to help students review, amend, synthesize, and extend what they know. The specific goals of this formative assessment lesson – to make sure that students see the relationships between the parameters of quadratic equations and their graphical manifestations – are reviewed in the next episode (episode 3), and they are put into use in the main part of the lesson (the Domino problem). So this review/synthesis lesson does have a specific focus on those mathematically important issues. But in our opinion, we have larger responsibilities as well. Mathematical thinking includes mathematical practices as well as content, with mathematical reasoning being critically important. So if opportunities to address incomplete understandings or to focus on mathematical reasoning arise, what do we do? It does take time to address them, and doing so might deflect us from the intended lesson plan.

This is an issue of values and priorities, which also depend on the instructional context and the time available. If we think only about the specific learning objectives of the FAL, it is unlikely we'd be thinking about whether quadratics have asymptotes, or about a quadratic function residing solely in one quadrant. But, if we think about our goals for the year rather than this lesson, those are important topics; and, addressing them (briefly!) could reinforce the class's sense-making habits. So, the challenge is how to do justice to such larger topics while working to achieve lesson goals. To put things in perspective, the class devoted 90 minutes to the FAL. About 6 minutes in total were devoted to the "how many quadrants" and "asymptotes" discussion.

The mathematics addressed in the asymptotes discussion provides a reminder of the big picture, and another opportunity for formative assessment and review. It's worth noting that the asymptotes of rational functions are vertical (where the denominator would be zero), and that's what people tend to think of. So, the references to the asymptotes in the case of $y = \ln(x)$ (the y-axis) and $y = e^x$ (the x-axis) are a good way to check student knowledge. A lot of important mathematics was referenced in this brief discussion!

4.3.3 Episode 3 (from 19:30 to 30:15 in the Lesson)

Overview

The class now turns to two tasks that highlight the relationship between the parameters of quadratic equations that are written in "standard," "root," and "vertex" forms, and the attributes of the quadratics' graphs that can be readily identified from those forms. These tasks provide an opportunity for the students to demonstrate and refine the skills that they will employ in the "Dominos" part of the lesson, which the students will work next. The give-and-take over the examples provides an opportunity to tie together some of the understandings about the forms and what one can "see" in them.

Narrative: (Starting at 19:30)

Ms. Smith says "Alright!" as she transitions to the next phase of the lesson. The task in Figure 4.22 is projected on the screen in front of the class:

Ms. Smith continues,

> I want you to copy those three functions on your page, and I want you to write a coordinate of a key feature of one of the functions. Maybe one of these things [she points to the whiteboard, which shows the list of terms she had transcribed while the students were presenting] describes a key feature. I don't know.

"And let's be clear. I don't want you to do any mathematical manipulation of those functions. For my people who love to compute. No, no computations. OK?"

"Using the markers [for the smartboard] for each function write a coordinate – a coordinate, an ordered pair[7] – of a key feature of the following functions. So copy them down, I'll give you a minute to do that."

As Ms. Smith circulates through the class she notes what students are doing at their desks, counting something (we can't tell what) and saying "hmm, that's interesting…"

Walking by Anthony she whispers, "Anthony, write a coordinate for one of the graphs up on the whiteboard." Looking to her for clarification, he says "huh?"; she says "write one of your answers up there [pointing to the board] please." "Up there?" "Yeah yeah yeah, on the whiteboard. Pick a function, just one function." Anthony stands in front of the board as the teacher says "now pick a function…"

Ms. Smith continues walking until she reaches Neo and makes a similar request: "Neo, pick a function up there. Write a coordinate for f or g because he's doing h."

While she has been circulating around the classroom Anthony has written "vertex" under the equation for h. Following suit, Neo writes "roots" under the equation for g. See Figure 4.23.

Write a coordinate of a key feature of the following quadratic functions:

$$f(x) = x^2 - 10x + 24 \qquad g(x) = (x - 6)(x - 4) \qquad h(x) = (x-5)^2 - 1$$

Figure 4.22 The coordinates assignment

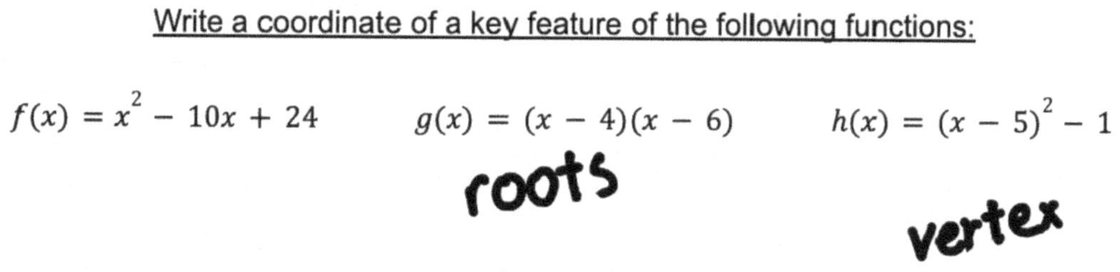

<u>Write a coordinate of a key feature of the following functions:</u>

$$f(x) = x^2 - 10x + 24 \qquad g(x) = (x - 4)(x - 6) \qquad h(x) = (x - 5)^2 - 1$$

roots

vertex

Figure 4.23 Student work on the coordinates task

Noting this, Ms. Smith says "Make sure you're following instructions... Make sure you're following instructions. I'll say that to Anthony and then to Neo."

Anthony returns to the board and, without hesitation, writes (5, –1) under the word "vertex." Neo follows, writing (4,0) and (6,0) under the word "roots." As this takes place the teacher is chatting off-camera with a student. She then says, "Estelle, can you go up and do f please? Ah, she's so excited."

Seeing Estelle waiting at the board for Neo's black marker, she says "you can use the red one." Estelle starts writing on the board.

Off camera, Ms. Smith says to a student, "Oh you weren't sure, so let's see what they have."

In the meantime, Estelle has written "y-intercept" (0,24) under equation f on the board.

Off camera, Brandon gets the teacher's attention. She asks, "You have another one?" and when he says "yeah" she announces, "Brandon has another one, let's see." Brandon goes to the board, clearly doing some calculating in his head as he points to different parts of equation f. He adds "root (4,0)" underneath the equation for f. The board now looks like Figure 4.24:

Noting Brandon's addition to the board, the teacher says, "Hmm, that's interesting." Then she starts the whole-class review: "First, let's do this... we'll start with... we'll go left to right with the first three. y-intercept, Estelle has (0,24). Do we agree?"

There is a chorus of yeses, to which she says "How do we know, how do we know?"

There are some mumbles in response, but Ms. Smith wants a clear answer. "How do we know? We agreed with her, right? How do you know?... She knew it, you knew it, how do you know it? What's that based on, Mr. Lamar?" How do we know that the y-intercept of $f(x)$ is 24?

Lamar says, "Because... we know that c is the y-intercept and it's 24..."

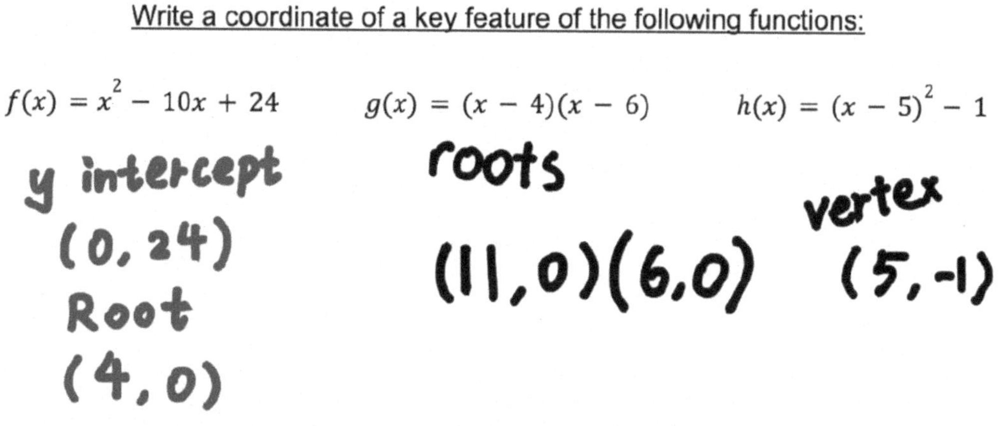

<u>Write a coordinate of a key feature of the following functions:</u>

$$f(x) = x^2 - 10x + 24 \qquad g(x) = (x - 4)(x - 6) \qquad h(x) = (x - 5)^2 - 1$$

y intercept
(0, 24)
Root
(4, 0)

roots
(11, 0)(6, 0)

vertex
(5, -1)

Figure 4.24 More student work on the coordinates task

The teacher asks, "What do you mean by *c*?" Lamar responds, "Well, [pointing at equation f] that's in, I think, it's standard form…"

Ms. Smith interrupts, "He thinks it's in standard form, is that in standard form?… Yes or No?" Students call out "yes" and she continues, "so what does that mean?"

Lamar continues, "it would be $ax^2 + bx + c$, and we know that *c* is always the *y*-intercept, so that's how you guys got [the value?]."

The teacher asks the class, "are we good?" and is met by a bunch of yeses in return.

She then turns to Neo and says, "Neo, they were giving you a hard time up there. Why were they giving you a hard time?" He says and then repeats more loudly, "because I forgot to write the coordinates."

The teacher chuckles and says, "Oh, OK. OK, So how do you know those are the roots? (4,0), (6,0)?" Neo says "because it's written in… [in a questioning tone] intercept form?…"

Ms. Smith asks the class "Do you guys hear that?" and there are some "no"s in response. Neo repeats himself without hesitation, "it's written in intercept form." The teacher asks, "Is he correct?" and some yeses are called out in response.

Ms. Smith proceeds: "$h(x)$. (5, –1) How do we know that?" Anthony says,

> Because it was in the vertex formula so… *h* [here Anthony is referring to the *x*-coordinate of the vertex, (h, k)] is the one in $(x–5)$ and –5 is the form but since it's always "minus *h*" in the equation, so rather a 5, and since [there's a] –1, that would be *k*.

The teacher then calls on Brandon, who had written the "root (4,0)" under the function f. Brandon says

> So to find the roots, we can find it if we know what we're doing, so you can use 24 and find the factor and see what number can add up to negative 10 so I find 4 and 6 that can equal to zero if you can solve two equations.[8]

Ms. Smith calls on Kenneth, who says, "I believe we weren't supposed to do math." The teacher repeats this, and Brandon says "well, I did it in my head." The teacher laughs as Brandon says, "It said no calculations and I just did it in my head."

Turning to the class, the teacher says, "Do we agree with the calculations?" There are yeses in response, and some more casual back-and-forth between the teacher and Brandon about what constitutes "calculation."

Ms. Smith asks the whole class, "Do you notice anything else about these equations f, g, and h?… Yes, Imani?"

Imani says, softly, "Isn't $g(x)$ basically the factored form of $f(x)$?"

In response Ms. Smith says "Who heard her," and when only a few hands go up, Imani loudly says "I think $g(x)$ is the factored form of $f(x)$."

There are some "oohs" in the class and the teacher repeats "g is f in factored form…. Look, I don't know, is it?" A student says, "are they all the same?" and the teacher echoes this loudly: "are they all the same?" Some students say yes, and the teacher walks toward the smartboard.

"So g: if we factor f we get g, yes? So it makes sense they have the same roots. Is *h* the same?" After a small pause she repeats, "Is *h* the same? All right, we could expand that, what do we get if we expand $(x–5)^2$?" She starts writing on the board as she says, "We get *x* squared minus 10 x plus 25 minus 1, which is…"

The board looks like Figure 4.25. as Ms. Smith turns to the class. She says, "They're all the same, right?"

They're all the same. Different forms, obviously they give us different information… but they're all the same. Very good, very good. Somebody had a question, I thought I saw…"

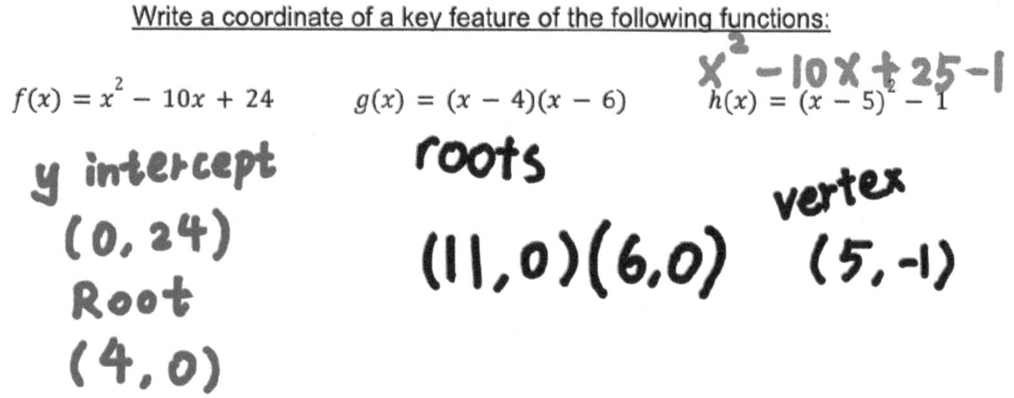

Write a coordinate of a key feature of the following functions:

$$f(x) = x^2 - 10x + 24 \qquad g(x) = (x - 4)(x - 6) \qquad h(x) = (x - 5)^2 - 1$$

[Handwritten annotations:]

$x^2 - 10x + 25 - 1$

y intercept
(0, 24)
Root
(4, 0)

roots
(11, 0) (6, 0)

vertex
(5, -1)

Figure 4.25 The fully annotated coordinates task

Pointing to the right side of the board, Brandon says, "Well, I'd say, That's the vertex form. And [pointing left] we also have the standard form over there."

Ms. Smith finishes up "All 3. So good, we're good. This is the good part.... The better part, it's all good."

She quickly puts up the next task (Figure 4.26.):

In the interests of time, she says "We'll do that together, I think you can kinda see... What do they have in common, guys? What do these two functions have in common?" She calls on Richard. He says, "They're both negative." The teacher asks "What do you mean by that?" and Richard replies "They're both traveling... going downwards." In response to the teacher's "How do you say that?" the class choruses "Concave downwards." She responds "Concave down, good, what else?"

There are loads of hands in the air. The teacher calls on Joanna, who says "They have the same roots." Echoing this with emphasis, Ms. Smith says "they have the same roots. OK, what are the roots?" Sean responds, "negative 4 and 5." The teacher repeats this with a question mark in her voice and the class confirms the roots. She goes on, "OK, anything else they have in common?"

Hilary says, "$f(x)$, isn't it wider than $g(x)$?"

Deferring this, the teacher says, "So, not in common, this is something different. Wait. Do we have anything else in common, first?"

Someone says "Yes"

"What?"

"They are both parabolas."

There is scattered clapping in humorous response and another student calls out, "Oh wow, they are both parabolas." The teacher asks, "Anything else besides that?"

Compare and contrast the following functions:

$$f(x) = - (x + 4) (x - 5) \qquad g(x) = -2 (x + 4) (x - 5)$$

Figure 4.26 The final task in the lesson segment

Turning back to Hilary, the teacher says "What's different, Hilary?" Hilary responds, "$f(x)$ is wider than $g(x)$." The teacher asks the class, "Is that true?" and there is a chorus of yeses. She asks, "Why is $f(x)$ wider than $g(x)$? Because, cause, cause...

A student says "the lead coefficient... a" as the camera fades out.

Focus Questions for Episode 3

 This third set of questions explores the ways in which two straightforward content-related questions round out the review of the properties of quadratic equations. Our primary goals are to compare the design properties of these two questions with the questions in episodes 1 and 2, and to focus on the ways in which the teacher makes use of formative assessment.

Focus Question 1
The first task in episode 3,

<u>**Write a coordinate of a key feature of the following quadratic functions:**</u>

$f(x) = x^2 - 10x + 24$ $\qquad$ $g(x) = (x - 6)(x - 4)$ $\qquad$ $h(x) = (x-5)^2 - 1$

is very precise, in contrast to the open task from episodes 1 and 2, in which students were asked to draw two "very different" quadratic functions. What are the key mathematical ideas you would hope to see emerge from working on this task? What do you see as the advantages and challenges of using such a narrowly defined task, in comparison to the advantages and challenges of using the task from episodes 1 and 2?

Focus Question 2
The second task in episode 3

<u>Compare and contrast the following functions:</u>

$f(x) = -(x + 4)(x - 5)$ $\qquad$ $g(x) = -2(x + 4)(x - 5)$

appears to be more open than the first task in this episode. What are the key mathematical ideas you would hope to see emerge from working on this task? Can you identify some advantages to making this task somewhat more open than the previous task?

Focus Question 3
As always, tasks provide opportunities for formative assessment; those opportunities play out in classroom dynamics. Where in episode 3 do you see the teacher soliciting student thinking, or assessing student thinking

and probing further? Where do you see these acts of formative assessment playing out in the service of rich mathematics (TRU dimension 1), a seemingly appropriate level of cognitive demand (TRU dimension 2), equitable access (TRU dimension 3), and opportunities for the development of agency and initiative, ownership, and mathematical identities (TRU dimension 4)?

Our Commentary

1. The first task in Episode 3,

<u>**Write a coordinate of a key feature of the following quadratic functions:**</u>

$f(x) = x^2 - 10x + 24$ $g(x) = (x - 6)(x - 4)$ $h(x) = (x-5)^2 - 1$

is very precise, in contrast to the open task from Episodes 1 and 2, in which students were asked to draw two "very different" quadratic functions. What are the key mathematical ideas you would hope to see emerge from working on this task? What do you see as the advantages and challenges of using such a narrowly defined task, in comparison to the advantages and challenges of using the task from episodes 1 and 2?

As we wrote in Section 4.1, learning to "read" properties of the graphs of quadratic functions from their algebraic forms is an important curricular goal. This task gets at those skills directly. In addition, the skills highlighted in this task are necessary for the "Dominos" task the students will work on in the subsequent part of this Formative Assessment Lesson. The hope is that they enter the Dominos problem knowing that:

- the coordinates of the y-intercept of the "standard form" of the function $y = f(x) = ax^2 + bx + c$ are $(0,c)$;
- the coordinates of the roots of "root form" of the function $y = f(x) = a(x - p)(x - q)$ are $(p, 0)$ and $(q, 0)$;
- the coordinates of the vertex of the "vertex form" of the function $y = f(x) = a(x - h)^2 + k$ are (h, k).

This task highlights those facts. As such, the specificity is useful.

Two additional possibilities inherent in the task are worth discussing. First, there is always the opportunity to have students explain why the three properties just discussed are true. The students should be able to explain that:

- the y-intercept of any function is the point where its graph crosses the y-axis – in other words, it's the y-value when $x = 0$. If you plug $x = 0$ into the standard form $y = f(x) = ax^2 + bx + c$, you get $y = c$, so the y-intercept of $f(x)$ is $(0,c)$.
- If $x = p$ or $x = q$ when a function is written in the root form, then $y = f(x) = a(x-p)(x-q)$ has a factor whose value is 0. So, $f(p) = f(q) = 0$.

- When $x = h$ in the vertex form $y = f(x) = a(x - h)^2 + k$, the value of $a(x - h)^2$ is 0, and $y = a(x - h)^2$ is a parabola with vertex at $(h, 0)$. Adding the value k to $a(x - h)^2$ translates the graph of $a(x - h)^2$ by k units, placing the vertex of the function $y = f(x) = a(x - h)^2 + k$ at the point (h, k).

In fact, it's good if the student explanations can go back to the fundamentals. Years from now the students will forget which values are called h or k, for example. But if they have a basic understanding of why

- replacing (x) by $(x - ɣ)$, where $ɣ$ could be any number, results in translating the graph of f(x) by $ɣ$ units to the right, and
- the graph of $g(x) = f(x) + ɣ$ is obtained by taking the graph of $f(x)$ and translating it up or down $ɣ$ units, depending on the sign of $ɣ$, then they can re-generate the information they need.

The second possibility for further discussion was highlighted when the teacher closed the discussion of the problem. The three functions $f(x) = x^2 - 10x + 24$, $g(x) = (x-4)(x-6)$, and $h(x) = (x-5)^2 - 1$ are all algebraically equivalent. That means that you can "read" all of the graphical properties of the function directly from the different algebraic forms in which it is expressed. That makes graphing the function a cinch. At a strategic level, it's good for students to know that re-expressing a function in an equivalent algebraic form can sometimes make it easier to find information that they're looking for.

In comparing the task that begins episode 3 with the task at the core of episodes 1 and 2, we see the advantages and limitations of both kinds of tasks. The episode 3 task highlights essential skills, the pieces of information that are fundamental aspects of fluency in graphing. Without them, students are pretty much stuck plotting functions point-by-point. It's important to have such skills at your disposal.

At the same time there's the issue of seeing the big picture, of getting a feel for what matters when you sketch the graphs of functions. You can ask fine-grained questions all day – in fact, that's what most curricula do! – without getting to the issues that underlie intuitions about what the graphs of functions look like.

Asking students to draw two "very different" quadratic functions provided them with the opportunity to develop such intuitions. They had to think about which properties of parabolas matter, and how to characterize them. Further, their conversations about those properties indicated what they thought the key features of quadratics actually are, and opened up conversations that would have been unlikely to happen with focused questions like the one boxed above.

In short, if you want to check that your students see the forest *and* the trees, you need to provide opportunities of both kinds. (There may or may not be reasons to address one before the other. In this FAL, we appreciated the fact that the open question gave students the opportunity to reveal "big picture" understandings and misunderstandings before the more focused problems prepped them on the specific skills they'd need for the Dominos problem.)

In terms of openness, this task occupies a middle ground between the "sketch two very different quadratics" task that opened the lesson and the task discussed in Focus Question 1. It focuses on the root form for quadratics, but it allows students to raise what's important to them. This more specific focus allows for some visualization and making connections. The functions f and g have the same roots, at $(-4,0)$ and $(5,0)$; these serve as "anchors" for the visualization. Both parabolas open downward. And, since the lead coefficient of g is larger in magnitude than the lead coefficient of f, g is "narrower."

How much more can you say? There's a lot to say, if you want to devote some time and energy to the task. We'll point it out, just for completeness.

2. The second task in Episode 3,

Compare and contrast the following functions:

$$f(x) = - (x + 4) (x - 5) \qquad g(x) = -2 (x + 4) (x - 5)$$

appears to be more open than the first task in this episode. What are the key mathematical ideas you would hope to see emerge from working on this task? Can you identify some advantages to making this task somewhat more open than the previous task?

Since g is actually twice f, there's an informal sense that g grows (downward) twice as fast as $f(x)$. But what does this look like?

Because they have the same roots, f and g also have the same axis of symmetry. The axis of symmetry is the vertical line that hits the x-axis half-way between the roots – in this case half-way between (–4) and (+5), or $x = 1/2$. The vertex of each curve lies on its axis of symmetry, so both curves have $x = 1/2$ as the x-coordinate of their vertex. Note that g's vertex is twice as "tall" as f's vertex, since $g(x) = 2f(x)$.

The other "low hanging fruit" is the y-intercept: $f(0) = 20$ and $g(0) = 40$.

Let's build a mental image of the two functions from this information. Both f and g "start" at their vertices and then dive down, intersecting each other at the two roots, (–4,0) and (5,0). Both vertices, with x-coordinates $x = 1/2$, are a bit higher than the y-intercepts of 20 (for f) and 40 (for g). [In fact, you can calculate them without too much trouble: $f(1/2) = -(4.5)(-4.5) = 20.25$ and $g(1/2) = 2f(1/2) = 40.5$.] So, f arches down from its vertex at (1/2,20+) and crosses the x-axis (–4) and (+5). g starts twice as high, at (1/2,40++), zooms down twice as fast, and crosses the x-axis at the same points as f. Then g continues diving down at twice the rate that f does.[9]

How did we do? Figure 4.27 shows the graphs.

That's not bad, all things considered!

But now, back to reality. In every lesson, the teacher has to make decisions about how much time and attention can be devoted to any particular activity. In this particular class, Ms. Smith had devoted extra time, in episode 2, to pursuing two unexpected questions: whether a parabola can "live" in just one quadrant, and whether parabolas can have asymptotes. Those were useful excursions – and they used some valuable class time. The clock was ticking and the class needed time to work on the Dominos problem, the main activity of the lesson. So, we come once again to a question of decision-making in the moment, based on context and goals. The fine points discussed above can wait for another day.

This decision reflects one of many decisions that teachers make every day. Teaching is always a matter of juggling goals and setting priorities. When unexpected events come up – and they always do! – we are compelled to make choices in the moment, based on what we know about our students and what we think will be most helpful.

Formative assessment begins with the tasks. The tasks in episode 3 are aimed at finding out whether students have specific understandings about the graphical properties of the standard, root, and vertex forms of quadratic functions. But, the teacher asks for more than just answers. When discussing function f, Ms. Smith could have simply confirmed the value of the y-intercept and moved on, but instead she used the answer as a launching point. She asked if the class

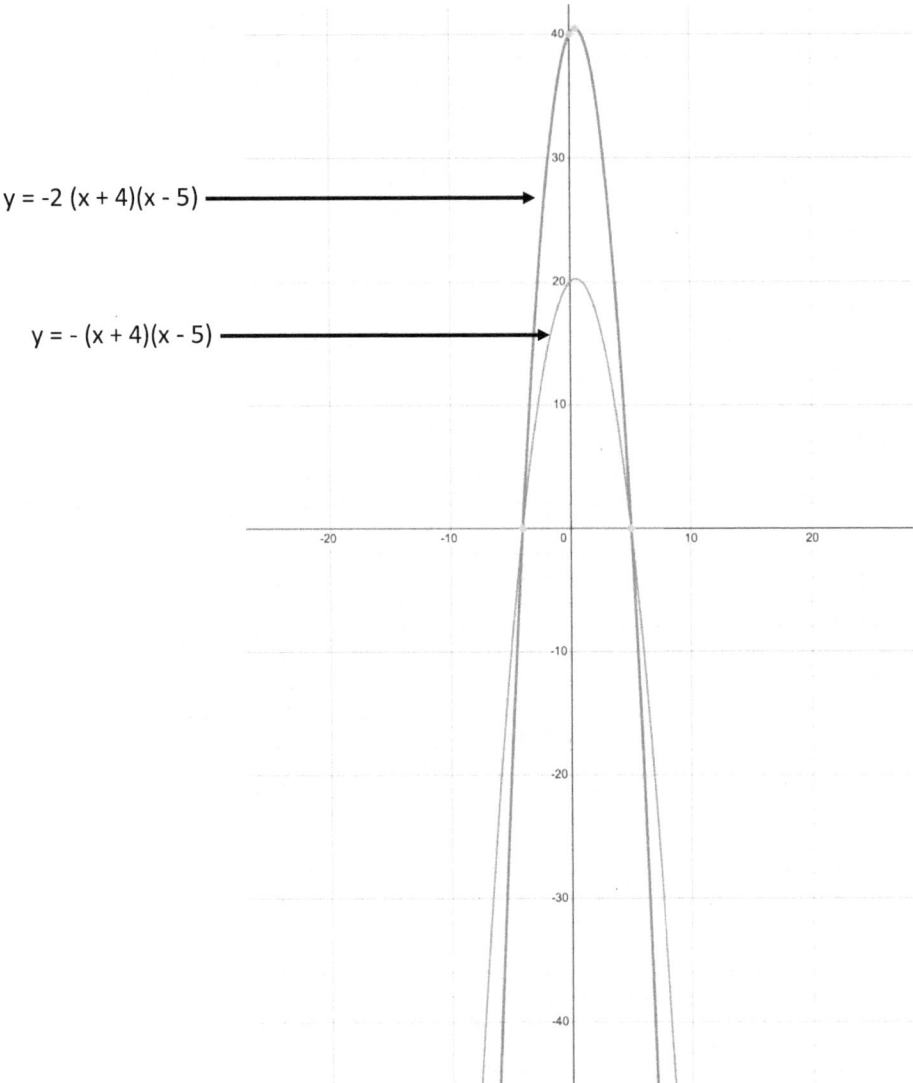

$y = -2 (x + 4)(x - 5)$

$y = - (x + 4)(x - 5)$

Figure 4.27 The two graphs

agreed that the y-intercept is (0,24). When they did, she asked "how do we know?" and pressed on a student's explanation until it was clear that the equation was written in standard form and that the constant c in the standard form (in this case, 24) was the y-intercept. Only when there was a chorus of yeses to her question "are we good?" did she proceed.[10]

3. As always, tasks provide opportunities for formative assessment; those opportunities play out in classroom dynamics. Where in Episode 3 do you see the teacher soliciting student thinking, or assessing student thinking and probing further? Where do you see these acts of formative assessment playing out in the service of rich mathematics (TRU dimension 1), a seemingly appropriate level of cognitive demand (TRU dimension 2), equitable access (TRU dimension 3), and opportunities for the development of agency and initiative, ownership, and mathematical identities (TRU dimension 4)?

Similarly, when discussing function h Ms. Smith asks Anthony "how do we know" that the vertex of function h is the point $(5,-1)$. And, rather than summarizing herself, she gives Brandon the opportunity to explain how he obtained the root $(4,0)$ for function f. Each of these actions goes beyond the task itself, with the teacher leveraging student answers and student thinking to make important ideas public.

Even though the class marches through the second problem at a pretty rapid pace, the teacher also makes sure the students make the relevant connections – when Hilary says "$f(x)$ is wider than $g(x)$" the teacher asks the class, "Is that true?" and, in response to a chorus of yeses, asks "Why is $f(x)$ wider than $g(x)$? Because, cause, cause..." In sum (as we also saw in episodes 1 and 2), formative assessment consists of the ongoing solicitation of student thinking and in-the-moment adjustments to classroom discourse.

Let's run quickly through TRU dimensions 2 through 4. We don't know how challenging the tasks in episode 3 were, although it seems that the level of cognitive demand was in the right ballpark (for review – for a learning activity, you might want to stretch the students more). The teacher constantly pressed for explanations, which took some time for the students to produce; and there was some genuine surprise when students noted that the functions f, g, and h in the first task were the same. This surprise indicates that they were "stretched" by the problem.

It's worth noting that having students work problems in small groups provides the teacher with the opportunity to circulate through the classroom, and to judge for herself how challenging the students are finding the tasks. In this sense, she has a much better sense than we do of the level of cognitive demand! (And, as we saw in episodes 1 and 2, noting challenges that the students faced led her to bring some issues to the whole class for discussion.)

Let's turn to issues of equitable access and AOI. Throughout this episode we've seen the teacher asking questions of the whole class and surveying student work as the class works on the tasks she has assigned. When she poses a whole-class question, the teacher appears (as far as the camera lets us see) to call on students who have their hands raised. When she circulates through the room, she suggests that particular students present their work to the class. (For example, she suggested that the students Anthony, Neo, and Estelle present their work on the first problem.) In all three episodes in this case study we have seen volunteers present; their contributions were accepted or built on in ways that supported the development of their mathematical identities. But that's volunteers. As far as we know, Anthony, Neo, and Estelle had not yet volunteered. We find it interesting that the teacher asked them to go to the board to present their work. She did so after looking over their work, which means that she also knew that they would be presenting the correct answers. Expanding the group of presenters is one way of providing equitable access, and knowing that those students are on safe ground is also a way to ensure that the exchanges with their classmates will contribute positively to their sense of agency and to their mathematical identities.

4.4 Reflecting on "Graphing Quadratic Functions"

We'll use two figures to provide the focus for this concluding discussion. The first, in the tradition of case studies 1 and 2, is to observe instruction through the eyes of the student. We *always* want to do this when we think about instruction. In our reflections we'll address the questions found in Figure 4.28.

In addition, since the focus of this case is on formative assessment, we will use Figure 4.1 to highlight the role of formative assessment in relation to each of the other dimensions. The question for each of TRU Dimensions 1 through 4 is, in what ways can we see formative assessment supporting classroom interactions related to that dimension?

Observe the Lesson Through a Student's Eyes

The Content	• What's the big idea in this lesson? • How does it connect to what I already know?
Cognitive Demand	• How long am I given to think, and to make sense of things? • What happens when I get stuck? • Am I invited to explain things, or just give answers?
Equitable Access to Content	• Do I get to participate in meaningful math learning? • Can I hide or be ignored? In what ways am I kept engaged?
Agency, Ownership, and Identity	• What opportunities do I have to explain my ideas? In what ways are they built on? • How am I recognized as being capable and able to contribute?
Formative Assessment	• How is my thinking included in classroom discussions? • Does instruction respond to my ideas and help me think more deeply?

Figure 4.28 Access suggestions from the Conversation Guide

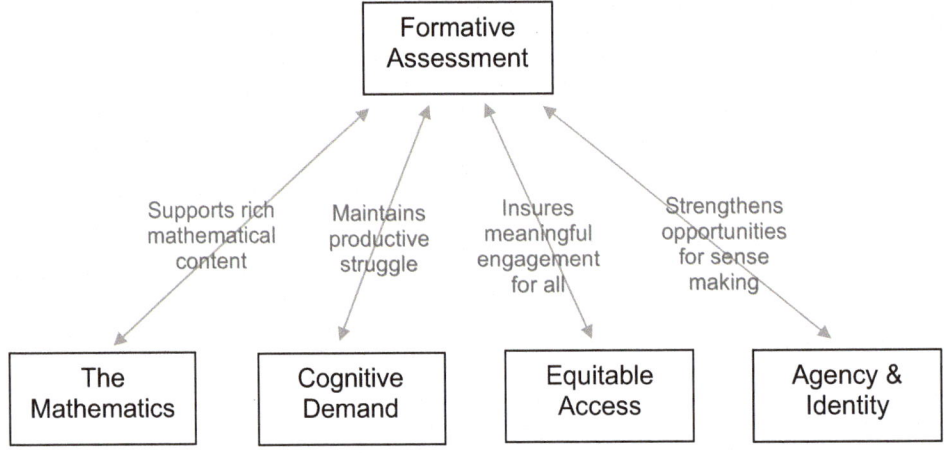

Figure 4.1 The key roles of formative assessment in powerful mathematics classrooms

Dimension 1, the Mathematics

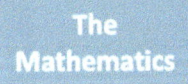

The Mathematics	• What are the big ideas in this lesson? • How do they connect to what I already know?

This case study reviews and synthesizes a substantial amount of material. It covers a lot more mathematical territory than the lessons in case studies 1 and 2 – there are both "big picture" issues to think about and lots of specific information.

The half-hour of instruction in this lesson segment addresses a substantial part of the quadratic functions overview that was discussed in Section 4.1. The big-picture items the students discussed include:

- the general shapes of quadratic functions
- the descriptive terms that characterize them, ranging from informal language that includes terms like "wide" and "narrow" to technical terms such as domain and range, concave up and concave down, vertex, and y-intercept,
- the three forms in which quadratics are typically expressed, the easily "readable" information from each of those forms, and the utility of re-expressing a quadratic in alternative forms,
- issues of domain and range, including the question of whether a quadratic function can "reside" in just one quadrant,
- questions of which categories of functions have asymptotes.

The more detailed knowledge includes the following:

- $f(x) = ax^2 + bx + c$ is concave up if a is positive and concave down if a is negative
- the y-intercept of $f(x) = ax^2 + bx + c$ has coordinates $(0,c)$
- the roots of $f(x) = a(x - p)(x - q)$ are $x = p$ and $x = q$, and the graph of $y = f(x)$ passes through $(p, 0)$ and $(q, 0)$
- the graph of $f(x) = a(x - h)^2 + k$ has its vertex at (h, k).

That's a lot of content to address in a half hour of instruction, even if most of it is review. It's interesting to see how much of that information emerges through formative assessment.

The task that opened the lesson, which asked students to sketch two "quite different" quadratics and to discuss how they were different, gave students the opportunity to select what they thought were the main features of quadratics. Almost all of the students focused on whether the quadratics were concave up or down, and how wide or narrow the functions were; some focused on how many roots the quadratics had.

Since the students were working at tables, the teacher had the opportunity to examine their work and to note some misunderstandings, for example the idea that a "narrow" parabola could live entirely in one quadrant. The subsequent (and obviously unplanned) discussion of this issue gave rise to conversations about limiting the domain of the function to non-negative values and about whether quadratic functions have asymptotes. Providing students opportunities to reveal their understandings and then responding nimbly when they do resulted in a lot of the big picture issues being addressed.

Two of the most important mathematical practices are reasoning and explaining. Ms. Smith's constant interjections, "How do we know? How do we know?" resulted in students providing mathematical justifications for most of their assertions, both in small groups and in whole-class discussions. In addition, the justifications that students provided often helped to reveal their more fine-grained content-related understandings, for example when they referred – unsolicited – to the parameters in the equations. The fact that the lesson began with a somewhat open task provided room for students to express their thinking about the big ideas. The students' responses pointed to some issues worthy of discussion that might not otherwise have emerged.

The follow-up questions in episode 3, which asked students to identify key features of the graphs of three equations *without doing computations,* and to compare the graphs of two functions, $f(x)$ and $g(x) = 2f(x)$, delved into detail. Again, the teacher's repeated question "How do we know?" and the students' responses made both emergent thinking and the correct answers publicly available.

In sum: soliciting student thinking via carefully designed tasks, picking up on what student thinking on those tasks reveals, and discussing those issues with the whole class serve to address both big picture issues and to review the particular skills that will be honed in subsequent parts of the lesson. Formative assessment played a central role in highlighting the important mathematics in this lesson.

Dimension 2, Cognitive Demand

Cognitive Demand	• How long am I given to think, and to make sense of things? • What happens when I get stuck? • Am I invited to explain things, or just give answers?

The fundamental question related to cognitive demand is: Are students being stretched – but not too far? Given that the quadratic functions lesson is a review/synthesis activity, "productive struggle" looks somewhat different than it did in case studies 1 and 2. In those lessons the students were grappling with new material, while here they were called on to use what they know. Yet, whether students are learning new material or solidifying their understanding of material they have studied, the underlying issue is the same. Can the students draw upon what they know to make connections and reinforce them?

When students engage with brand new content, the pace is often (appropriately!) slow as the students struggle to make connections for the first time. In a review/synthesis lesson, the pace may be a good deal faster – the students should have many of the resources at their disposal – but the students are still putting things together, sometimes in new ways. The challenges are to keep things moving, to identify places where the students need to do more thinking, and to make sure they are being stretched – even in review.

The students did bring a lot of knowledge to the discussions in this lesson. Overall, the conversations were pretty fast-paced. When that happens there is always the possibility that the coverage of content will be rote and superficial. That wasn't the case here. The pacing, the teacher's identification of various instances where the students were somewhat misdirected (not necessarily stuck), and her repeatedly asking "How do we know?" meant that the discussion was neither mechanical nor superficial.

Let's take the questions at the beginning of this subsection one at a time.

How long am I given to think, and to make sense of things?

Imagine if, before reading this case study, we'd presented you with the task at the heart of episodes 1 and 2,

> Have the students draw two quadratic curves that look quite different from each other, and discuss the differences.

If we had asked you, "How long do you think this discussion would take in a review lesson?", what would you have said?

We might have thought, "five to ten minutes; it's a warm-up." Yet, the discussion took nearly 20 minutes. That's because the small-group conversations among students raised interesting issues that kept the students talking with each other for some time. Then, the whole-class discussion in episode 2 veered into unexpected but valuable territory.

Part of the episode 2 discussion was triggered by the teacher's observation that one group of students had been discussing a parabola that appeared to stay entirely within the first quadrant. The teacher had observed the students discussing this issue and brought the topic to the whole class for consideration. That's formative assessment in action! The whole-class conversation addressed issues related to the domain and range of a quadratic function, which (given the fact that numerous students thought it *might* be possible for a thin, U-shaped parabola to reside in one quadrant) did help the students make relevant connections they hadn't previously made.

The second unexpected discussion was related to Sean's question about asymptotes – if a parabola has the *y*-axis as an asymptote, might that allow it to remain in the first quadrant? Here it's worth thinking about classroom norms. Sean's question was made in the spirit of sense-making – "is this possible?". There are many classrooms in which students wouldn't feel comfortable venturing such ideas. The fact that Sean felt safe enough to do so is important. And, as the discussion in episode 2 showed, his question led to a useful and informative discussion.

What happens when I get stuck?

In contrast to the situations in case studies 1 and 2, we didn't see students get stuck in this lesson segment. On occasions when students do get stuck, the challenge is to provide enough scaffolding to get them "unstuck," but to not remove cognitive demand in the process. Here that wasn't an issue, but there still were issues of cognitive demand.

A useful re-phrasing of the "stuckness" question above is, *Did students have an adequate amount of time for potentially problematic issues to arise, and to work through those issues?* In the small-group discussions in episode 1, we saw students explaining their own ideas and exploring each other's. It was in those conversations that the question of whether quadratic functions can reside in just one quadrant first arose. That issue and the question of whether quadratic functions can have asymptotes were given substantial air time in episode 2. As always, how much time to devote to any issue is a matter of judgment. We are not in a position to judge the ways in which the discussion took place, but rather to note that potentially problematic issues were noted and brought to the students for discussion.

Am I invited to explain things, or just give answers?

Ms. Smith's oft-repeated phrase, "How do we know?" indicates that providing explanations is a classroom norm. That students provide explanations on their own in the small-group discussions is an indication that the question is not simply rhetorical.

Dimensions 3 and 4, Equitable Access and Agency, Ownership, and Identity

Equitable Access to Content	• Do I get to participate in meaningful math learning? • Can I hide or be ignored? In what ways am I kept engaged?
Agency, Ownership, and Identity	• What opportunities do I have to explain my ideas? In what ways are they built on? • How am I recognized as being capable and able to contribute?

As always, Dimensions 3 and 4 are inextricably linked. A key point in the framing of equitable access is the word *meaningful* – every student should have access to the central content and practices of the lesson. Here, "access" doesn't simply mean that there is the potential to engage; it means that the activities provide significant ways for each student to engage with important ideas.

It goes without saying that classroom activities should invite engagement. Above and beyond that, however, is the question of ongoing classroom norms and culture. Do classroom interactions support the active engagement of every student? Do students feel safe enough to contribute their thoughts to classroom discussions? Once they do, how are those contributions taken up? An individual's participation and the classroom's reaction to it form the basis for agency, ownership, and identity.

Once again, we are not in the business of judging; moreover, to make any meaningful judgments on the basis of a half-hour visit to a classroom is both impossible and unfair. Our goal is to look for indicators of classroom climate and to think about issues they raise for us as teachers.

We discuss three sets of indicators in terms of their supporting access and AOI: small-group interactions, whole-class interactions, and discussions of not-fully-correct statements.

Small-group interactions

In episode 1 we saw snippets of conversation in most of the small groups. What struck us was the naturalness of the collaborations within each group. There was clear turn-taking and a substantial amount of careful listening and discussion.

In the first interaction, for example (Table 2, starting at 3:20), we saw Estelle describing her graphs to her tablemates. When she faltered briefly, looking for the right phrase, Damien jumped in to supply "double root." Estelle continued, with the others following intently. When she was done, Damien took his turn naturally: "What I would do is…" got his table's attention and he continued, "Alright, for mine, the *a* coefficient will be positive for one, and this one would have a negative *a* coefficient because it's more narrow." When he said this, Damien mis-spoke: the sign of the *a* coefficient determines the direction of concavity, not the "width" of the parabola. Without missing a beat – but naturally and gently – Estelle interjected, "It's not gonna be narrow because the *a* value…"

Most of the interactions had that quality. In both of the conversations recorded at Table 6, for example, the students interrogated each others' approaches. In conversation 1 (4:40–6:05), Richard's reference to the parameters p and q led Sean to ask about the standard forms. In conversation 2, the students had a lively conversation after the teacher highlighted Sean's observation that two of his parabolas "were small enough to fit entirely in one quadrant."

Obviously these few minutes offer a very small sampling of interaction patterns. Not everybody spoke when we looked at their tables, but all students leaned in to observe (and may well have spoken at other times).

Whole-Class Interactions

Before proceeding, you might want to leaf through the descriptions of episodes 2 and 3 to see if any patterns of student participation strike you. Here are some of the things we noticed.

Most of the time, the teacher asked for volunteers and then called on students who had raised their hands in response. (There were often quite a few). Making a habit of doing so provides a safe climate for students, assuming that their contributions are appreciated and that any mistakes they make are handled gently.

As far as we can tell, the teacher called on different volunteers each time. If a student was called on more than once, it was because the student raised their hand to comment or ask a question.

There was a different pattern at the beginning of episode 3. Ms. Smith had posed the task

<u>Write a coordinate of a key feature of the following quadratic functions:</u>

$$f(x) = x^2 - 10x + 24 \qquad g(x) = (x - 6)(x - 4) \qquad h(x) = (x-5)^2 - 1$$

and walked around the room looking at what the students had written as they worked the task. As she did, she suggested to three students that they write their answers at the board. Interestingly, none of those three students had presented earlier in the class. We don't know if they had raised their hands before, but we do know that the teacher had looked at their work and seen that the students had worked the task correctly. So, when Ms. Smith suggested that they present, she was giving them a safe space – knowing ahead of time that what they said would be correct.

In other words, not only did Ms. Smith provide what appears to be an equitable distribution of speaking opportunities, but she did it with a "safety net." That's an aspect of equitable access.

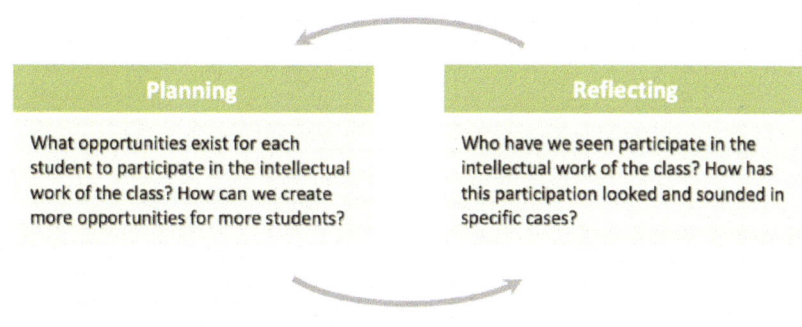

Planning

What opportunities exist for each student to participate in the intellectual work of the class? How can we create more opportunities for more students?

Reflecting

Who have we seen participate in the intellectual work of the class? How has this participation looked and sounded in specific cases?

Things to think about

- What is the range of ways that students can and do participate in the mathematical work of the class (talking, writing, leaning in, listening hard; manipulating symbols, making diagrams, interpreting text, using manipulatives, connecting different ideas, etc.)?
- Which students participate in which ways?
- Which students are most active, and when?
- In what ways can particular students' strengths or preferences be used to engage them in the mathematical activity of the class?
- What opportunities do various students have to make meaningful mathematical contributions?
- What are the language demands of participating in the mathematical work of this class (e.g., academic vocabulary, mathematical discourse practices)?
- How can we support the development of students' academic language?
- How are norms (or interactions, lesson structures, task structure, particular resources, etc.) facilitating or inhibiting participation for particular students?
- What teacher moves might expand students' access to meaningful participation (such as modeling ways to participate, holding students accountable, point out students' successful participation)?
- How can we support particular students we are concerned about (in relation to learning, issues of safety, participation, etc.)?
- How can we create opportunities for more students to participate more actively?

Figure 4.29 Access suggestions from the Conversation Guide

This came about because of ongoing formative assessment. The teacher was aware of who had spoken to the whole class and that these three students had not yet done so – and that in this context they could do so safely. Being aware of student thinking and adjusting classroom actions accordingly (in line with the goals of instruction) is the essence of formative assessment.

There are many other ways to think about equitable access; we have, here, simply highlighted one technique evident in the lesson segment. The big question is how we might support increasing opportunities for our students in general. There are many things to think about in that regard. Figure 4.29 reproduces some suggestions from the TRU Conversation Guide that are worth keeping in mind when thinking about equitable access.

Discussions of not-fully-correct statements

Equitable access provides every student with meaningful opportunities to participate – but then, the nature of that participation is critically important. In this concluding section we discuss two events and suggest their implications for agency, ownership, and identity.

Let's work through the first student presentation in episode 2. It began when the teacher asked for volunteers. Grace was among those who raised her hand. The teacher asked her to go to the front of the classroom.

As Grace spoke, the teacher wrote down the technical terms she used: vertex, y-intercept, x-intercept, concave up, concave down. Recording the terms in this way is a validation of the student's presentation; that the teacher wrote them on the board meant that Grace is saying things that are important. What happened after that was minor, but interesting. Here's what took place.

> Grace said "… my first graph is a parabola that goes downward, it's concaving down, it's vertex is the y-intercept, and it has a positive x-intercept and a negative x-intercept and it has a wide… [Grace looked somewhat unsure of herself and glanced at the teacher]… a wide curve." A student said "a what?" and Grace repeated "a wide curve".
>
> At that point the teacher asked, "a wide curve, what does Grace mean when she says 'a wide curve'?"
>
> There was some talking at the tables, after which the teacher called on Brandon who said, "that means that the lead coefficient, a, is closer to zero." In response the teacher asked, "Does everyone understand?" There was a chorus of yeses.
>
> When Grace concluded her presentation the teacher said "thank you" and the class applauded.

Grace appeared unsure when she said "a…wide curve" and had she not fared well in the ensuing interaction there might have been negative implications for her mathematical identity. The fact that the teacher took this issue as being worthy of discussion and asked the groups to talk about it took the pressure off Grace. The issue was dealt with, and Grace sat to her classmates' applause. As far as we can tell, the net impact of the conversation was positive.

This kind of discussion took place more than once. Later, when Marcus made a comment about a function having neither x- nor y-intercepts, the teacher repeated it and asked the class to think about it. Doing so signaled that the topic was worth talking about, and the give-and-take of the conversation deflected attention from the person who made the mathematically incorrect statement.

Likewise, in the discussion of whether a parabola could reside in one quadrant, Ms. Smith said

> I asked a question to one of the groups, I think it was that group, and I said, can a parabola be in one quadrant?… Only one quadrant?… Can a parabola be in only one quadrant? [3 sec pause] Think about that. Can a parabola… Maybe on the other side of your card, you can try to draw a parabola that's only in one quadrant. Can a parabola only reside in one quadrant? Can it, can it, can it… That's a good question…

Here Ms. Smith took responsibility for the question, which was generated by a student mistake. The result was a robust conversation – and no embarrassment for the student.

Once again, formative assessment served multiple purposes. Raising the issue of whether a parabola could reside in one quadrant, which arose from observing the students' work, advanced the class's mathematical conversations and understanding. The way in which the issue was raised maintained the safety of the environment, a point that is underscored by the fact that Sean felt comfortable raising the issue of asymptotes.

And, there's more than safety involved. In general, the teacher tended to take student questions and – rather than answering them herself – echo them back to the class for discussion. This has positive implications for agency ("we can figure it out"), ownership (it wasn't handed down "from above"), and identity ("we're mathematical sensemakers").

Once again, the broader issue for AOI is, how can our observations of student thinking be leveraged to provide positive opportunities for rich mathematical discourse? Figure 4.30 provides some useful suggestions from the TRU Conversation Guide (Figure 4.29).

In general, the goal of instruction is for every student to have rich mathematical experiences, in ways that build their understanding of important mathematics and provide opportunities for the development of positive mathematical identities. All five TRU dimensions are important – and they reinforce each other. Whether we're thinking about a lesson that introduces new

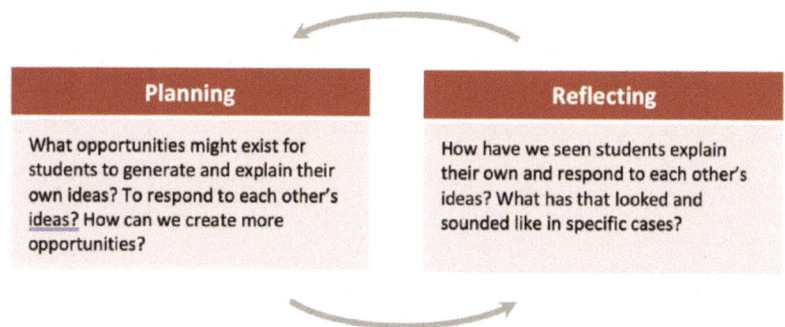

Agency, Ownership, and Identity

Core Questions: What opportunities do students have to see themselves and each other as powerful doers of mathematics? How can we create more of these opportunities?

Planning

What opportunities might exist for students to generate and explain their own ideas? To respond to each other's ideas? How can we create more opportunities?

Reflecting

How have we seen students explain their own and respond to each other's ideas? What has that looked and sounded like in specific cases?

Things to think about

- Who generates the ideas that get discussed?
- What kinds of ideas do students have opportunities to generate and share (strategies, connections, partial understandings, prior knowledge, representations)?
- Who evaluates and/or responds to others' ideas?
- How deeply do students get to explain their ideas?
- How does (or how could) the teacher respond to student ideas (evaluating, questioning, probing, soliciting responses from other students, etc.)?
- How are norms about students' and teachers' roles in generating ideas developing?
- How are norms about what counts as mathematical activity (justifying, experimenting, connecting, practicing, memorizing, etc.) developing?
- Which students get to explain their own ideas? To respond to others' ideas in meaningful ways?
- Which students seem to see themselves as powerful mathematical thinkers right now?
- How might we create more opportunities for more students to see themselves and each other as powerful mathematical thinkers?

Figure 4.30 AOI suggestions from the Conversation Guide

material as in case studies 1 and 2, or one that reviews and synthesizes previous study as in this lesson, the mathematics needs to be rich and coherent. Then, students need to be engaged in sense-making and productive struggle. By "students," we mean each and every student; and, the ways in which students are engaged needs to support the students' sense of who they are as mathematical thinkers. As this case study indicates, all this is supported by formative assessment – by attending to student thinking and adjusting instruction in ways that all students can engage productively with rich mathematics.

Notes

1 This sounds simple, and we often gloss over it. But, it's worth remembering – for example, the point $(3,9)$ is on the graph of the function $y = x^2$ because $9 = 3^2$, and the point $(3,10)$ is *not* on the graph of the function $y = x^2$ because $10 \neq 3^2$.

2 For completeness we'll note that expressions that are "quadratic in y" such as $x = y^2$ are also parabolic; their graphs are sometimes referred to as "horizontal" parabolas. In this discussion we'll limit our attention to expressions that are quadratic in x.

Reflecting on Images of Practice (The Case Studies)

3 We stress that the questions themselves are not enough. Equitable access and AOI depend on the teacher's having established norms that support all students feeling safe to contribute, and on ideas being refined in productive ways.

4 Throughout the lesson, Ms. Smith grabs and maintains students' attention by using a singsong voice ("Is that cleeeear?"), repeating something rapidly ("How do we know? How do we know? How do we know?), etc. This is in keeping with the fast but comfortable pace of classroom conversations.

5 In this instance we were able to get clear screenshots, which show the index cards produced by the students. For the sake of clarity, most of the other figures show reproductions.

6 It says "Asymptotes: None" on the poster, so this is clearly a review.

7 The authors are accustomed to a different use of the term "coordinate," in which the ordered pair (a,b) is said to have two coordinates, the x-coordinate "a" and the y-coordinate "b." However, It's clear the class is accustomed to the way the term is used here – a "coordinate" is an "ordered pair" – so we follow the classroom usage and transcribe directly.

8 Brandon is referring to the fact that $f(x) = (x - p)(x - q) = x^2 - (p + q)x + pq$. That means that the roots of $f(x)$ are factors of the constant term, and their sum is the coefficient of x. He worked backwards to factor 24 into 4 x 6, observing that their sum is 10. That means that $f(x) = x^2 - 10x + 24$ can be factored into $(x - 4)(x - 6)$.

9 Remember that "diving" is relative when seen on a graph. g dives down to 0 from a height slightly above 40 as x goes from 1/2 to 5. That's pretty steep.

10 Is it possible that the students would respond "yes" to the question "are we good" when they didn't see why the y-intercept of $f(x)$ was (0,24),? Yes it's possible. But, given the aspects of classroom culture we've observed, it seems reasonable to believe that if the students didn't follow, at least some would say so.

Conclusions and Next Steps

Introduction to Part III

We hope you've found the classroom visits in the three case studies to be interesting, useful, and generative as sources of reflection – and we hope that you've come to see how thinking with TRU can enrich your understanding of the factors that help students become powerful mathematical thinkers. Now it's time to think about applying those understandings to your own work as an educator.

Our goal in Part III is to provide you with the tools and perspectives that distill the key ideas in TRU and enable you to use them to support your teaching, in cycles of planning, teaching, and reflection. The first part of Chapter 5, "Looking back and looking forward: TRU in the case studies, TRU in our classrooms," looks back over the way we conducted our classroom observations. It addresses the ways that TRU shaped our discussions of the case studies and highlights some of the main "take-aways" that emerged for us about how TRU helps us to think about what happened in those three classrooms. The second part of the chapter, looking forward, turns to practicalities. Now that you have a sense of what TRU can offer, how can you use those ideas to shape your own teaching practice and work with your colleagues? We describe a number of models that individual teachers, small groups of teachers, and professional learning communities have used to make TRU-based planning and reflection central to their own teaching.

Chapter 6 provides two of the tools that enable you to bring TRU into your classroom, the *TRU Conversation Guide* and the *TRU observation Guide*. The Conversation Guide opens up thinking about each of the five dimensions. It offers sets of questions that we can ask before, during, and after teaching, to sharpen the ways we use TRU with our students. The Observation Guide helps frame the observations and conversations you might have before and after someone sits in on your class. The key questions it addresses are, what activities are students engaged in, and how do they support students in becoming more powerful and empowered mathematical thinkers? What actions by the teacher might help scaffold the students in engaging with the mathematics and each other in powerful ways?

Beyond these two fundamental tools, we point to a collection of other tools and approaches that various groups we've worked with have found to be helpful. With these at your disposal, you're positioned to take next steps on the lifelong journey of refining your teaching in ways that support students to engage deeply with rich mathematics.

DOI: 10.4324/9781003375197-7

Looking Back and Looking Forward

TRU in the Case Studies, TRU in Our Classrooms

It's time for us to take stock – to think about what we've learned from the case studies and about how we can apply what we've learned in our teaching. The first part of this chapter, Sections 5.1 and 5.2, uses the five dimensions of TRU to reflect on the three cases. In those sections we highlight differences, similarities, and unifying themes across the three classroom stories. We identify the take-aways that we hope you, the reader, have come to understand from working through them. The goal is to learn to see things through the lens of TRU – to plan, teach, and reflect on instruction in ways that make the learning experience richer and more empowering for students.

The second part of this chapter, Sections 5.3 and 5.4, addresses the question, "now what?" Whether you're a pre- or in-service teacher reading this book by yourself, part of a small group of teachers who are working together to share ideas and teaching practices, or engaged in a more formally organized professional or teacher learning community, there are some tools and practices that can deepen your experience and that apply directly to your teaching. Section 5.3 describes two tools we've made use of throughout this book, the *TRU Conversation Guide* and the *TRU observation Guide*. Section 5.4 begins with a set of principles regarding powerful professional development, both in general and with regard to TRU. It then provides a range of models for digging more deeply into TRU and using it to shape your teaching. By the end of the chapter (and with the help of the resources offered in Chapter 6) you should be ready to take next steps in applying the core ideas in TRU to your teaching.

5.1 Reflecting on the Three Cases

5.1.1 TRU in the Case Studies – Differences, Similarities, and Unifying Themes

Just as no two snowflakes are the same, no two classrooms are the same. Although each of the three classrooms we visited in the case studies dealt with a topic in high school algebra, the classrooms were far more different than similar – indeed, they were chosen for that reason. In what follows we discuss differences, similarities, and unifying themes. In this section we organize the discussion by the five TRU dimensions, in order.

DOI: 10.4324/9781003375197-8

Dimension 1: The Mathematics

We begin with the mathematics. In "The Car Values Problem" the class was asked to grapple with a modeling problem that was set in a meaningful context, and to extend what they knew about exponential functions in a new way. The mathematical connections explored in that case, including the importance of the problem context as a check on students' ongoing work, were deep and complex. In "Where is the Ten?" the students, working on what might seem on the surface to be a simple problem of adding algebraic terms, were confronted with some very subtle issues related to connections between the visual and algebraic representations of the object's perimeter. The students were also held to a very high standard of mathematical explanation, a key mathematical practice. In "Graphing Quadratic Functions" the students were asked to review and synthesize a large number of concepts central to understanding quadratic functions and their graphs – and to confront some of their own incomplete understandings. Looked at "up close," the math in the three case studies differs in type and focus.

What the cases have in common, however (beyond the superficial fact that they all deal with some aspect of algebra) is that each case focuses on robust understanding – on mathematical concepts and practices that fit together in important ways. Students should make deep connections between different ways of conceptualizing and expressing exponential functions, and they should make meaningful use of the contexts in which problems are set. Students should make sense of algebraic representations and the connections between symbols and the objects they represent; and they should learn to produce coherent and compelling explanations. Students should develop fluency with algebraic manipulation and in visualizing the impact of parameter changes on the graphs of functions. What these three vignettes exemplify is rich, deep, connected mathematics. Whether or not the students make all the connections they might have, or as richly as one might hope, isn't the point; no lesson is perfect. What matters here is that the mathematics in which the students engaged in all three cases was anything but cut-and-dried, mechanical answer producing. It was mathematics worth learning. That's what Dimension 1 is all about. The goal for all lessons should be for students to engage with important mathematical ideas.

Dimension 2: Cognitive Demand

Let's turn to Dimension 2, cognitive demand. The issues play out differently in all three cases. In the lesson on depreciating car values, the students in the focal group and in the class as a whole found the task to be quite challenging. In consequence the questions faced by the teacher were: How much challenge is too much challenge? When does struggle become unproductive? There's no right answer to these questions, especially when there's a whole class of students who have different understandings and different tolerances for frustration. The car value lesson provides us with a close look at the tensions involved.

The tensions in "Where is the ten?" are different. In the focal group the students all had responsibility for each other. Their goal was to make sure that Alicia, the group representative, could provide the teacher with a clear explanation of where the 10 is. The challenges and frustrations were high at times, but the context for their interactions was different: The teacher was monitoring group dynamics, and the students were working – although not always smoothly – as a team. The question for the teacher was how much, if at all, to scaffold the group dynamics and explanations. The goal for the students was to work as a team, focusing on the big ideas; each had some learning to do. That's a complex question of productive struggle.

It might appear at first that there's very little challenge in "Graphing quadratic functions." The students seemed to be on top of the material. Would this turn out to be a rote review? Well, no – in part because the FAL's well designed tasks pushed the students to do some synthetic thinking and in part because the teacher observed various misunderstandings in student conversations and turned some of those misunderstandings into topics for small group and then whole-class conversations. The dialogue was fast paced and for the most part is grounded in solid understandings, but there was a level of challenge.

What we see, then, are three very different kinds of instructional context and cognitive demand. That's no surprise. Every group of students is different. More importantly, at different times students will find themselves struggling in different ways. The overarching questions are, Are the students engaged in activities in which there's something meaningful to learn? Are they challenged at least some of the time, as they're engaged in sense-making? On the one hand, they shouldn't be bored because things are too easy; on the other hand, they shouldn't be frustrated to the point where their efforts are counterproductive. The three lessons, as different as they are, show a range of ways in which such tensions can arise. There are infinitely different ways – the issue is always there, in some way or another.

Dimensions 3 and 4: Equitable Access and AOI

Because issues of equitable access and student agency, ownership, and identity are so deeply intertwined, we will again discuss Dimensions 3 and 4 together. Once again we see significant variation but common underlying themes. During the focal group interactions in the car value lesson there were different levels of verbal participation and engagement. Some students said much more than others, although all seemed engaged. If you multiply that diversity of participation by the number of small groups in the class, the magnitude of the management issues related to equitable access becomes clear. Moreover, the degrees and kinds of participation of different students have implications for AOI. In every classroom, supporting small groups so that all students participate in meaningful ways with the core content is an ongoing challenge. Similarly, there are significant risks when students present incorrect work to the whole class. Unless it is managed carefully, the ensuing feedback could have a negative impact on students' identities.

In "Where is the ten," Alicia's struggles and their impact were apparent: Access was not an issue, but issues of AOI were front and center. In fact, in all three case studies the tensions related to cognitive demand (how hard is too hard?) had profound implications for students' mathematical identities. In the quadratic functions lesson, some very clear errors in understanding emerged. Here too, how they were dealt with made a difference. There were issues of norms (do students feel comfortable raising questions, or will they remain quiet for fear looking dumb?), and of teacher moves to normalize the ways in which issues are raised and addressed.

What the three case studies show is that issues of equitable access and AOI are always present, as they are in every human interaction. They differ in every classroom, in every exchange between students – and they're important. The point is that we need to be aware of them, and to try to address them. Here too, there are no easy answers. Awareness helps, as does knowing a range of tools and techniques for addressing such issues. Those are discussed in Section 5.3.

Dimension 5: Formative Assessment

As with the first four dimensions, there are distinct but overlapping examples of formative assessment (Dimension 5) in the three case studies. In the car value lesson, the teacher constantly monitored student progress in both small groups and the whole class. She provided different kinds of feedback, depending on the circumstances. Early on, for example, she simply pointed out where a group had gone wrong and left it to the group to figure out how to recoup. Toward the end of the lesson, as the students were tiring, she pointed out and discussed a flawed approach with the whole class and provided much more direction to the students than earlier. In "Where is the ten?" the teacher paid close attention to student understanding, and set and maintained very high standards. This case study highlights a critical aspect of formative assessment – that students are powerful resources in instruction, and that the burden of identifying and addressing misconceptions or other partial understandings does not lie solely on the teacher's shoulders.

Finally, the quadratic functions lesson highlights multiple aspects of formative assessment. To begin, there's design. The carefully crafted tasks featured in the FAL provide students with opportunities to reveal what they think is important about quadratic equations and their graphs in general, and then pinpoint specific understandings that the students should have developed regarding the different forms of quadratic expressions and corresponding features of their graphs. There is, as in the first two cases, the use of small groups as a mechanism for making student thinking public (sometimes within the group, sometimes brought to the whole class), and for addressing issues that arise. In addition there's the unplanned but central use of formative assessment, when the teacher hears a student comment that merits full discussion by the class. More broadly, the way the lesson played out allowed us to show how (by design!) formative assessment serves as a mechanism for enriching all four of the other TRU dimensions.

5.1.2 What Lessons can we Take from these Three Deep Dives?

Perhaps better, the question is, what do the authors hope that you'll take from your reflections on the events described in these three classrooms?

We hope you've appreciated our close examination of what took place in these three classrooms. The more is known about student thinking, the more is known about classroom interactions, the more is known about powerful learning environments … the more we can do to help our students become agentive and powerful thinkers.

As we wrote in the "welcome" that started this volume, we selected the three case studies for a number of reasons. They're examples of real, everyday practice, in which teachers are faced by interesting pedagogical dilemmas. The vignettes are not intended to be exemplars or "model instruction"; our goal is to think through and reflect on interesting and challenging situations. The cases take place in a range of classroom contexts: Small group, whole class, and a mix of both. "Where is the ten" shows students learning something brand new. The "Car Value" problem shows students building on what they know. And the quadratic Functions FAL is a review/synthesis lesson. What they have in common is that they highlight some of the interesting challenges we face during instruction, and that they provide opportunities for reflecting on the five TRU dimensions. Reflecting on instruction in the ways we've done here can and should be an ongoing, lifelong process.

We also hope that, in working through the case studies, TRU's focus on the student's experience has become increasingly natural for you. When your focus is on the students and their learning, the key question is not "What is the teacher doing?" (or more personally, "What am I doing as a teacher?") but, "How is each and every student experiencing instruction?"

It's essential to put ourselves in the student's shoes. That's why Figure 5.1 (= Figure I.2), "Observe the lesson through a student's eyes," is so important. The issue with regard to content (Dimension 1) is not simply whether the mathematics is rich and connected. The issue is how the mathematics is connected to what the students know, so the students can see and experience mathematical ideas in ways that are rich, connected, and powerful.

The students' engagement with mathematical content and practices is the substance of Dimensions 2 through 5. Our goal is for students to be mathematical sensemakers. That means they need to be stretched (but not too far!) when learning, so that they can build connections in ways that last. Do they have time to think, to work things through? Do they explain, or simply memorize? These are key aspects of Dimension 2, Cognitive Demand, as reflected in Figure 5.1. Dimension 3 means that each and every student builds those connections. We have to be concerned about making sure that all students participate in meaningful ways, and that they can' hide or manage to be ignored. And, of course, access is the gateway to issues of Agency, Ownership, and Identity (Dimension 4). What's important at the top level is the students' capacity and willingness to be mathematical sensemakers. These capacities need to be continually

Observe the Lesson Through a Student's Eyes	
The Content	• What's the big idea in this lesson? • How does it connect to what I already know?
Cognitive Demand	• How long am I given to think, and to make sense of things? • What happens when I get stuck? • Am I invited to explain things, or just give answers?
Equitable Access to Content	• Do I get to participate in meaningful math learning? • Can I hide or be ignored? In what ways am I kept engaged?
Agency, Ownership, and Identity	• What opportunities do I have to explain my ideas? In what ways are they built on? • How am I recognized as being capable and able to contribute?
Formative Assessment	• How is my thinking included in classroom discussions? • Does instruction respond to my ideas and help me think more deeply?

Figure 5.1 What's important, from the student perspective

nurtured in classroom interactions, with opportunities for participation and recognition of the students' contributions. Finally, Dimension 5 (formative assessment) is the glue that binds all this together. Formative Assessment is fundamentally concerned with making student thinking accessible, refining it in discussions, and using what's been learned about student thinking to optimize what's happening in Dimensions 1 through 4.

5.2 Reviewing the Big Ideas about TRU, in Preparation for Applying it to Our Teaching

What we've tried to show throughout this book is how complex – and how important! – each of the five TRU dimensions is. In Sections 5.3 and 5.4 we put those understandings in context and think about possible next steps. Here we want to pave the way by looking at the big picture, in Q&A form. In doing so we draw upon a large body of research about TRU and our own experience.

- Why are there five dimensions? Why these five?

  Here's the story in a nutshell. Over a period of years the Teaching for Robust Understanding Project surveyed the literature. We found literally hundreds of different factors that various studies said were important. That's far too many. Even if you cut the list down to 20 or 30 of the "most important" classroom suggestions, that list would be impossible to work with. It's impossible to keep tabs on 20 or more different things while you're engaged in something as complex as teaching. (In fact, there's a very famous paper titled "The Magical Number Seven, Plus or Minus Two: Some Limits on Our Capacity for Processing Information" by George Miller (1956), which implies that keeping 7 things in mind can be a stretch.) So, we set out to distill the list into a small number of categories of related ideas.

Some of the criteria we used to organize the categories were:

- The categories (which, as you know, we call the dimensions of powerful classrooms) should make sense from a practical point of view as well as from a theoretical perspective.

  A research technique called factor analysis is often used to separate collections of studies into theoretically coherent subcollections. The problem is that those subcollections often don't correspond to anything meaningful in practice – in which case they can't be used to improve practice.

- Everything important is still contained within the dimensions.

  There can't be anything important missing. It's essential to be able to say, "if things go well along these key dimensions, students will emerge from instruction being powerful and empowered learners."

  It's impossible to list everything that counts in a figure such as Figure I.1 – in that kind of representation each dimension is characterized by just a few dozen words. But, everything should fit *conceptually* with the short descriptions we've provided, in ways that feel natural. For example, one colleague said after hearing about TRU that their major concern was that classrooms need to be safe environments for students to think and learn. Was classroom safety in TRU?

  Absolutely! A feeling of safety is a necessary precondition for AOI: If students don't feel safe venturing ideas, then there's little chance they will develop positive mathematical identities. In putting TRU together we had to choose the "headers" for the five dimensions. The decomposition into 5 dimensions isn't unique; someone else might produce a different set of dimensions, with "safety" as one dimension. Any choice of dimensions will bring some aspects of each dimension into the foreground and will background others. That said, we're pretty happy with the dimensions we've chosen.

- Each dimension is somewhat independent from the others.

  As we've pointed out, the dimensions interact, and all the dimensions are in play all the time. But, each dimension makes sense by itself – you can talk about it, and you can work on it in meaningful ways. Our case studies exemplify this. "The car value problem" provided a close look at cognitive demand (admittedly the most abstract of the five dimensions); "Where is the ten?" provided a comparably close look at issues of equitable access and AOI; and the entire collection of Formative Assessment Lessons (which includes the quadratic functions lesson) highlights issues of formative assessment. The fact that you can work on each dimension by itself is fundamentally important, because:

- A major goal is for the 5 TRU dimensions to serve as a solid foundation for professional development.

  As we'll see in Section 5.3, it's possible to focus on each dimension in productive ways. That makes a *big* difference.

- What does it mean to "understand" each dimension?

  There isn't a neat and clean answer to this question! You know, and we know, that teaching is extremely complex. We never understand it fully, but we do get better at it, over time, with reflection.

  We can *always* get better at mathematics (Dimension 1), seeing new connections or developing deeper understandings. This past semester a student in one of our classes produced a solution to the "Where is the ten?" problem we'd never seen before. It was marvelous!

  Similarly with regard to Dimension 2, you don't simply "master" cognitive demand – there are no algorithms for determining the "right" level of challenge. There are no such rules because every class and every student is different; what seems to be right at any particular time will depend on our knowledge of each particular student, the mathematics they're studying and what they seem to find difficult, and maybe their mood that day. What happens is that as we observe and reflect, we become more perceptive. We develop a deeper sense of what to look for and our repertoire of tools and techniques, including specific content knowledge that helps us to scaffold students more effectively, continues to grow.

The same is the case for the other dimensions. There are many techniques for providing the students in our classes with increasingly equitable opportunities for engaging key mathematical ideas. Some of those methods work better with a teacher's individual style than others. Over time we find out what works best for us – although things always depend on the student, the topic, and how people are feeling that day. That's especially the case for issues of agency, ownership, and identity. Everything depends on knowing our students and getting a sense of what works for them. Likewise for formative assessment. There are specific content-related ways in which students are likely to go wrong. That's why there are so many different Formative Assessment Lessons, each of which highlights "common issues" related to student understandings of the content in that lesson.

So where does this leave us?

In summary, TRU provides a comprehensive but telegraphic view of the dimensions of classrooms from which students are likely to emerge as mathematically powerful thinkers and learners. Each of the TRU dimensions is important, and each is something we can get better at. What matters over the long run is thoughtful attention to those five dimensions, each by itself and in interaction with the others. There are no "magic bullets" – anyone who tells you there is a simple way to improve teaching is either a liar or a fool – yet there are productive ways we can work on those dimensions, individually and with partners.

5.3 Tools that Can Help and Possible Next Steps

The key to getting better at teaching is to pay ongoing attention to what happens in our classrooms and to reflect on it. TRU has some tools to help. You'll find two tools, the *TRU conversation Guide* and the *TRU Observation Guide,* in Chapter 6. A third, synthetic tool, *On Target* (Schoenfeld, Fink, Sayavedra, Weltman, & Zuñiga-Ruiz 2023), is book length. All of these tools can be used by individuals for cycles of planning, teaching, and reflection – but it's so much better to work on these issues with colleagues. We'll discuss a number of ways that teachers or teacher learning communities can use these and other tools.

5.3.1 The TRU Conversation and Observation Guides

The lessons in Chapters 2, 3, and 4 differed in significant ways – as do all lessons. The focal questions in each case study resembled each other in some ways, but they differed too. It's reasonable to ask: Where did those questions come from? As we plan our lessons, teach them, and reflect on them, what kinds of questions can we ask? That's the issue that the two tools in this section are designed to help with.

We know the five dimensions of TRU: The Mathematics; Cognitive Demand; Equitable Access, Agency, Ownership, and Identity; and Formative Assessment. We want the learning environment to be as rich as possible along those five dimensions. So, let's consider some questions from the *TRU Conversation Guide* that ask how the learning environment can be enriched along each dimension. See Figure 5.2.

Figure 5.2 provides a start. It turns *criteria* for powerful learning environments into a set of questions we can ask in order to make our teaching more powerful.

The Five Dimensions of Mathematically Powerful Classrooms

The Mathematics	How do mathematical ideas from this unit/course develop in this lesson/lesson sequence? How can we create more meaningful connections?
Cognitive Demand	What opportunities do students have to make their own sense of mathematical ideas? To work through authentic challenges? How can we create more opportunities?
Equitable Access to Content	Who does and does not participate in the mathematical work of the class, and how? How can we create more opportunities for each student to participate meaningfully?
Agency, Ownership, and Identity	What opportunities do students have to see themselves and each other as powerful mathematical thinkers? How can we create more of these opportunities?
Formative Assessment	What do we know about each student's current mathematical thinking? How can we build on it?

Figure 5.2 Questions to enrich planning in the five TRU dimensions. From the *TRU Conversation Guide*

Those questions are still pretty general, however. Let's get more specific. As an example, take Dimension 4, AOI. The overarching AOI question in Figure 5.2 is,

> What opportunities might exist for students to generate and explain their own ideas? To respond to each other's ideas? How can we create more opportunities?

In more detail we might ask,

- Who generates the ideas that get discussed?
- What kinds of ideas do students have opportunities to generate and share (strategies, connections, partial understandings, prior knowledge, representations)?
- Who evaluates and/or responds to others' ideas?
- How deeply do students get to explain their ideas?
- How does (or how could) the teacher respond to student ideas (evaluating, questioning, probing, soliciting responses from other students, etc.)?
- How are norms concerning students' and teachers' roles in generating ideas developing?
- How are norms concerning what counts as mathematical activity (justifying, experimenting, connecting, practicing, memorizing, etc.) developing?
- Which students get to explain their own ideas? To respond to others' ideas in meaningful ways?
- Which students seem to see themselves as powerful mathematical thinkers right now?
- How might we create more opportunities for more students to see themselves and each other as powerful mathematical thinkers?

This set of questions gives us lots to work with. As we plan our lessons, we can examine the lesson plans to see how well the plans anticipate such issues. And after we've taught, we can think about how we might do things differently the next time.

We've drawn the bulleted points above from the *TRU Conversation Guide*. The core idea is to "problematize" our teaching – to ask questions about our plans that will help us engage the

Vstudents more deeply. The *TRU Conversation guide* expands each of the enrichment questions in Figure 5.2 to a family of questions to consider. You can find the guide in Chapter 6.

A quick note: There are a lot of questions in the *Conversation Guide*! But, they're organized under the five dimensions, so they're more manageable. At the top level, the questions on the right hand side of Figure 5.2 are all you need to think about. They're the "question forms" of the five dimensions. When you're planning or revising, however, more specific detail helps. The questions given in the bullets above are laid out for use in the *Conversation Guide*.

We can tell you from experience that the more you work with these questions the more they become second nature. We didn't actively need to keep the guide in front of us when we wrote the case studies. We'd pretty much internalized the questions, and our thoughts were stimulated by what we saw.

The *TRU Conversation Guide* is especially useful for planning and revising lessons. Now we introduce a second tool, the *TRU Observation Guide,* that's designed for observing lessons and reflecting on them (with an eye toward revision). It will feel familiar in spirit – the core ideas in the *TRU Observation Guide* have structured our commentaries on all three case studies.

Imagine that you have a partner (a fellow teacher, a coach, or an administrator) visiting your class, or that you've videotaped the lesson and now have a chance to review it. What we'd like for the observer to do is *not* to evaluate the teacher, but rather, to put themself in the position of being a student in the class. That is, we'd want them to observe the lesson through a student's eyes. What do the students see and experience, how does it feel, what does it suggest?

Sound familiar? We hope so; that's Figure 5.1. As in the case of the *Conversation Guide,* each of the questions in Figure 5.1 can be thought of as representing an area for investigation. We can expand the questions that accompany each dimension in two ways. We can ask, (1) what kinds of student activities indicate that things are going well along that dimension? (2) what kinds of moves by the teacher might facilitate productive activities in that dimension?

Once again, these are specific questions designed to be used in context. The responses to them depend on the students and how things play out in that lesson. Under ideal circumstances, the teacher has at least one partner; the teacher and partner(s) might agree beforehand that the observer will attend to specific issues.

Figure 5.3 reproduces the page in the *TRU Observation guide* that corresponds to an agreed-upon focus on Dimension 4, agency, ownership, and identity. Our experience has been that using structured observations of this type fleshes out and contributes to the review process, and results in increasingly rich planning and implementation of lessons. The *Conversation and Observation Guides* together contribute to a program of ongoing professional development. Various models for such programs are discussed in Section 5.4.

5.3.2 Mathematics Teaching On Target

The TRU team has also produced a tool called *Mathematics Teaching On Target: A Guide to Teaching for Robust Understanding at All Grade Levels* (Schoenfeld et al. 2023), which helps with more detailed planning. Here's the basic idea. Imagine that you have a particular task or activity in mind – possibly one from the curriculum or textbook, or one you're creating. Can you identify aspects of the task or activity that are not as rich as they might be, or possible modifications that are more aligned with TRU and offer richer opportunities for students? For example, suppose the syllabus calls for having students work a series of similar exercises. Are there ways that you might vary the exercises to make them more mathematically rich, to introduce some strategic thinking, or to make connections to other mathematical ideas or real world contexts? Can you make the task more interactive? etc. There are lots of specific ways such improvements can be done, and lots of more pointed questions than those in the conversation and observation guides.

AGENCY, OWNERSHIP, AND IDENTITY	
The extent to which every student has opportunities to explore, conjecture, reason, explain, and build on emerging ideas, contributing to the development of agency (the willingness to engage academically) and ownership over the content, resulting in positive disciplinary identities.	
Each student…	Teachers…
• Takes ownership of the learning process in planning, monitoring, and reflecting on individual and/or collective work • Asks questions and makes suggestions that support analyzing, evaluating, applying and synthesizing ideas • Builds on the contributions of others and help others see or make connections • Holds classmates and themselves accountable for justifying their positions, through the use of evidence and/or elaborating on their reasoning	• Provide time for students to develop and express their ideas • Work to make sure all students have opportunities to have their voices heard • Encourage student-to-student discussions and promote productive exchanges • Assign tasks and pose questions that call for marshaling, analyzing and synthesizing evidence, and for students to explain their reasoning • Employ a range of techniques that attribute ideas to students, to build student ownership and identity
• Other focal points for observation:	
What opportunities do all students have to see themselves and others as proficient disciplinary thinkers, to grapple with challenges and construct new understandings, to build on others' ideas, and demonstrate their understandings? How can more of these opportunities be created?	
Goal: All students build productive disciplinary identities through taking advantage of opportunities to engage meaningfully with the discipline and share and refine their developing ideas.	

Figure 5.3 A sample page (Agency, Ownership, and Identity) from the *TRU Observation Guide*

Our basic metaphor is a target that lists possible attributes of a task or activity. Attributes on the outer ring of that target can stand some significant improvement; those moving toward the center are increasingly rich. So, if you consider any task, you can ask, what attributes does the current task have? Where do those attributes "land" on the targets? Are there natural modifications to the task that would result in the attributes landing more toward the center of the targets?

Mathematics Teaching On Target contains 15 targets, three for each dimension. Figure 5.4 shows the first mathematics target. For fun, you might take a typical task or activity and think about concrete improvements suggested by the target.

5.4 Where Might You Go from Here?

At this point you're familiar with the core ideas underlying TRU and you've had some experience looking at instruction through the lens of TRU. This concluding section of the chapter explores some possible ways to deepen your engagement with TRU-related ideas.

We start by outlining the big picture in Section 5.4.1, describing six general principles related to professional learning. None of these will surprise you, but they're good to keep in mind as you contemplate next steps. Then we provide three TRU-related principles that supplement them. In Section 5.4.2 we offer a range of models for digging more deeply into TRU.

5.4.1 Principles Related to Professional Learning

We've said a lot about TRU, and we've explored ways TRU can be used to think deeply about instruction. Here we think about key principles related to people's growth as teachers, a.k.a. their professional learning and development. We start with six general principles.

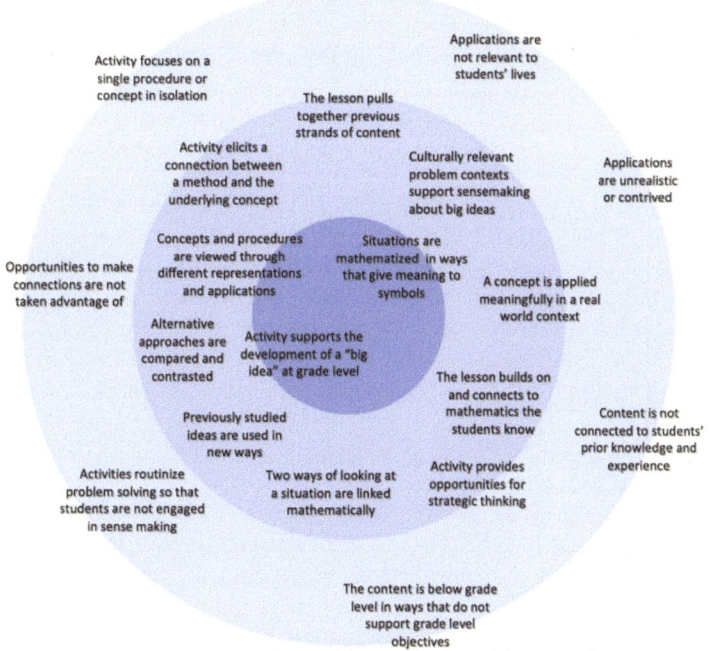

Figure 5.4 The first of 15 targets in *Mathematics Teaching on Target*

General Principle 1. Teaching is complex. Getting better at teaching takes time and effort, even when you have solid support from administration and are part of a robust teacher learning community.

The three case studies in this book indicate just how knowledge-intensive, context-dependent, and challenging the process of teaching really is. There are no quick fixes or tricks to improve teaching, despite the fact that many people are happy to offer them.

We say this for two reasons. The first is that if you're an administrator, it's natural to look for quick results – usually, improved test scores. That's the wrong place to look. The first place to look is at what students are doing. (See the *TRU Observation Guide.*) Are the students engaged in sense-making? Are they grappling with ideas in meaningful ways? If they are getting better at this and the teacher is getting better at supporting students in doing so, improved test scores will follow. So will a lot else, such as meaningful changes in AOI.

The second is that if you're a teacher, you can expect to engage in a lot of – guess what? – productive struggle in order to evolve as a teacher. Teaching for Robust Understanding is challenging – but the rewards in the development of classroom community, in students' thinking more deeply and mathematically, and in their coming to see themselves as mathematical thinkers, are more than worth it.

General Principle 2. Planning, thoughtful observation, and reflection are the keys to ongoing improvement.

Most of us know the phrase "the wisdom of practice." It refers to the fact that, as we teach, we learn new things about mathematics and about our students. For example, a very common error is for students to write "$(a + b)^2 = a^2 + b^2$." The first time we see a student say or write this, we may be taken aback. After a while we recognize it as a common occurrence. After some time, we have a number of ways to address the misconception. We learn about student thinking and how better to address it as we teach.

That kind of learning is important, but it's haphazard. We learn a lot more if our learning is deliberate. If we plan and experiment deliberately, track what happens, and reflect on the results, we accelerate the process of our own learning. It's possible to do that individually, but...

General Principle 3. Reflecting with colleagues is better than trying to do it by yourself.

When you're busy teaching, you're ... busy teaching. You can only observe so much. And, of course, you're limited to your own observations and reflections. We find that, whether it's with a peer, a coach, or a professional learning community, having multiple sets of eyes and minds to plan, observe, and reflect is tremendously valuable. We can "test run" ideas during planning – and we may think more deeply about our plans when a colleague asks why we intend to take any particular approach. Whether they're in the room with you or looking at a video, other eyes may notice things you didn't catch or have different interpretations of how or why things happened. Moreover, having sounding boards for reflections can enrich your thinking substantially.

General Principle 4. Professional development is much more effective when it builds on teachers' everyday experiences and concerns.

Knowledge of all sorts is good – it adds to the set of resources we can draw upon while planning and teaching. However, if the things we're learning are abstract and somewhat removed from what we do in the classroom, it can be hard to make the connections – and, if participants don't see the connections or feel that the topics have been imposed on them, they may not be motivated to learn. Providing teachers meaningful choices is important (see Section 5.4.2.2 for an example). Connections are most meaningful when teachers share videos of their own practice for discussion.[1]

General Principle 5. Feeling safe to share ideas and artifacts is essential for professional learning.

Professional learning calls for stretching oneself, and for taking risks. Just as students need to feel safe before sharing their work with a class, teachers need to feel safe before sharing examples or videos of their teaching with colleagues or others. Growth comes from honest and respectful conversations in which participants feel safe to ask for help, to show and discuss examples where things didn't go as well as hoped, and to contribute to each other's growth by addressing such issues. Norms and practices that support trust and safety are a must.

General Principle 6. There is no single "right way" to engage in professional development or to support professional growth.

Just as there is no one "right way" to teach, there is no best model of professional development. We'll elaborate on this point in Section 5.4.2, in which we document a number of different ways that teachers have worked together to enhance teaching and learning using TRU. What is essential, as discussed above, is that teachers feel empowered to work on problems that are meaningful to them. Activities need to be structured in ways that provide participants opportunities to reflect deeply and safely on aspects of teaching that really matter.

The six general principles above hold in all professional development. Our hope, of course, is that if you've come this far, you'll want to engage in some form of learning that has TRU at its core. The following three principles are essential for TRU-related efforts.

TRU-Related Principle 1. TRU does not tell you how to teach. It embodies a perspective on teaching and learning, and a language for talking about them. Learning to "see" with TRU, and teach accordingly, is what makes it effective.

TRU doesn't tell you how to teach because there's not one right way to teach. Think of the best teachers you know. The odds are they have different styles and different classroom routines. We're willing to bet, however, that their classrooms do well along all five dimensions of TRU.

TRU is a way of thinking and talking about teaching. The district math coach in Chicago, who introduced TRU city-wide (See Schoenfeld et al. 2019a), said that for the first time in her experience, second-grade teachers had a natural way to discuss what was taking place in their classrooms with their principals, district administrators, and middle and high school teachers. With a common language, it's much easier to focus on what counts.

TRU-Related Principle 2. The main goal of TRU is to "problematize" – to raise questions for planning and reflection.

The idea behind Figure 5.2 (in Section 5.3.1) is that thinking about each of the five TRU dimensions can be enriched by asking the kinds of questions suggested there – and then, expanding those questions as indicated in the text that follows Figure 5.2. That's what the *TRU Conversation Guide* is all about. (*The TRU Observation Guide* comes at the same issues from a slightly different angle, and *Mathematics Teaching on Target* from a third.) This approach fits with the general idea that TRU focuses on what matters, and on general principle 2: Once you've identified what matters, planning, thoughtful observation, and reflection are the keys to ongoing improvement.

TRU-Related Principle 3. TRU lives in synergy with all meaningful professional development. If you're already engaged in some form of PD, seeing it through a TRU lens can make it richer.

You could probably name a dozen or more initiatives that teachers in your school or district have been asked to implement in recent years. There's complex instruction, a range of equity initiatives, Lesson Study, professional development on the Math Practices, Problem-Based Learning, SEAL, GLAD, and lots more. What we often hear, from administrators and teachers alike, is "My plate's full. There's not room on it for any more initiatives."

That would be the case if TRU were another, separate entity. But that's not the way that TRU works. Suppose you're committed to implementing Initiative X. As you do so, you can ask the following kinds of questions:

- What can I do to enrich the mathematics content and practices students are engaging with as I implement Initiative X?
- What can I do to enrich productive struggle as I implement Initiative X?
- What can I do to support more equitable access for students as I implement Initiative X?
- What can I do to provide greater opportunities for students to develop agency, ownership, and productive mathematical identities as I implement Initiative X?
- What can I do to make student thinking more public, so that I have more opportunities to enhance activities along the lines of the four previous bullets?

In other words, once you come to view classroom issues through the lens of TRU (that's TRU-Related Principle 1) and problematize those issues using TRU (that's TRU-Related Principle 2), you have a mechanism for enhancing all of the work you do with other initiatives. Thinking with TRU doesn't add work. You do it as part of the natural work you do within those initiatives.

5.4.2 Models of TRU-related Professional Development

This section describes a wide range of ways that TRU has been implemented over the past decade. Our purpose is to show how, whatever the nature of your learning context, there is a natural form of TRU-related activity that can be adapted to fit that context. For obvious reasons, these descriptions are brief – we could write a book fleshing out each of the models described here. More detail on specific approaches can be found in "Learning with and from TRU" (Schoenfeld et al. 2019a) and other references described below.

5.4.2.1 One Path to Getting Started

Disston (2019) discusses three structures he uses to introduce pre-service teachers to the ideas in TRU. The three structures described below provide a quick introduction to viewing instruction through a TRU lens and using that perspective productively to enhance planning and teaching. They were developed for a quick intervention, such as a student teaching seminar, but they can be modified for alternative circumstances.

A VIDEO JIGSAW

The TRU Video Jigsaw provides a quick introduction to TRU to any group of (say) 15 or more participants. Participants discuss the math task in the video they are about to watch. They then watch a 5-minute classroom video and comment on it. At that point they are introduced to the *TRU Observation Guide* and *TRU Conversation Guide*. They then break into five groups, with each group focusing on one TRU dimension. After reviewing the norms for respectful discussions, each group discusses the video with an eye toward their chosen dimension. Then the whole group rearranges into new groups, each of which contains a representative from each of the five dimensions. In those discussions, every student gets to hear about and discuss all five dimensions. After that there is a whole group discussion of the issues. This particular approach provides a very quick introduction to the key ideas in TRU.

A TRU CLASSROOM OBSERVATION FORM

Figure 5.5 provides a condensed version of a form used by an observer to provide lesson feedback. The form addresses all five TRU Dimensions. This broad focus (in contrast to the single dimension focus in Figures 5.3 and 5.4) is useful for people who are beginning to teach or who are just being introduced to TRU. Using the form to reflect on the lesson with the student

Observation Form

Teacher: _____   School: _____
Observer: _____   Subject: _____
Date: _____   Period: _____

Goal and/or focus of the observation:
- *(Space expands as necessary...)*

Link to Lesson Plan or Lesson Plan Summary:
-

Lesson segment, initial focus of the debrief, and rationale for focusing on this segment:
-

Agenda:
-

Content: *As a student: What do I think the big ideas are? What's my main science/math focus for this lesson?*
-

Cognitive Demand: *As a student: What opportunities do I have for sense making? How much time do I spend:* <ul><li>*On small tasks?*</li><li>*Grappling with ideas?*</li><li>*At sea?*</li></ul> •	**Equitable Access:** *As a student: Are there consistent opportunities for me to be engaged meaningfully with content? Can I hide, will I be ignored?* •
Assessment: *As a student: Do the classroom discussions relate to my thinking? Does the teacher know what I understand or don't?* •	**Agency/Ownership/Identity:** *As a student: Do I have the opportunity to explain my ideas, have them recognized? How do I feel when the spotlight lands on me?* •

On separate pages there is room for:
- General notes by the observer
- Reflections on the teacher's specific goals and their implementation
- The teacher's responses to the observer's notes.

Figure 5.5 A form used for observing lesson observations

teacher helps build the use of TRU as a language for thinking about and communicating ideas about teaching, as well as reinforcing the use of TRU as a mechanism for planning and reflection.

TRU AS A FOCUS FOR WEEKLY REFLECTIVE JOURNALS – A QUICK "TOUR" OF THE TRU DIMENSIONS

Participants in ongoing weekly meetings wrote journals that were first posted online and reacted to electronically by other participants. The posts and reactions then served as fodder for collective discussions. The journal assignments focused on each of the TRU dimensions for a week or two, providing participants opportunities to "become experienced in looking through a specific lens to identify the moments in a lesson that align to that dimension. [Students are asked] to focus on connections between the dimensions, and aspects of teaching and learning that fall within the intersection between dimensions, as a way to explore how certain lesson structures or teacher moves might be leveraged to achieve particular teaching practice goals." (Disston 2019, p. 280).

5.4.2.2 TRU as the Main Focus of a Teacher Collective

Over the past decade we have collaborated with a number of groups that made TRU their primary mechanism of shared professional development. In some cases a few teachers banded together to exchange ideas about their teaching. In others there were school-wide or district-wide efforts, or special professional development communities. There were varied models of implementation. We begin with some common threads.

OVERVIEW

Given our view that professional growth is a life-long process, our goal is to create a form of professional development that lasts – one that teacher collectives make their own, working through relevant issues of practice on an ongoing basis. The idea isn't to "do" professional development and move on to something else; it's to help build a self-sustaining community.

The general sequence of events consists of the following components: Introducing TRU; selecting the particular model the group wants to use; supporting deep dives into the TRU dimensions (and lessons as a whole), and transferring ownership to the group of teachers.

INTRODUCING TRU

We've discussed one model of introducing TRU, the "TRU video jigsaw" described in Section 5.4.2.1. That's an effective method to use if you're time-constrained. Here we describe two more expansive versions, which take 90 minutes or so to implement. They provide deeper immersion, under the assumption that some version of TRU will be an ongoing focus of the collective's work for some time.

Version 1 proceeds as follows. The facilitator discusses norms for commenting on videos, and then shows a short video. Participants discuss what they notice about the video in small groups, and then report their observations to the whole group. The facilitator writes down their statements in what may seem a strange way – the statements are written in different places on the board, without there being any apparent rationale for where the statements are written.

This process is repeated with a second video, and then a third. It becomes apparent that the facilitator has placed all of the comments in five columns. At that point the facilitator reveals the reason for the placement: The columns are labeled "The mathematics; Cognitive Demand; Equitable Access; Agency, Ownership, and Identity; Formative Assessment." The facilitator goes on to explain that every time a group has done this exercise, their comments fall into these five categories – they are comprehensive and correspond to the PLC's observations. These are the dimensions of TRU, which:

- cover everything important in classrooms
- are validated by a substantial amount of research
- make a big difference in student performance
- have just been validated by the PLC on their own.

It's a powerful learning experience for participants when their own observations fit naturally into a theoretical framework.

The videos for Version 1 can be found at https://www.mathnic.org/tools/01_tru.html, along with a collection of materials (including a complete set of slides and support materials for the facilitator) that support the facilitator in running the whole session.

In *Version 2* the participants are provided brief descriptions of the TRU dimensions and their importance before they watch the videos in the link given above. They comment on each video in small groups and then discuss how their comments are related to the TRU dimensions, ultimately placing their comments under the most relevant dimension. In a collective discussion the facilitator helps the participants refine their preliminary understandings of the dimensions and makes the four points listed two paragraphs above.

Version 2 lacks the drama that accompanies Version 1 – the "aha" experience in Version 1 of seeing that the group's comments all fit into the framework is impressive. In compensation, the participants have more direct experience thinking about the five TRU dimensions. (And, since Version 1 calls for making snap decisions about where to place the comments when participants make them, Version 2 is less taxing on the facilitator.) Which version to use, then, depends on the preference of the facilitator. In either version, participants end the session with a clear sense that their observations of three very different classroom episodes all fit comfortably within the TRU Framework. They are positioned to dig more deeply into TRU, and to select their pathway through the TRU dimensions.

MODELS OF SEQUENCING

Once a teacher collective has a sense of what the TRU Framework offers, they discuss how they would like to proceed. Some groups decide they would like to do a series of deep dives into the TRU dimensions. Some decide to spend a year or two learning more about TRU, after which they will engage in some form of TRU-inflected Lesson Study, or Complex Instruction, or the continued exploration of problems of practice (e.g., formative assessment), using TRU as the mechanism for reflecting on their ongoing work. We briefly describe such models.

DEEP DIVES

Various groups of teachers have, over a period of two or three years, conducted a series of deep dives into individual TRU dimensions. Part of the reason for doing this is that there's a lot to notice in TRU: When you're beginning, it's hard to think about all five dimensions at once. That's one of the reasons each of the three case studies in this book has just one or two focal dimensions (plus the math, of course). As you're first settling into TRU, focusing on particular issues related to cognitive demand, or equitable access and AOI, or formative assessment, is more than enough to wrap your head around in a session that lasts 60–90 minutes. And, of course, there's a lot more to each dimension than can be dealt with in one session.

The decision point comes after the first few sessions, when the teacher collective has developed a general sense of TRU. Then it's time to pose the question, which dimensions would we like to focus on for our first deep dive? The two most popular choices are AOI and formative assessment. Many teacher learning communities choose AOI because they have focused on issues of equity and AOI is a natural extension. Many choose formative assessment because it's an intriguing topic and the Mathematics Assessment Project's collection of FALs (see https://www.map.mathshell.org/) provide resources for engagement.

For the early sessions in a deep dive it's good to choose episodes that, like the case studies in this book, support rich discussions of mathematics and feature opportunities to focus on the particular dimension(s) of interest. Once a safe working environment has been established (see General Principle 5 above), group members can be encouraged to bring their own problems of practice for discussion. This should be made as easy as possible. When we're working with a group, a teacher identifies something to work on. We record the video and identify things we think might be interesting to discuss. The teacher has veto power over the choice of focus, and the choice to either lead the discussion or have us lead it. Most often the teacher has us lead the session, at least at first – but, with the understanding that they can stop any discussions that veer into delicate territory. If possible, it's good for a partner teacher to make the video and help select the clip collaboratively. Things feel a lot less risky in general when the teacher who's being videoed has feelings of ownership and control.

Since circumstances and interests vary, there is no one script for conversations about a video, even when focusing on just one dimension. The idea, in general, is that the *TRU Conversation Guide* provides a general set of questions to keep in mind when looking at a video. When focusing on AOI, for example, participants turn to the AOI page of the *Conversation Guide*. The group might look at a slide with a few key questions, such as in Figure 5.6.

These are often modified for the specific clip the teacher has chosen, as in Figure 5.7.

Questions to Consider:
Agency, Ownership, and Identity

- Who generates the mathematical ideas that get discussed?
- Who evaluates and/or responds to others' ideas? How does the teacher respond?
- Which students respond? How do they respond?
- How deeply do students get to explain their ideas?
- How are norms developing around students' and teachers' roles in generating mathematical ideas?

Figure 5.6 Key questions related to AOI

To what extent do classroom activities provide access to "voice" for students? To what extent do students explain their thinking and make sense of other students' thinking?

Specifics for whole class or small group:

- How much "room" is there for student thinking and exploration?
- How is correctness determined? By looking to external authority (teacher or text), or working things through?
- Who initiates conversations?
- How long are students' speech turns?
- Do student ideas get built on? By whom?

Figure 5.7 Questions tailored to a particular video clip

Alternatively, if the teacher was working on a specific teaching goal, the teacher and/or facilitator(s) may craft reflection questions tied to that goal – e.g., "I was trying to encourage the students to build on each other's thinking. What's your impression of what the students did? What else could I try?"

Finally, although the emphasis in such sessions typically is on the mathematics (Dimension 1) and one or two other dimensions (in the case above, AOI), it's good for every session to end with a brief discussion of how all of the dimensions played out and interacted. (That's what we did in the three case studies in this book.) It's important not to lose sight of the fact that TRU is an organic whole, and that classroom interactions are best reviewed through the lenses of all five TRU dimensions.

Our experience has been that a deep dive into any of the dimensions can last for a semester or more, depending on how frequently the teacher collective meets. At some point the group decides it wants to move on to another dimension.

TRU-WITH-SOMETHING ELSE

An alternative model is to spend a year focusing on TRU and then dig deeply into another form of professional development. It is entirely possible to condense a solid introduction to TRU into a year, with brief but substantial deep dives into each dimension, e.g., in the ways described in Section 5.4.2.1. Afterward, having an understanding of TRU can be used to enhance whatever additional program of professional development a PLC decides to explore. We will say more below, describing a way to conduct TRU-with-Lesson-Study.

TRANSFERRING OWNERSHIP (OR BUILDING IT)

As noted above, our long-term goal is for teacher groups using TRU to be self-sustaining. We often play a role in helping groups get started – it's part of our mission – but we also need to fade out of the picture. It's worth re-emphasizing one of the fundamentals, both for groups starting afresh and for groups that begin with the help of external facilitators.

The key words (recall General Principle 5) are *trust* and *feeling safe*. Bringing your own work to a group for discussion can be scary. People will only allow themselves to ask for help if they feel that the community is respectful and supportive. That's one reason it's essential to emphasize the norms for respectful interactions (e.g., Section 2.2). We find that if our questions and comments in the opening sessions – which make use of external videos – model the kind of supportive, non-judgmental, and evidence-based problematizing we'd like to see, that teacher collectives settle into that mode of conversation within two or three sessions. At that point, it's usually possible to ask one participant in private if they'd be willing to bring a video of one of their classes to the next session. When they do, and the discussion turns out to be helpful, it's easy to move on from there. (There is nothing more gratifying than to have a teacher announce, "I tried what you suggested last time, and it made a big difference.")

TRU-LESSON STUDY

There are multiple ways for TRU and Lesson Study to live in synergy. Here we provide a cursory summary of Lesson Study and two descriptions of the ways in which TRU and Lesson Study have been integrated.

Lesson Study is a highly refined form of professional development practiced extensively in Japan and China, with increasing uptake in other nations. (See Huang, Takahashi, & Ponte 2019, for an international review.) Here is a brief description of one generic version. The core idea is that Lesson Study is a form of collaborative practice as described in Table 5.1.

Given space limitations, we can only point briefly to two ways in which TRU and Lesson Study have been integrated. The first is a deep integration. A substantial set of resources can be

Table 5.1 A somewhat generic description of the Lesson Study Process

---

In Lesson Study, a teacher collective:

- identifies a topic of interest and sets goals for student learning,
- conducts research into what is known about the relevant mathematics, student understanding of it that mathematics, and effective lessons for teaching it,
- designs a lesson aimed at the learning goals the PLC has identified,
- develops observational mechanisms to examine the efficacy of the lesson as it takes place, and how students learn,
- has a member of the PLC teach the lesson on behalf of the design team, while multiple observers take notes,
- has a formally structured debriefing session in which the teacher, designers, and observers provide detailed commentaries on how the lesson played out, vis-à-vis the learning goals,
- refines the lesson on the basis of the debriefings,
- re-teaches the lesson, repeating and refining the bullet points above.

---

found at the *TRU+LessonStudy* website, https://www.tru-lessonstudy.org/home. That website contains materials that support two years of monthly sessions for high school math teachers. In accord with the *TRU-with-Something Else* model discussed above, the website offers a set of materials for a first year learning about TRU and ways to explore it through a series of mini-cycles of inquiry. The second year focuses on a TRU-inflected version of Lesson Study that, much like classic Lesson Study, explores questions about mathematics instruction in great depth. One of the key features of this integrated model is that the goals that the team sets for investigation are TRU-related – they are aimed at some of the challenges that teachers typically encounter when trying to use Lesson Study in the instructional context in the US. Examples of such goals include "finding the sweet spot of productive struggle," "fostering dynamic group work in which meaningful sense-making occurs in ways that students own it," and "keeping mathematical big ideas front and center so students make more connections." A more "researchy" description of the process that led to the TRU+LessonStudy website can be found in Schoenfeld et al. 2019b ("Teaching for Robust Understanding with Lesson Study").

A second model, once a teacher learning community has enough knowledge of TRU for "thinking with TRU" to seem natural, doesn't require recasting Lesson Study in TRU's image. Here we briefly describe how it can be used to enhance the process described in Table 5.1. This example shows how TRU can work in synergy with any established professional development program.

Consider the set of bullets in Table 5.1. Now ask, which of those bullets can be enriched if you know about TRU? Here are some possibilities.

- If you know about TRU, you can set TRU-related goals.

- Your TRU-related understanding of what's important for student learning might affect your thinking about lesson design. But even if not...

- Once your lesson is designed, you can enrich the design by asking the problematizing questions discussed in the section *TRU-Related Principle 3*.

- The observational tools and the lesson debrief can be shaped to address the same TRU-related questions.

This approach to enrichment can enhance the standard practices of Lesson Study without calling for any significant redesign. The same kind of approach can be used for any approach a teacher collective plans to adopt, for example, complex instruction or video clubs.

5.4.3 Some Final Thoughts

The universe of professional development is vast, and this discussion has barely scratched its surface. We hope to have shown that TRU is, first and foremost, a way of thinking about teaching and learning. Once it becomes comfortable to plan, observe, and reflect on classroom events through the lens of TRU, you have a natural way of deepening your understanding and teaching practice.

There are a number of ways to do this. For starters, take a look at the approaches various teams have taken to enrich their own work, in Schoenfeld et al. (2019a). Above and beyond the approaches briefly described here, you will find descriptions by David Foster and Tracy Sola of the ways that the Silicon Valley Mathematics Initiative built professional development networks across the US; by Ruth Haumersen, Alanna Mertens, Lynn Narasimhan and Nicole Louie of the ways they supported significant TRU-related reform across the Chicago schools; and by Michael Driskill, Eileen Murray, and David Wilson from Math for America characterizing their work with multiple teacher learning communities.

One main point about these TRU-related efforts is that they all differ, in major ways. That's because the contexts in which the approaches were developed differ. TRU adapts to context because it is a way of thinking, not a set of procedures to be implemented. A second main point is that our understanding evolves. Michael Driskill has observed that the discussions among Math for America's teachers got much richer when *MfA* started looking at videos of participating teachers when they were implementing the Formative Assessment Lessons. When the lessons themselves provided lots of material for students to grapple with, the conversations about them got correspondingly deeper. (Case Study 3 came from the *MfA* team.)

There is no endpoint to learning about teaching. The more we reflect, the deeper our thinking; the more we try things and refine them, the better prepared we are to help our students. This volume offers ideas that we have found useful. We hope that you do too, and that you and your colleagues can use them in ways that take you further in your journeys as teachers.

Note

1 This is now easy to arrange. Teachers by themselves can use a smartphone mounted on a small tripod. For fancier videos there's technology such as Swivl, and partners or aides can help with video, monitor small groups, etc.

Chapter 6

Resources and Tools
for Moving Forward

6.0 Overview

This chapter offers resources and tools to help you move forward with TRU, whether you do so by yourself, as part of a small study-and-action group, or as part of a professional learning community.

Section 6.1 describes a collection of resources that our partner teachers and teacher educators have found useful. These include general information about the Teaching for Robust Understanding project; the Formative Assessment Lessons and other resources available from the Mathematics Assessment Project; the Mathematics Improvement Network; background information on Lesson Study, Complex Instruction, and video clubs; and pointers to the professional development efforts organized by AIM-TRU, the Silicon Valley Mathematics Initiative, and VIDEO-LM.

The TRU project's two main tools are the *TRU Conversation Guide* and *the TRU Observation Guide*. You're familiar with aspects of them, since we've referred to them throughout this book. You'll find them in full in Sections 6.2 and 6.3.

With access to all these resources, you're poised to take next steps.

6.1 Resources

TRU has profited greatly from partnerships with a number of national and international groups, who have provided wisdom, tools, and varied resources – both for TRU directly and for ways to support teachers' professional growth that live in synergy with TRU. As we indicated in Chapter 5, TRU can be a focus of professional development, or it can enhance other approaches. Here is a sampling that may be of use to you.

6.1.1 The TRU Website, at https://truframework.org/

If you want to know more about TRU, the TRU website is the place to go. The *TRU Conversation Guide* and *the TRU observation guide* are available for downloading, if you'd like easy access to the planning and observation sheets you saw in this chapter.

 DOI: 10.4324/9781003375197-9

The "publications" page of the website offers a continuously updated collection of research papers that provide the substance for all the claims we've made about TRU. You can learn about how TRU was developed, and see more descriptions of TRU in action. We won't list the references here, because a fully annotated list of papers is on the TRU website.

While we're mentioning TRU resources, we also want to note *Mathematics Teaching On Target: A Guide to Teaching for Robust Understanding at All Grade Levels* (Schoenfeld et al. 2023). See Section 5.3.2 for an introduction to *On Target*. We view that book and this one as being complementary and synergistic, each being coherent on its own and yet extending the other – this book going into depth to study classroom reality through the lens of TRU, and *On Target* providing similar depth with an eye toward improving classroom activities.

6.1.2 The Mathematics Assessment Project, at https://www.map.mathshell.org/

The Mathematics Assessment Project (MAP) offers a range of tools that enhance teaching and professional development. The Formative Assessment Lesson at the heart of Chapter 4 was developed by MAP; it's one of 100 such lessons available at no cost from the MAP website. In fundamental ways, the values underlying the MAP work and TRU are the same. The website offers a range of tasks, tests, and professional development that are well worth looking at.

6.1.3 The Mathematics Improvement Network, at https://www. mathnic.org/

The Mathematics Network of Improvement Communities (Math NIC) was a joint project of MAP and the TRU team. It produced a collection of resources and workshops that schools and districts could use for ongoing improvement. Among the tools developed is a 90-minute workshop for introducing TRU, which was briefly discussed in Section 5.4.2.2. The workshop, available at https://www.mathnic.org/tools/01_tru.html, features three video clips and a full set of slides for organizing and orchestrating the conversations involving the clips. There are a number of other valuable resources at the Math NIC website as well.

6.1.4 Lesson Study

As we indicated in Section 5.4.2.2, there is a natural synergy between TRU and Lesson Study – in fact, there are various forms of TRU-Lesson Study. For example, you can begin by learning about TRU and then build a form of lesson study in which the research questions are TRU-related. That's the approach taken by the *TRU+LessonStudy* project, which provides materials that support a year of TRU followed by a year of TRU-Lesson Study. Their website is https://www.tru-lessonstudy.org/home.

The synergy between TRU and Lesson Study comes from the parallels in their underlying conceptualizations of what matters in helping students become powerful mathematical thinkers. There is, of course, a great deal of complexity to lesson study – some of the references below will unpack it – but at core, two aspects of lesson study are fundamental. The first is making sure that the mathematics that the students encounter is rich and meaningful. The second is a focus on understanding student thinking and building on it in ways that help the students engage in mathematical sense-making. Those fundamentals correspond to Dimension 1 and Dimensions 2–5 of TRU, respectively.

Those parallels explain why it was easy to build the second model of TRU-Lesson Study, which consists of TRU-based enhancements of the lesson study process. As we described in

Section 5.4.2.2, it's straightforward to look at various stages of the lesson study process through the lens of TRU. One can: use the questions in the Conversation Guide to help enrich the mathematics in the lesson; review the lesson plan with an eye toward the TRU dimensions; and similarly, conduct formal lesson commentaries using TRU as the mechanism for organizing those commentaries.

These two papers provide short and easy-to-read introductions to Lesson Study:

Lewis, C., Perry, R., Hurd, J., & O'Connell, M. P. (2006). Lesson Study Comes of Age in North America. *Phi Delta Kappan*, 88(04), 273–281.

Takahashi, A., & Yoshida, M. (2004). How Can We Start Lesson Study?: Ideas for establishing lesson study communities. *Teaching Children Mathematics*, Volume 10, Number 9., pp. 436–443.

Professional developers and school districts often make use of this handbook to conduct Lesson Study:

Lewis, C., & Hurd, J. (2011). *Lesson Study step by step: how teacher learning communities improve instruction. Heinemann.*

The following book is helpful for teachers who are interested in trying Lesson Study.

Takahashi, A., McDougal, T., Friedkin, S., & Watanabe, T. (Eds.). (2022). *Educators' Learning from Lesson Study: Mathematics for Ages 5–13.* Routledge.

6.1.5 Complex Instruction

You may recall that the classroom we visited in Chapter 3 ("Where is the Ten?") was a Complex Instruction classroom. CI, as it's known, is an equity-focused instructional approach employing group work. It aims to support teachers in ways that ensure "that all group members are active and influential participants and that their opinions matter to their fellow-students" (Cohen et al. 1999, p. 80). As you've seen in Chapter 3, a major aspect of Complex Instruction involves the use of "group-worthy problems" – tasks that are mathematically important, have multiple entry points, and support conversations in which students engage in sense-making and are held accountable for it. CI emphasizes that there are multiple ways of contributing to a group's mathematical progress, all of which are aspects of mathematical proficiency. Thus CI aims directly at aspects of rich mathematics (Dimension 1) and offers mechanisms for supporting equitable access and AOI (Dimensions 3 and 4). The best teachers attend to issues of formative assessment and cognitive demand, but these concerns are not embedded in the formal structure of CI. In consequence the kinds of reviewing and attention to all five dimensions of TRU in planning, teaching, and reflecting on lessons can work in synergy with CI, just as they do with lesson study.

An article that lays out the basics of CI is

Cohen, E.G., Lotan, R.A., Scarloss, B.A., & Arellano, A. R. (1999). Complex instruction: Equity in cooperative learning classrooms. *Theory into Practice, 38,* 80–86.

More detail on group work can be found in

Cohen, E.G., Lotan, R.A. (2014). *Designing Groupwork. Strategies for the heterogeneous classroom. (3rd edition).* New York: Teachers College Press.

The CI website is at http://cgi.stanford.edu/group/pci/cgi-bin/site.cgi?page=index.html. An annotated bibliography of CI papers is at http://cgi.stanford.edu/group/pci/cgi-bin/site.cgi?page=research.html.

Two books that work in the same general arena are

Horn, I. S. (2012). *Strength in Numbers: Collaborative Learning in Secondary Mathematics*. Reston, VA: National Council of Teachers of Mathematics.

Horn, I., & Garner, B. (2022). *Teacher learning of ambitious and equitable instruction*. New York: Routledge.

An example of a teaching partnership in which a teacher describes how she came to understand and appreciate her students' mathematical strengths is

Skinner, A., Louie, N., & Baldinger, E. (2019). Learning to see students' mathematical strengths. *Teaching Children Mathematics, 25*(6), 339–344.

6.1.6 Video Clubs

Video clubs are (often facilitated) teacher collectives that gather to discuss problems of practice that emerge from or are deliberately sought in videos of the members' classrooms. Major themes in video club work involve developing a particular kind of professional vision or "noticing" (Sherin & Dyer 2017, Sherin & Linsenmeier 2011) and the importance of reflecting deeply on aspects of student work and teaching, specifically taking advantage of videos for purposes of preparing to notice, noticing in action, and reflective noticing (Sherin 2000, Sherin, Richards, & Altshuler 2021).

If all this sounds familiar, it should. Video clubs are grounded in the general principles discussed in Section 5.4.1: long-term commitment; cycles of planning/ observation/ reflection; collective reflection; building on everyday experience; and trust. If you add in a focus on particular TRU-related issues, you have the TRU deep dives. For further information on teacher noticing and video clubs, see the following papers.

Sherin, M. G., Richards, J., & Altshuler, M. (2021). Recording one's classroom: A new source for teacher learning. *Kappan, 103*(2), 44–48.

Sherin, M. G. & Dyer, E. B. (2017). Teacher self-captured video: Learning to see. *Kappan, 98*(7), 49–54.

Sherin, M. G. & Linsenmeier, K. (2011). Pause, rewind, reflect: Video clubs throw open the classroom doors. *Journal of Staff Development, 32*(5), 38–41.

Sherin, M. G. (2000). Viewing teaching on videotape. *Educational Leadership, 57*(8), 36–38.

van Es, E. A., & Sherin, M. G. (2021). Expanding on prior conceptualizations of teacher noticing. *ZDM Mathematics Education, 53*, 17–27.

6.1.7 AIM-TRU

AIM-TRU (Analyzing Instruction in Mathematics using the Teaching for Robust Understanding Framework) is a research and development collaboration between Montclair State University, SUNY Buffalo, Math for America and DePaul University. It is a model for professional development that gives teachers the opportunity to understand high-quality instructional resources and their use through video cases.

In AIM-TRU, Professional Learning Teams of teachers meet on a regular basis to engage in the collaborative investigation of video cases utilizing a shared repertoire that includes questioning protocols adapted from TRU. The video cases are centered around videos of teachers using the Formative Assessment Lessons. (Sounds familiar? The case study at the center of Chapter 4, "Graphing Quadratic Functions," was developed from a video shared with us by AIM-TRU.)

Each AIM-TRU video case has a set of materials that can be adapted to fit any professional development context or group of collaborating teachers. These materials include the lesson itself, information about the mathematical context, supplementary information about the school and lesson, a transcript of the video, facilitation slides (for both in-person and remote facilitation), and remote versions of lesson resources (if available). AIM-TRU has also created facilitation guides, slide decks, and orientation materials for teachers or teacher educators who would like to investigate the video cases for professional development or in teacher education courses.

Descriptions of AIM-TRU and access to case materials can be found at https://tle.soe.umich.edu/MFA, https://www.aimtru-buffalo.org/, and https://csh.depaul.edu/academics/stem-studies/research/aim-tru/Pages/default.aspx, and https://www.montclair.edu/mathematics-education-phd/2018/10/04/aim-tru/.

6.1.8 The Silicon Valley Mathematics Initiative

For more than a quarter century the Silicon Valley Mathematics Initiative has partnered with school districts, providing a broad spectrum of professional development support. Participating districts receive year-round professional learning, a formative and summative performance assessment system, funding to support district mathematics coaching, and a network including meetings and workshops with mathematics teachers, leaders, and administrators.

SVMI has been a long-term partner of the TRU project, the MAP project, and their antecedents. Many of the Formative Assessment lessons were field tested in SVMI-affiliated classrooms. SVMI was an "early adopter" of the TRU Framework, contributing immeasurably to its refinement. For more detail, see https://svmimac.org/.

6.1.9 VIDEO-LM

The VIDEO-LM project (Viewing, Investigating and Discussing Environments of Learning Mathematics) is aimed at enhancing mathematics teachers' reflection on their professional practice, through watching and discussing videotaped lessons of unfamiliar teachers. The project has created a large data bank of videotapes and support materials (most in Hebrew, with a small subset of the materials in English).

While there are significant differences between TRU and VIDEO-LM, both programs are very much in the same part of the professional development space. Both are aimed fundamentally at problematizing teaching, empowering teacher collectives to inquire into key aspects of their practice. The discussions in VIDEO-LM are guided by the use of an analytic framework, comprised of six detailed "viewing lenses" that focus on (1) mathematical and meta-mathematical ideas, (2) goals for lessons and instruction; (3) the mathematical and conversational affordances of tasks; (4) classroom interactions; (5) the dilemmas and issues of decision-making raised by the video; and (6) underlying beliefs held by the participants, which might shape the decisions that were made, or might be made.

Video-LM uses videos of unfamiliar teachers as ongoing stimuli for discussion, while TRU tries to move toward participants bringing in their own videos. There are surface differences in language and focus between the five dimensions of TRU and the six viewing lenses of VIDEO-LM. The library of videos and support materials for VIDEO-LM is far larger than that of TRU. But the similarities between the two programs far outweigh the differences: both programs aim at empowering teachers in reflecting on what best helps their students engage powerfully and productively with mathematics. Learning about one helps to enrich one's understanding of the other.

The VIDEO-LM website is https://stwww1.weizmann.ac.il/en/?page_id=326. For more detail, see

Arcavi, A., & Karsenty, R. (2018). Enhancing mathematics teachers' reflection and knowledge through peer-discussions of videotaped lessons: A pioneer program in Israel. In N. Movshovitz-Hadar (Ed.), *K-12 Mathematics Education in Israel - Issues and Challenges* (Chapter 33, pp. 303–310). Series on Mathematics Education (Volume 13). Singapore: World Scientific.

Karsenty, R. (2018). Professional development of mathematics teachers: Through the lens of the camera. In G. Kaiser, H. Forgasz, M. Graven, A. Kuzniak, E. Simmt, & B. Xu (Eds.), *Invited Lectures from the 13th International Congress on Mathematical Education* (pp. 269–288). Cham, Switzerland: Springer.

Karsenty, R., & Arcavi, A. (2017). Mathematics, lenses and videotapes: A framework and a language for developing reflective practices of teaching. *Journal of Mathematics Teacher Education, 20,* 433–455.

Karsenty, R., & Arcavi, A. (2021). "Life trajectory" of a professional development project: The case of VIDEO-LM. In A. Hofstein, A. Arcavi, B. Eylon, & A. Yarden (Eds.), *Long-term research and development in science education: What have we learned?* (pp. 306–332). Brill.

We now turn to two of the main TRU tools, the *TRU Conversation Guide and the TRU Observation Guide.*

6.2 The TRU Conversation Guide[1]

Whether we speak as teachers, teacher educators, or researchers, we can say that our most meaningful learning has occurred when we have interacted with others, developing and sustaining relationships that simultaneously challenge and support us. These relationships push us to expand our vision of teaching and learning. They offer perspectives on our work that differ from our own. They respect our intelligence, skill, and intentions—as well as our need to continually grow. These supportive relationships help us to alter our practice and to deepen our understanding of the complex work we are undertaking. Here, we have tried to create a professional development tool that builds on what teachers, coaches, and professional learning communities know. Our intention is for you to engage in conversation with us and with those around you.

This Conversation Guide (Baldinger, Louie, and the Algebra Teaching Study and Mathematics Assessment Project, 2016) represents our best efforts to use research to support teacher learning and growth in a way that accounts for both how people learn and the complexity of teaching practice. Instead of prescribing instructional techniques or tricks, we offer a set of questions organized around five dimensions of teaching identified by research as critical for students' mathematics learning.

The dimensions are summarized in Figure 6.1. Together, they offer a way to organize some of the complexity of teaching so that we can focus our learning together in deliberate and useful ways. They include attention to content, practices, and students' developing identities as thinkers and learners. There is necessarily some overlap between dimensions; rather than capturing completely distinct categories, each dimension is like a visual filter, highlighting different aspects of the same phenomena in everyday classroom life. We encourage you to think about interactions between dimensions when it is useful for you. The questions on subsequent pages of the Guide will also direct your attention to particular kinds of overlap.

The Five Dimensions of Mathematically Powerful Classrooms	
The Mathematics	How do mathematical ideas from this unit/course develop in this lesson/lesson sequence? How can we create more meaningful connections?
Cognitive Demand	What opportunities do students have to make their own sense of mathematical ideas? To work through authentic challenges? How can we create more opportunities?
Equitable Access to Content	Who does and does not participate in the mathematical work of the class, and how? How can we create more opportunities for each student to participate meaningfully?
Agency, Ownership, and Identity	What opportunities do students have to see themselves and each other as powerful mathematical thinkers? How can we create more of these opportunities?
Formative Assessment	What do we know about each student's current mathematical thinking? How can we build on it?

Figure 6.1 The five dimensions of mathematically powerful classrooms

6.2.1 What the Conversation Guide is for

The purpose of this Conversation Guide is to facilitate coherent and ongoing discussions in which teachers, administrators, coaches, and others *learn together*. We hope that the questions in the Conversation Guide will support educators with different experiences, different expertise, and different strengths to work together to develop a common vision, common priorities, and common language, to collaboratively improve instruction and better support students to develop robust understandings.

The Conversation Guide can be used to support many different kinds of conversations, including (but not limited to):

- conversations to develop common vision and priorities across groups of teachers (within the same school and/or across different schools),
- conversations between teachers and administrators and instructional coaches around classroom observations (see also the TRU Observation Guide, available at http://map.mathshell.org/trumath.php),
- conversations between teachers around peer observations,
- conversations around video recordings of mathematics teaching and learning,
- conversations about planning a particular unit or lesson,
- conversations about a particular instructional strategy or set of strategies,
- ongoing individual reflection.

We have found that the Conversation Guide can be useful for facilitating a one-time conversation. *But its real power lies in its support for creating coherence across conversations.* The Guide can help individuals as well as groups of educators to set an agenda and work on it consistently over time. For example, a teacher team (such as a math department) might decide to spend a semester focusing on issues of Equitable Access to Content (Dimension 3). Meeting time might then be spent reflecting on the kinds of access that are currently available to students and planning lessons with the goal of monitoring and expanding access in mind, using the Equitable Access to Content questions and prompts in this Guide. Members of the team might observe each other's classrooms focusing on these same questions and prompts. The principal might find ways to support teachers to attend workshops related to the theme of Equitable Access to Content, rather than supporting a series of disconnected trainings.

In the remainder of this document we provide an overview of each dimension; discussion questions for each dimension, for your use in reflecting on and planning instruction; and a set of suggestions for how to use the discussion questions.

We hope you will find the Conversation Guide useful. Happy teaching and learning!

6.2.2 How to Use this Conversation Guide

Our field tests and experiences as instructional coaches have led to a few suggestions that may help you make the most of this Conversation Guide. In this section, we share these suggestions and give some examples of how conversations using the guide might look.

1. *Set a long-term learning agenda.*

    Complex learning—like learning how to teach for robust student understanding—has so many facets that it is easy to jump from one thing to another, without making clear progress on anything. Setting a long-term learning agenda can help us focus our energies, whether we're full-time classroom teachers or people who support classroom teachers. Opportunities to have deep conversations about practice are few and far between. Nonetheless, if we have a core learning agenda that we can return to again and again, we stand a better chance of leveraging all our strengths to learn together about something that matters.

The process of setting an agenda can unfold in many different ways. Various stakeholders may come in with clear (and perhaps competing) ideas about what they want to focus on, or it may happen that no one has a particular preference. Whatever the case may be, it is important that all participants, especially classroom teachers, feel connected to the learning agenda. Our learning is much more powerful when we get to learn about things that trouble or inspire us.

Some examples of long-term learning agendas might be, "This semester, I want to focus on getting students to share their reasoning, not just answers or steps," or "This year, I want to get better at engaging students who get frustrated and give up easily." As you set your own learning agenda, it may be useful to read through the dimensions and discussion questions, to see if anything jumps out as particularly important or exciting.

2. *Use the discussion questions like a menu. Pick and choose.*

You might have noticed that there are a lot of questions in this Guide! Our design assumes that you WILL NOT try to discuss every bullet, one by one, each time you use the guide. Instead, we hope you will identify areas of the guide that are appropriate for your learning agenda and return to these areas often. We expect that some of the questions will be difficult to answer—and that by discussing them together you will find new ways of understanding teaching and learning and come up with ideas for things to try in order to improve both.

3. *Ground discussion in specific, detailed evidence.*

We've all made statements like, "My kids seem to really get linear equations" or "They're really struggling with fractions." While these statements convey a picture of student understanding in a quick and concise way, they need to be followed up with more detailed information. Otherwise, it is difficult to make instruction responsive to student thinking and easy to miss opportunities to build on students' strengths or address their misconceptions. One way to make our observations more specific is to talk about content with as much detail as possible; for example, instead of saying "My kids are really struggling with fractions," you might observe that "Even though I've seen my kids do just fine with finding equivalent fractions and even adding them, they just seem to shut down every time they see a fraction," or "they're reducing fractions in a mechanical way, but they don't seem to *see* that 4/6 of a chocolate bar and 2/3 of a chocolate bar represent the same amount."

Pressing for specific examples also makes observations more accurate and concrete, helping us get away from our general impressions and closer to actual student thinking. Talking about specific students—and ways that their thinking is or isn't typical of the class— is another strategy. Not only does this strategy give us a more detailed and accurate picture of the thinking that is going on in our classrooms, but it also opens up instructional possibilities. For example, noticing that today, Jessica drew a really helpful picture to represent fractions could lead you to invite Jessica to share her method with the rest of class, creating a learning opportunity that is invisible in "They're really struggling with fractions."

Finally, attending to particular students can help us think about patterns of marginalization in society at large (e.g., fewer resources for ELLs, or stereotypes that link race, gender, and mathematics ability), and how our classrooms might work to replicate or counter those patterns for our own students.

If you are able to ground your conversations in shared experiences of the same classroom (from peer observations, co-teaching, instructional coaching, etc.), you will benefit from more eyes and more perspectives on the details of classroom activity. But even if this isn't possible in any particular conversation, working with evidence of specific students' thinking and understanding will make your conversation a richer resource for your own learning.

4. *If you are conducting a classroom observation, pre-brief.*

Classroom observations are generally accompanied by a debrief conversation. Pre-brief conversations can be just as important. If you can, have a conversation prior to each observation. In this conversation, clarify the goals not just for the lesson, but also for the observation of the lesson. Talk about goals for students, so that observers can understand what the teacher is trying to accomplish. Also remind each other of the teacher's learning agenda so that you can discuss how the observer can be most helpful. We have found this question especially useful: "What do we want to be able to talk about in our debrief conversation?" From there, you might discuss what the observer should be looking for (e.g., recording the questions the teacher asks, or focusing on a particular student), and what kinds of interactions (if any) the observer should have with students.

The pre-brief conversation is one way of capitalizing on the focus and organization that a learning agenda offers. Without it, it's easy to get distracted during the observation. It's also easy for the observer to notice things that are not interesting or important to the teacher, which are less likely to help the teacher learn and grow.

The Conversation Guide includes prompts for planning, which can be used for planning observations as well as for planning lessons. Discussing these prompts should bring to the surface ideas about what is likely to happen in the lesson, given the tasks students will be given, the participation structures that will be used, and so on. This kind of anticipatory thinking might lead to tweaks in the lesson plan, but just as important, it can establish common focus between the teacher and observer. This adds richness to the debrief after the lesson; everyone can then reflect on the ways that things worked out the way they were intended to, ways they were surprising, and next steps in light of that information.

5. *Link planning to reflection and vice versa.*

This Guide includes prompts for "planning" and prompts for "reflecting." We do not mean to suggest that you restrict each conversation to a focus on one or the other; rather, it will be useful to connect these perspectives in many conversations. Reflection is most practical when it leads to next steps, and next steps (planning) should be firmly grounded in reflection on what has already happened. It is worthwhile to make space for thinking about what has already happened without jumping to next steps too quickly, however. Reflecting on the details of what we have observed opens up possibilities for future action that might otherwise remain hidden (as described above). In addition, different people see different things, and sharing our observations can enrich everyone's understanding of what students have been doing, thinking, and learning.

6. *Work from teachers' strengths.*

Our culture often prompts us to focus on our weaknesses, and on the areas where we need improvement. But our *strengths* are huge assets when it comes to learning and improving our practice. Knowing our strengths supports us to engage with challenges, giving us a starting point to work from and a reason to believe that we can be successful. Identifying teachers' strengths, making them explicit, and using them as authentic resources for growth can therefore support teachers to think deeply and critically about their practice, to strive for improvement, to actually improve by building on their strengths, and to develop productive relationships with supportive others, all at once.

In practice, this might mean prompting teachers (not just supervisors) to share *their* observations, interpretations, and ideas for moving forward; creating diverse opportunities to identify what teachers already do well, including planning together, reflecting together, and observing various kinds of interactions with students (e.g., leading discussions, intervening at small groups, and building rapport with individual students); and building next steps around strengths instead of deficits (e.g., working on supporting students who have been reluctant to participate by building on a teacher's skill at noticing something that each student is good at).

The planning/reflecting sheets for the five dimensions are on the next five pages.

The Mathematics

Core Questions: How do mathematical ideas from this unit/course develop in this lesson/lesson sequence? How can we create more meaningful connections?

Students often experience mathematics as a set of isolated facts, procedures and concepts, to be rehearsed, memorized, and applied. Our goal is to instead give students opportunities to experience mathematics as a coherent and meaningful discipline. This requires identifying the important mathematical ideas behind facts and procedures, highlighting connections between skills and concepts, and relating concepts to each other—not just in a single lesson, but also across lessons and units. It requires engaging students with centrally important mathematics in an active way, so that they can make sense of concepts and ideas for themselves and develop robust networks of understanding. And it requires engaging students in authentic performances of important disciplinary practices (e.g., reasoning abstractly and quantitatively, constructing mathematical arguments and critiquing the reasoning of others).

Planning

How will important mathematical ideas and practices develop in this lesson and unit? How can we connect the ideas and practices that have surfaced in recent lessons to this lesson and future lessons?

Reflecting

How have we seen students engage with important mathematical ideas and practices? How has this engagement looked and sounded in specific cases?

Things to think about

- What are the mathematical goals for the lesson?
- What connections exist (or could exist) between important ideas in this lesson and important ideas in past and future lessons?
- How do important mathematical practices develop in this lesson/unit?
- How are facts and procedures in the lesson justified?
- How are facts and procedures in the lesson connected with important ideas and practices?
- How do we see/hear students engage with important ideas and practices during class?
- Which students get to engage deeply with important ideas and practices?
- How can we create opportunities for more students to engage more deeply with important ideas and practices?

Cognitive Demand

Core Questions: What opportunities do students have to make their own sense of mathematical ideas? How can we create more opportunities?

We want students to engage authentically with important mathematical ideas, not simply receive knowledge. This kind of learning requires that students engage in *productive struggle*, grappling with difficult concepts and challenging problems. As teachers, we must support students in ways that maintain their opportunities to do this grappling for themselves. Our goal is to help students understand the challenges they confront and persist in solving them, while leaving them room to make their own sense of those challenges.

Planning

What opportunities might students have to make their own sense of important mathematical ideas? How can we create more of these opportunities?

Reflecting

How have we seen students make their own sense of important mathematical ideas? How has this sense-making looked and sounded in specific cases?

Things to think about

- What opportunities exist for students to struggle with important mathematical ideas?
- How are students' struggles supporting their engagement with important mathematical ideas?
- How does (or how could) the teacher respond to students' struggles, and how do (or how could) these responses maintain students' opportunities to develop their own ideas and understandings?
- What resources (other students, the teacher, notes, texts, technology, manipulatives, various representations, etc.) are available for students to use when they encounter struggles? Are there more resources we can make available?
- What resources are students actually using, and how might they be supported to make better use of resources?
- Which students get to engage deeply with important mathematical ideas?
- How can we create opportunities for more students to engage more deeply with important mathematical ideas?
- What community norms seem to be evolving around the value of struggle and mistakes?

Equitable Access to Content

Core Questions: Who does and does not participate in the mathematical work of the class, and how? How can we create more opportunities for each student to participate meaningfully?

All students should have access to opportunities to develop their own understandings of rich mathematics, and to build productive mathematical identities. For any number of reasons, it can be extremely difficult to provide this access to everyone, but that doesn't make it any less important! We want to challenge ourselves to recognize who has access and when. There may be mathematically rich discussions or other mathematically productive activities in the classroom—but who gets to participate in them? Who might benefit from different ways of organizing classroom activity?

Planning

What opportunities exist for each student to participate in the intellectual work of the class? How can we create more opportunities for more students?

Reflecting

Who have we seen participate in the intellectual work of the class? How has this participation looked and sounded in specific cases?

Things to think about

- What is the range of ways that students can and do participate in the mathematical work of the class (talking, writing, leaning in, listening hard; manipulating symbols, making diagrams, interpreting text, using manipulatives, connecting different ideas, etc.)?
- Which students participate in which ways?
- Which students are most active, and when?
- In what ways can particular students' strengths or preferences be used to engage them in the mathematical activity of the class?
- What opportunities do various students have to make meaningful mathematical contributions?
- What are the language demands of participating in the mathematical work of this class (e.g., academic vocabulary, mathematical discourse practices)?
- How can we support the development of students' academic language?
- How are norms (or interactions, lesson structures, task structure, particular resources, etc.) facilitating or inhibiting participation for particular students?
- What teacher moves might expand students' access to meaningful participation (such as modeling ways to participate, holding students accountable, point out students' successful participation)?
- How can we support particular students we are concerned about (in relation to learning, issues of safety, participation, etc.)?
- How can we create opportunities for more students to participate more actively?

Agency, Ownership, and Identity

Core Questions: What opportunities do students have to see themselves and each other as powerful doers of mathematics? How can we create more of these opportunities?

Many students have negative beliefs about themselves and mathematics, for example, that they are "bad at math," or that math is just a bunch of facts and formulas that they're supposed to memorize. Our goal is to support all students—especially those who have not been successful with mathematics in the past—to develop a sense of mathematical agency and ownership over their own learning. We want students to come to see themselves as mathematically capable and competent—not by giving them easy successes, but by engaging them as sense-makers, problem solvers, and creators of mathematical ideas.

Planning

What opportunities might exist for students to generate and explain their own ideas? To respond to each other's ideas? How can we create more opportunities?

Reflecting

How have we seen students explain their own and respond to each other's ideas? What has that looked and sounded like in specific cases?

Things to think about

- Who generates the ideas that get discussed?
- What kinds of ideas do students have opportunities to generate and share (strategies, connections, partial understandings, prior knowledge, representations)?
- Who evaluates and/or responds to others' ideas?
- How deeply do students get to explain their ideas?
- How does (or how could) the teacher respond to student ideas (evaluating, questioning, probing, soliciting responses from other students, etc.)?
- How are norms about students' and teachers' roles in generating ideas developing?
- How are norms about what counts as mathematical activity (justifying, experimenting, connecting, practicing, memorizing, etc.) developing?
- Which students get to explain their own ideas? To respond to others' ideas in meaningful ways?
- Which students seem to see themselves as powerful mathematical thinkers right now?
- How might we create more opportunities for more students to see themselves and each other as powerful mathematical thinkers?

Formative Assessment

Core Questions: What do we know about each student's current mathematical thinking? How can we build on it?

We want instruction to be responsive to students' actual thinking, not just our hopes or assumptions about what they do and don't understand. It isn't always easy to know what students are thinking, much less to use this information to shape classroom activities—but we can craft tasks and ask purposeful questions that give us insights into the strategies students are using, the depth of their conceptual understanding, and so on. Our goal is to then use those insights to guide our instruction, not just to fix mistakes but to integrate students' understandings, partial though they may be, and build on them.

Planning

What do we know about each student's current thinking, and how might this lesson or unit build on that thinking? How can we learn more about each student's thinking?

Reflecting

What have we learned in recent lessons about each student's thinking? How did this thinking look and sound in specific cases? How was this thinking built upon?

Things to think about

- What opportunities exist (or could exist) for students to develop their own strategies, approaches and understandings of mathematics?
- What opportunities exist (or could exist) for students to share their ideas and reasoning and to connect their ideas to others'?
- What different ways do students get to share their mathematical ideas and reasoning (writing on paper, speaking, writing on the board, creating diagrams, demonstrating with materials/artifacts, etc.)?
- Who do students get to share their ideas with (a partner, a small group, the whole class, the teacher)?
- What opportunities exist to build on students' mathematical thinking, and how are teachers and/or other students taking up these opportunities?
- How do students seem to be making sense of the mathematics in the lesson, and what responses might build on that thinking?
- How can activities be structured so that students have more opportunity to build on each other's ideas?
- What might we try (what tasks, lesson structures, questioning prompts, etc.) to surface student thinking, especially the thinking of students whose ideas we don't know much about yet?

A NOTE: What We Mean By "Important Mathematical Ideas and Practices"

"Important mathematical ideas" are notoriously hard to define. Which ideas are important? Which are not? What even counts as an "idea"? Who should have the authority to decide? Our intention with the Conversation Guide is to support discussions about these questions rather than to offer answers. To us, it is much more important to work together to push our students and ourselves as educators toward more interconnected and fundamental understandings of mathematics than to decide exactly which ideas are most important. This pushing is crucial, because traditional views of school mathematics—and many of today's textbooks and standards documents—define mathematics in terms of isolated topics, skills, and sub-skills. Thinking about the progression of mathematical ideas as "Day 1: Add and Subtract Fractions With Like Denominators; Day 2: Multiply Fractions; Day 3: Divide Fractions; Day 4: Add and Subtract Fractions with Unlike Denominators" (a typical textbook progression) makes it difficult to develop conceptual understanding and a sense of meaning behind all of the mechanics. This is both untrue to mathematics as a discipline and alienating for many students.

One way of finding connections among apparently isolated topics is to focus on core mathematical practices. For example, *constructing an argument* is one such practice. Creating opportunities for students to develop skill in constructing mathematical arguments can bridge the otherwise disparate topics that math courses are typically supposed to cover. (Note the differences between a skill like constructing an argument and a skill like adding fractions.) Yet a focus on core practices does not eliminate the need to identify important mathematical ideas and use these ideas to organize instruction.

We find the questions below useful for shifting our focus from facts and procedures to important mathematical ideas. We hope they will be helpful for you as well.

- What do we want students to understand about the relevant mathematical objects (fractions, negative numbers, the coordinate plane, triangles, etc.) in this lesson? In this unit?
- What mathematical relationships, patterns, or principles do we want students to understand in this lesson? In this unit?
- How might students connect math ideas in this lesson/unit with ideas that came before or will come later? Are there overarching principles or relationships or patterns that they might work toward understanding?
- What are different ways of representing the math in this lesson/unit? How might different representations be connected to each other and how might these connections deepen our students' understanding?
- How do the ideas we're considering develop across multiple lessons/units?
- What are some ways to make connections to this idea in different lessons/units/content areas?

Some examples of math ideas that might be considered "important":

- Area and perimeter are fundamentally different measurable attributes of two-dimensional shapes. It is possible to change shapes such that neither, one, or both of these attributes change. For some families of shapes, there are interesting relationships between them.
- Relationships between two variables can be represented using equations, tables, graphs, and verbal descriptions. Parameters of the relationship between the variables (e.g., the rate of change) can be identified in each of these representations and connected across representations.
- Right triangles have special properties that are different from the properties of other triangles. These properties give us special access to information about things like angle measures and side lengths in particular right triangles.
- Many sets of changing quantities are proportionally related. This means that certain aspects of the relationship are constant and unchanging, which allows us to use the relationship to determine one quantity given the other.

One characteristic of all of these ideas is that they go beyond naming topics and skills. For example, we might know that we want to "cover proportional relationships" in a particular unit, or that we want students to be able to solve proportions. However, without consideration of the important underlying ideas that we want our students to make sense of, we are likely to get lost in facts and procedures. We are likely to miss opportunities to support students to build conceptual understandings, to make connections, and to develop a sense of themselves as powerful learners and thinkers.

Our hope is that as teachers and others think together about teaching, they can continuously push each other to think about the mathematics that students need to learn in bigger, deeper, richer, and more interconnected ways. So while our discussion questions frequently refer to "important mathematical ideas" as though there were a set list of such ideas somewhere that you could simply consult, we hope that you will instead find ways to explore and interrogate what "important mathematical ideas" means to you.

6.3 The TRU Observation Guide[2]

The TRU Observation Guide (Schoenfeld & The Teaching for Robust Understanding Project 2016b) is designed to support teachers, coaches, administrators, and professional learning communities in planning, conducting, and reflecting on observations in mathematics classrooms. The key idea behind TRU is that the five dimensions of classroom activity described in Figure 6.2 are central in determining the degree to which students will emerge from the classroom being proficient mathematical thinkers and problem solvers.

This Observation Guide is part of a support system for collaborative partnerships between teachers and observers. Optimally, each observation is one of a series of classroom visits contributing to teacher growth. There should be ample time to plan observations, to observe lessons, and to discuss the observations, over the course of a term or a year.

Prior to an observation, it is useful for the teacher and observer to discuss the lesson plan and decide on the main points of focus for the observation. The observation might be general; it is possible for a practiced observer to take notes on all dimensions. Alternatively, the teacher and observer might agree to focus on one or two areas the teacher wants to address in detail. Either

The Five Dimensions of Powerful Mathematics Classrooms				
The Mathematics	Cognitive Demand	Equitable Access to Mathematics	Agency, Ownership, and Identity	Formative Assessment
The extent to which classroom activity structures provide opportunities for students to become knowledgeable, flexible, and resourceful mathematical thinkers. Discussions are focused and coherent, providing opportunities to learn mathematical ideas, techniques, and perspectives, make connections, and develop productive mathematical habits of mind.	The extent to which students have opportunities to grapple with and make sense of important mathematical ideas and their use. Students learn best when they are challenged in ways that provide room and support for growth, with task difficulty ranging from moderate to demanding. The level of challenge should be conducive to what has been called "productive struggle."	The extent to which classroom activity structures invite and support the active engagement of all of the students in the classroom with the core mathematical content being addressed by the class. Classrooms in which a small number of students get most of the "air time" are not equitable, no matter how rich the content: all students need to be involved in meaningful ways.	The extent to which students are provided opportunities to "walk the walk and talk the talk" – to contribute to conversations about mathematical ideas, to build on others' ideas and have others build on theirs – in ways that contribute to their development of agency (the willingness to engage), their ownership over the content, and the development of positive identities as thinkers and learners.	The extent to which classroom activities elicit student thinking and subsequent interactions respond to those ideas, building on productive beginnings and addressing emerging misunderstandings. Powerful instruction "meets students where they are" and gives them opportunities to deepen their understandings.

Figure 6.2 The five dimensions of powerful mathematics classrooms

Observe the Lesson Through a Student's Eyes

The Content	• What's the big idea in this lesson? • How does it connect to what I already know?
Cognitive Demand	• How long am I given to think, and to make sense of things? • What happens when I get stuck? • Am I invited to explain things, or just give answers?
Equitable Access to Content	• Do I get to participate in meaningful math learning? • Can I hide or be ignored? In what ways am I kept engaged?
Agency, Ownership, and Identity	• What opportunities do I have to explain my ideas? In what ways are they built on? • How am I recognized as being capable and able to contribute?
Formative Assessment	• How is my thinking included in classroom discussions? • Does instruction respond to my ideas and help me think more deeply?

Figure 6.3 Observing a mathematics lesson from the student perspective

way, reflecting beforehand on goals for the lesson and for the observation is a good way to make the most of the observation. The TRU Conversation Guide can serve as a resource for thinking about the plan, and what the teacher is trying to achieve.

Our primary focus in observing any lesson is what the classroom experience looks and feels like from the perspective of a student – students, after all, are the ones experiencing the instruction! The questions in Figure 6.3 provide an orientation that helps in seeing the lesson from the student perspective.

The form of the observation guide and the way to use it are both straightforward. Each observation sheet focuses on one dimension and looks like Figure 6.4.

The top and bottom parts of each observation sheet provide concise descriptions of the relevant dimension and goals for it. Beneath the description of the dimension are some examples of "look fors" – actions on the part of students and the teacher that are indicators that things are going well. They are things to aim for in general, and over time – they are NOT a list of things to be checked off in any particular lesson. We imagine teacher and observer discussing these prior to a lesson and deciding which, if any, might be things to focus on in the upcoming observation. The list is not meant to be comprehensive; teacher and observer may decide on another focus and write it in the space provided. The center of the observation sheet provides space for writing down observations.

There are many possible goals for classroom observations. Teacher and observer may decide to focus on one or two issues, or they may agree that the observer will provide a systematic run-through of all the dimensions. It is useful, and typically most comfortable, for the post-lesson conversation to start with the main focal points – on agreed-upon foci, along with events in the lesson that were particularly interesting and salient. But, even if particular foci have been chosen for the observation, it is valuable to run briefly through all of the dimensions – the Teaching for Robust Understanding Framework is intended as a way of seeing and talking about instruction, and it provides a language for thinking about it. After a few such conversations, teachers, coaches, administrators, and professional learning communities find that TRU is an easy and natural way to talk about teaching.

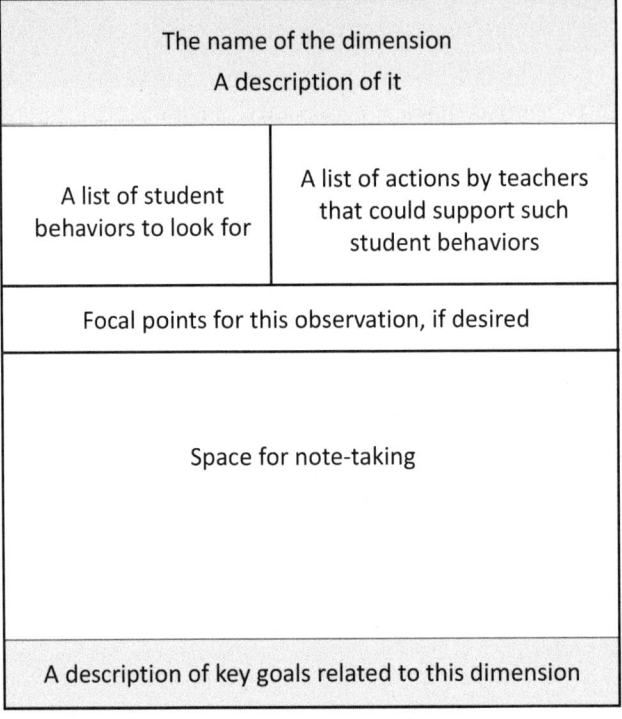

The name of the dimension A description of it	
A list of student behaviors to look for	A list of actions by teachers that could support such student behaviors
Focal points for this observation, if desired	
Space for note-taking	
A description of key goals related to this dimension	

Figure 6.4 The structure of an observation sheet

Here are a few important points about the TRU framework and its use. The framework highlights five dimensions of classroom mathematical activity. They are described separately because each can be the object of coherent focus, as part of ongoing professional development. In the classroom, however, they are all deeply interrelated. In particular:

- Issues related to mathematical content permeate all five dimensions – and all classroom activities. Dimension 1 focuses on the quality of the content *per se*. If the mathematics isn't rich, there is nothing meaningful for the students to learn. But what matters in addition is the set of opportunities that each student has to engage with and make sense of the mathematics. Thus Dimension 2, Cognitive Demand, should be conceived of as opportunities for productive struggle with core mathematical concepts and practices. Issues of access (Dimension 3) and opportunities to develop Agency, Ownership, and Identity (Dimension 4) concern the ways in which every student relates to the big ideas of the discipline. And, of course, the purpose of Formative Assessment (Dimension 5) is to facilitate meaningful access to the mathematics.
- Similarly, issues of equity also permeate all five dimensions and should be central at all times. The key point is that *every* student should be supported in developing a positive mathematical identity (Dimension 4) through meaningful access and participation (Dimension 3) to rich mathematical content (Dimension 1). That participation can only be meaningful for a student if the level of cognitive demand is right for sense-making (Dimension 2), something achieved by formative assessment (Dimension 5). At the same time, Dimension 3, Equitable Access to Mathematics, does need specific, focused attention: teaching in ways that provide meaningful opportunities for all students to engage with central mathematical content, and to build productive mathematical identities, is extremely challenging.

- As noted above, each observation sheet has room for specific observational goals established by the teacher and observer. One place where this will be essential is for Dimension 1, the content. The "look fors" on the first observation sheet are general across mathematics, and they should be supplemented by specifics for the lesson being observed.

The observation sheets follow.

THE MATHEMATICS
The extent to which central mathematical content and practices, as represented by State or he Common Core State Standards, are present and embodied in instruction. Every student should have opportunities to grapple meaningfully with key ideas and, in doing so, to become a knowledgeable, flexible, and resourceful mathematical thinker and problem solver. Teachers should have opportunities to consider and discuss how each lesson's activities connect to the concepts, practices, and habits of mind they want students to develop over time.

Each Student...	Teachers...
• Engages with grade level mathematics in ways that highlight important concepts, procedures, problem solving strategies, and applications • Has opportunities to develop productive mathematical habits of mind • Has opportunities for mathematical reasoning, orally and in writing, using appropriate mathematical language • Explains their reasoning processes as well as their answers.	• Highlight important ideas and provide opportunities for students to engage with them • Use materials or assignments that center on key ideas, connections, and applications • Explicitly connect the lesson's big ideas to what has come before and will be done in the future • Support the purposeful use of academic language and of representations (e.g., graphs, tables, symbols) central to mathematics • Support students in seeing mathematics as being coherent, connected, and comprehensible

• Other focal points for observation:

What are the big ideas in this lesson? How do they connect to what has come before, and/or establish a base for future work? How do the ways students engage with the material support the development of conceptual understanding and the development of mathematical habits of mind?

Goal: All students work on core mathematical issues in ways that enable them to develop conceptual understandings, develop reasoning and problem solving skills, and use mathematical concepts, tools, methods and representations in relevant contexts.

COGNITIVE DEMAND
The extent to which students have opportunities to grapple with and make sense of important mathematical ideas and their use. Students learn best when they are challenged in ways that provide room and support for growth, with task difficulty ranging from moderate to demanding. The level of challenge should be conducive to what has been called "productive struggle."

Each Student...	Teachers...
• Engages individually and collaboratively with challenging ideas • Actively seeks to explore the limits of their current understandings • Is comfortable sharing partial or incorrect work as part of a larger conversation • Reasons and tests ideas in ways that connect to and build on what they know • Explains what they have done so far before asking for help • Continues to wrestle with an idea after the teacher leaves	• Position students as sense makers who can make sense of key conceptual ideas. • Use or adapt materials and activities to offer challenges that students can use, individually or collectively, to deepen understandings • Build and maintain classroom norms that support every student's engagement with those materials and activities • Monitor student challenge, adjusting tasks, activities, and discussions so that all students are engaged in productive struggle • Supports students without removing the challenge from the work they are engaged in

• Other focal points for observation:

What opportunities do students have to make sense of mathematical content and practices? How are they supported in sense making so that they are not lost – yet real challenge has been maintained, so that they have opportunities to grapple with important ideas?

Goal: All students have opportunities to make their own sense of important mathematical ideas, developing deeper understandings, connections, and applications by building on what they know.

EQUITABLE ACCESS TO MATHEMATICS

The extent to which classroom activities invite and support the meaningful engagement with core mathematical content and practices by all students. Finding ways to support the diverse range of learners in engaging meaningfully is the key to an equitable classroom.

Each Student...	Teachers...
• Contributes to collective sense making in any of a number of different ways (e.g., proposing ideas, asking questions, creating diagrams...) • Actively listens to other students and builds on their ideas • Supports other students' developing understandings • Explains, interprets, applies and reflects on important mathematical ideas • Participates meaningfully in the mathematical work of the class	• Create safe environments • Use tasks and activities that provide multiple entry points and support multiple approaches to the mathematics • Provide opportunities for students to see themselves, and their personal and community interests, reflected in the curriculum • Validate different ways of making contributions • Build and maintain norms that support every student's participation in group work and whole class activities • Support particular needs, such as those of language learners, for full participation • Expect and support meaningful mathematical engagement from all students, helping them contribute and build on contributions from others

• Other focal points for observation:

In what ways does each student engage in the work of the class? How can more opportunities for every student to participate in meaningful ways be created?

Goal: All students are supported in access to central mathematical content, and participate actively in the work of the class. Diverse strengths and needs are built on through the use of various strategies, resources, and technologies that enable all students to participate meaningfully.

AGENCY, OWNERSHIP, AND IDENTITY

The extent to which every student has opportunities to explore, conjecture, reason, explain, and build on emerging ideas, contributing to the development of agency (the willingness to engage academically) and ownership over the content, resulting in positive mathematical identities.

Each Student...	Teachers...
• Takes ownership of the learning process in planning, monitoring, and reflecting on individual and/or collective work • Asks questions and makes suggestions that support analyzing, evaluating, applying and synthesizing mathematical ideas • Builds on the contributions of others and help others see or make connections • Holds classmates and themselves accountable for justifying their positions, through the use of evidence and/or elaborating on their reasoning	• Provide time for students to develop and express mathematical ideas and reasoning • Work to make sure all students have opportunities to have their voices heard • Encourage student-to-student discussions and promote productive exchanges • Assign tasks and pose questions that call for mathematical justification, and for students to explain their reasoning • Employ a range of techniques that attribute ideas to students, to build student ownership and identity

• Other focal points for observation:

What opportunities do all students have to see themselves and others as proficient mathematical thinkers, to grapple with challenges and construct new understandings, to build on others' ideas, and demonstrate their understandings? How can more of these opportunities be created?

Goal: All students build productive mathematical identities through taking advantage of opportunities to engage meaningfully with the discipline and share and refine their developing ideas.

FORMATIVE ASSESSMENT

The extent to which classroom activities elicit all students' thinking and subsequent interactions respond to that thinking, by building on productive beginnings or by addressing emerging misunderstandings. High quality instruction "meets students where they are" and gives them opportunities to develop deeper understandings, both as shaped by the teacher and in student-to-student interactions.

Each Student...	Teachers...
• Explains their thinking, even if somewhat preliminary • Sees errors as opportunities for new learning • Consistently reflects on their work and the work of peers • Sees fellow students as resources for their own learning • Provides specific and accurate feedback to fellow students • Makes use of feedback in revising their work	• Create safe climates in which students feel free to express their ideas and understandings • Use materials that elicit multiple strategies, and have students explain their reasoning, in order to gain information about student' emerging understandings • Flexibly adjust content and process, providing students opportunities for re-engagement and revision • Provide timely and specific feedback to students, as part of classroom routines that prompt students to make active use of feedback to further their learning • Create opportunities for students' individual and collaborative reflection on their knowledge and learning

• Other focal points for observation:

What opportunities exist for all students to demonstrate their understandings? What opportunities exist to build on the thinking that is revealed? How do teachers and/or other students take up these opportunities? Where can more be created?

Goal: Every student's learning is continually enhanced by the ongoing strategic and flexible use of techniques and activities that allow students to reveal their emerging understandings, and that provide opportunities both to rethink misunderstandings to build on productive ideas.

6.4 A Brief Coda

We've reached the end of this volume – but we hope that this is just the beginning of your adventures with the TRU Framework. TRU is a living entity; it improves as people use it and share their understanding of how to make it richer. Many of the insights that led to the current volume came from exchanges with partners who used TRU and shared their thoughts with us. Please feel free to do the same!

Notes

1 Evra Baldinger and Nicole Louie were the primary developers of the *Conversation Guide*. Please see the acknowledgments and the TRU website, https://truframework.org/.
2 We are indebted to the San Francisco Unified School District for its development of "Observational Tool for LEAD," which provided the inspiration for the design format of this *Guide*.

References

Aguirre, J., Mayfield-Ingram, K., & Martin, D.B. (2013). *The impact of identity in K-8 mathematics. Rethinking equity-based practices.* Reston, VA: NCTM.

American Association of University Women. (1992). *How schools shortchange girls.* Washington, DC: AAUW and NEA.

Ansalone, G. (2003). Poverty, tracking, and the social construction of failure: International perspectives on tracking. *Journal of Children and Poverty*, 9(1), 3–20, https://doi.org/10.1080/1079612022000052698

Arcavi, A., & Karsenty, R. (2018). Enhancing mathematics teachers' reflection and knowledge through peer-discussions of videotaped lessons: A pioneer program in Israel. In N. Movshovitz-Hadar (Ed.), *K-12 mathematics education in Israel - Issues and challenges* (Chapter 33, pp. 303–310). Series on Mathematics Education (Volume 13). Singapore: World Scientific.

Baldinger, E. Louie, N., & the Algebra Teaching Study and Mathematics Assessment Project. (2016). *TRU Math conversation guide: A tool for teacher learning and growth (mathematics version).* Berkeley, CA & E. Lansing, MI: Graduate School of Education, University of California, Berkeley & College of Education, Michigan State University. Retrieved from: http://truframework.org

Berry, R., Conway, M, Lawler, B., & Staley, J. (2020). *High school mathematics lessons to explore, understand, and respond to social injustice.* Thousand Oaks, CA, & Reston, VA: Corwin and NCTM.

Black, P., Harrison, C., Lee, C., Marshall, B., & Wiliam, D. (2003). *Assessment for learning: Putting it into practice.* Buckingham, UK: Open University Press.

Black, P., & Wiliam, D. (1998a). Inside the black box: Raising standards through classroom assessment. *Phi Delta Kappan, 80*(2), 139–147.

Black, P., & Wiliam, D. (1998b). Assessment and classroom learning. *Assessment in Education*, 5(1), 7–74.

Black, P.J., & Wiliam, D. (2009). Developing the theory of formative assessment. *Educational Assessment, Evaluation and Accountability* (21), 5–31.

Boaler, J. (2008). Promoting relational equity in mathematics classrooms – important teaching practices and their impact on student learning. *Proceedings of the 10th International Congress of Mathematics Education (ICME X)*, 2004, Copenhagen.

Boaler, J., & Staples, M. (2008). Creating mathematical futures through an equitable teaching approach: The case of Railside School. *The Teachers College Record, 110*(3), 608–645.

Burkhardt, H., & Schoenfeld, A.H. (2019). Formative assessment in mathematics. In R. Bennett, H. Andrade, & G. Cizek (Eds.), *Handbook of Formative Assessment in the Disciplines* (pp. 35–67). New York: Routledge. ISBN 9781138054363.

References

Burkhardt, H., & Schoenfeld, A. (2022). Assessment and Mathematical Literacy: A Brief Introduction. In: Tierney, R.J., Rizvi, F., Erkican, K. (Eds.), *International Encyclopedia of Education, 4th Edition, vol. 13.* Amsterdam: Elsevier. https://dx.doi.org/10.1016/B978-0-12-818630-5.09007-2. ISBN: 9780128186305.

Cohen, E.G. (1994). *Designing groupwork: Strategies for heterogeneous classrooms* (Revised edition). New York: Teachers College Press.

Cohen, E.G. & Lotan, R.A. (Eds.). (1997). *Working for equity in heterogeneous classrooms: Sociological theory in practice.* New York: Teachers College Press.

Cohen, E.G., & Lotan, R.A. (2014). *Designing groupwork. Strategies for the heterogeneous classroom* (3rd edition). New York: Teachers College Press.

Cohen, E.G., Lotan, R.A., Scarloss, B.A., & Arellano, A. R. (1999). Complex instruction: Equity in cooperative learning classrooms. *Theory into Practice, 38,* 80–86.

Common Core State Standards Initiative. (2010). http://www.corestandards.org/. See specifically the Common Core State Standards for Mathematics, http://www.corestandards.org/Math/

Darling Hammond, L. (2010). *The flat world and education: How America's commitment to equity will determine our future.* New York: Teachers College Press.

Daro, V. (2021). Growing student language in math class. https://envisionlearning.org/wp-content/uploads/2022/01/ELP-Growing-Student-Language-in-Math-Class-2021.pdf.

DIME (Diversity in mathematics education) center for learning and teaching. (2007). Culture, race, power, and mathematics education. In F. Lester (Ed.), *Handbook of research on mathematics teaching and learning* (2nd edition), pp. 405–434. Charlotte, NC: Information Age Publishing.

Disston, J. (2019). TRU in the Masters and Credential in Science and Mathematics Education program. In K. Beswick (Ed.), *International handbook of mathematics teacher education, Volume 4, The mathematics teacher educator as a developing professional,* pp. 278–282. Rotterdam, the Netherlands: Sense publishers.

Dweck, C. (2007). *Mindset: The new psychology of success.* New York: Ballantine.

Engle, R.A. (2011). The productive disciplinary engagement framework: Origins, key concepts, and continuing developments. In D.Y. Dai (Ed.), *Design research on learning and thinking in educational settings: Enhancing intellectual growth and functioning* (pp. 161–200). London: Taylor & Francis.

Fernandez, C., & Yoshida, M. (2004). *Lesson study: A Japanese approach to improving mathematics teaching and learning.* Mahwah, NJ: Erlbaum.

Fink, H. (2022). Centering students' voices: A multifocal mixed methods investigation of participatory equity in a distance learning calculus class. Ph.D. Dissertation, University of California, Berkeley.

Freeman, D., Freeman, Y., & Gonzalez, G. (1987). Success for LEP students: The Sunnyside Sheltered English program. *TESOL quarterly, 21,* 361–367.

Gawande, A. (2007). *Better: A surgeon's notes on performance.* New York: Picador.

Gawande, A. (2009). *The checklist manifesto: How to get things right.* New York: Picador.

Greeno, J. G. (2006). Authoritative, accountable positioning and connected, general knowing: Progressive themes in understanding transfer. *Journal of the Learning Sciences, 15,* 537–547.

Gutierrez, R. (2002). Beyond essentialism: The complexity of language in teaching mathematics to Latina/o students. *American educational research journal, 39*(4), 1047–1088.

Gutiérrez, R. (2009). Embracing the inherent tensions in teaching mathematics from an equity stance. *Democracy & Education, 18*(3), 9–16.

Gutstein, E. (2006). *Reading and writing the world with mathematics: Toward a pedagogy for social justice.* New York: Taylor & Francis.

Gutstein, E., & Peterson, B. (Eds.). (2005). *Rethinking mathematics: Teaching social justice by the numbers.* Milwaukee: Rethinking Schools.

Henningsen, M., & Stein, M.K. (1997). Mathematical tasks and student cognition: Classroom-based factors that support and inhibit high-level mathematical thinking and reasoning. *Journal for Research in Mathematics Education, 28*(5), 524–549.

Herman, J., Epstein, S., Leon, S., La Torre Matrundola, D., Reber, S., & Choi, K. (2014). *Implementation and effects of LDC and MDC in Kentucky districts* (CRESST Policy Brief No. 13). Los Angeles: University of California, National Center for Research on Evaluation, Standards, and Student Testing (CRESST).

Hess, K. (2006). Exploring cognitive demand in instruction and assessment. Downloaded April 1, 2015 from http://www.nciea.org/publications/DOK_ApplyingWebb_KH08.pdf.

Hess, K. (2013). A guide for using Webb's depth of knowledge with common core state standards. Retrieved April 1, 2015, from https://education.ohio.gov/getattachment/Topics/Teaching/Educator-Evaluation-System/How-to-Design-and-Select-Quality-Assessments/Webbs-DOK-Flip-Chart.pdf.aspx

Hodge, L.L., & Cobb, P. (2019). Two views of culture and their implications for mathematics teaching and learning. *Urban Education, 54*(6), 860–884.

Holland, D., Lachiotte, W., Jr., Skinner, D., & Cain, C. (1998). *Identity and agency in cultural worlds*. Cambridge, MA: Harvard University Press.

Horn, I., & Garner, B. (2022). *Teacher learning of ambitious and equitable instruction*. New York: Routledge.

Horn, I.S. (2012). *Strength in numbers: Collaborative learning in secondary mathematics*. Reston, VA: National Council of Teachers of Mathematics.

Huang, R., Takahashi, A., & Ponte, J.P. (Eds.). (2019). *Theory and practices of lesson study in mathematics: An international perspective* (pp. 136–162). New York: Springer. ISBN 978-3-030-04031-4.

Institute for Learning. (2016). *Accountable talk*. Retrieved from http://ifl.pitt.edu/index.php/educator_resources/accountable_talk.

Karsenty, R. (2018). Professional development of mathematics teachers: Through the lens of the camera. In G. Kaiser, H. Forgasz, M. Graven, A. Kuzniak, E. Simmt, & B. Xu (Eds.), *Invited Lectures from the 13th International Congress on Mathematical Education* (pp. 269–288). Cham, Switzerland: Springer.

Karsenty, R., & Arcavi, A. (2017). Mathematics, lenses and videotapes: A framework and a language for developing reflective practices of teaching. *Journal of Mathematics Teacher Education, 20*, 433–455.

Karsenty, R., & Arcavi, A. (2021). "Life trajectory" of a professional development project: The case of VIDEO-LM. In A. Hofstein, A. Arcavi, B. Eylon, & A. Yarden (Eds.), *Long-term research and development in science education: What have we learned?* (pp. 306–332). Boston: Brill.

Katznelson, N., & Bernstein, K. A. (2017). Rebranding bilingualism: The shifting discourses of language education policy in California's 2016 election. *Linguistics and Education, 40*, 11–26.

Koole, T. (2003). The Interactive Construction of Heterogeneity in the Classroom. *Linguistics and Education, 14*(1), 3–26.

Kozol, J. (1992). *Savage inequalities*. New York: Harper Perennial.

Krashen, S. (1985). *The input hypothesis*. London: Longman.

Ladson-Billings, G.J. (1997). *The dreamkeepers: Successful teachers of African-American children*. San Francisco, CA: Jossey-Bass.

Lamb, E. (2021) *Mathematical modeling in K–16: Community and cultural contexts*. A report on the 16th Critical Issues in Mathematics Education (CIME) conference held at the Mathematical Sciences Research Institute. Berkeley, CA: MSRI. http://library.msri.org/cime/CIME-v15.pdf

References

Lewis, C., & Hurd, J. (2011). *Lesson study step by step: how teacher learning communities improve instruction.* Portsmouth, N.H.: Heinemann.

Lewis, C., Perry, R., Hurd, J., & O'Connell, M.P. (2006). Lesson study comes of age in North America. *Phi Delta Kappan, 88*(4), 273–281.

Martin, D.B. (Ed.) (2009a). *Mathematics teaching, learning, and liberation in the lives of Black Children.* New York: Routledge.

Martin, D.B. (2009b). Researching race in mathematics education. *Teachers College Record, 111*(2), 295–338.

Martin, D.B. (2013). Race, racial projects, and mathematics education. *Journal for Research in Mathematics Education, 44*(1), 316–333.

Mason, J., Burton, L., & Stacey, K. (1982). *Thinking mathematically.* London: Addison-Wesley.

McDougal, T. (Ed.). (2017). *Essential mathematics for the next generation: What and how students should learn.* Tokyo, Japan: Tokyo Gagukei University.

Miller, G.A. (1956). The magical number seven, plus or minus two: Some limits on our capacity for processing information. *Psychological Review 63,* 81–97.

Moll, L., Amanti, C., Neff, D., & Gonzalez, N. (1992). Funds of knowledge for teaching: Using a qualitative approach to connect homes to classrooms. *Theory into Practice, XXXI*(2), 132–141.

Moschkovich, J.N. (2012). *Mathematics, the Common Core, and language: Recommendations for mathematics instruction for ELs aligned with the Common Core.* Proceedings of the "Understanding Language" Conference. Stanford, CA: Stanford University. Retrieved from http://ell.stanford.edu

Moschkovich, J.N. (2013). Principles and guidelines for equitable mathematics teaching practices and materials for English Language Learners. *Journal of Urban Mathematics Education, 6*(1), 45–57.

Moses, R.P. (2001). *Radical equations: Math literacy and civil rights.* Boston MA: Beacon Press.

Nasir, N., Cabana, C., Shreve, B., Woodbury, E., & Louie, N. (Eds). (2014). *Mathematics for equity: A framework for successful practice.* New York: Teachers College Press.

Nasir, N., & Cobb, P. (Eds.) (2007). *Improving access to mathematics: Diversity and equity in the classroom.* New York: Teachers College Press.

Nasir, N., & Shah, N. (2011). On defense: African American males making sense of racialized narratives in mathematics education. *Journal of African American Males in Education, 2*(1), 24–45.

National Academy of Education/National Council of Teachers of Mathematics Committee on Civic Reasoning and Discourse. (2023). *Civic discourse in elementary, middle school, and secondary mathematics classrooms.* Washington DC and Reston, VA: NAEd and NCTM.

National Council of Teachers of Mathematics. (1989). *Curriculum and evaluation standards for school mathematics.* Reston, VA: NCTM.

National Council of Teachers of Mathematics. (2000). *Principles and standards for school mathematics.* Reston, VA: NCTM.

National Research Council. (2001). *Adding it up: Helping children learn mathematics.* J. Kilpatrick, J. Swafford, & B. Findell (Eds.). Mathematics Learning Study Committee, Center for Education, Division of Behavioral and Social Sciences and Education. Washington, DC: National Academy Press.

Oakes, J. (2005). *Keeping track: How schools structure inequality* (2nd edition). New Haven: Yale University Press.

Oakes, J., Joseph, R., & Muir, K. (2001). Access and achievement in mathematics and science. In J.A. Banks & C.A. McGee Banks (Eds.), *Handbook of research on multicultural education* (pp. 69–90). San Francisco: Jossey-Bass.

Pólya, G. 1945 [1957]. *How to solve it* (2nd edition). Princeton: Princeton University Press.

Pólya, G. (1954). *Mathematics and plausible reasoning* (Volume 1, *Induction and analogy in mathematics;* Volume 2, *Patterns of plausible inference*). Princeton: Princeton University Press.

246

Pólya, G. 1962 [1965/1981]. *Mathematical discovery* (Volume 1, 1962; Volume 2, 1965). Princeton: Princeton University Press. Combined paperback edition, 1981. New York: Wiley.

Reinholz, D., Johnson, E., Andrews-Larson, C., Stone-Johnstone, A., Smith, J., Mullins, B., Fortune, N., Keene, K., & Shah, N. (2022). When active learning is inequitable: Women's participation predicts gender inequities in mathematical performance. *JRME, 53*(3), 204–226.

Reinholz, D.L., & Shah, N. (2018). Equity analytics: A methodological approach for quantifying participation patterns in mathematics classroom discourse. *JRME, 49*(2), 140–177.

Research for Action. (2015). *MDC's influence on teaching and learning*. Philadelphia, PA: Author. Retrieved March 1, 2015 from https://www.researchforaction.org/publications/mdcs-influence-on-teaching-and-learning/

Resnick, L., O'Connor, C., & Michaels, S. (2007). *Classroom discourse, mathematical rigor, and student reasoning: An accountable talk literature review*. Downloaded July 9, 2008 from http://einstein.pslc.cs.cmu.edu/research/wiki/images/f/ff/Accountable_Talk_Lit_Review.pdf

Schoenfeld, A.H. (1985). *Mathematical problem solving*. Orlando, FL: Academic Press.

Schoenfeld, A.H. (1992). Learning to think mathematically: Problem solving, metacognition, and sense-making in mathematics. In D. Grouws (Ed.), *Handbook for research on mathematics teaching and learning*, pp. 334–370. New York: MacMillan. https://doi.org/10.1177/002205741619600202

Schoenfeld, A.H. (2003). Making mathematics work for all children: Issues of standards, testing, and equity. *Educational Researcher, 31*(1), 13–25.

Schoenfeld, A.H. (2013). Classroom observations in theory and practice. *ZDM, the International Journal of Mathematics Education, 45*: 607–621. https://doi.org/10.1007/s11858-012-0483-1

Schoenfeld, A.H. (2014). What makes for powerful classrooms, and how can we support teachers in creating them? *Educational Researcher, 43*(8), 404–412. https://doi.org/10.3102/0013189X1455

Schoenfeld, A.H. (2015). Thoughts on scale. *ZDM, the International Journal of Mathematics Education, 47*, 161–169. https://doi.org/10.1007/s11858-014-0662–3

Schoenfeld, A.H. (2022). Why are learning and teaching mathematics so difficult? In M. Danesi, (Ed.), *Handbook of cognitive mathematics*. New York: Springer Nature. https://doi.org/10.1007/978-3-030-44982-7_10-1

Schoenfeld, A.H. (2023). On Problems, Problem-Solving, and Thinking Mathematically. In: Leikin, R. (Eds.), *Mathematical challenges for all: Research in mathematics education*. Springer, Cham. https://doi.org/10.1007/978-3-031-18868-8_29.

Schoenfeld, A.H., Baldinger, E., Disston, J., Donovan, S., Dosalmas, A., Driskill, M., Fink, H., Foster, D., Haumersen, R., Lewis, C., Louie, N., Mertens, A., Murray, E., Narasimhan, L., Ortega, C., Reed, M., Zuñiga-Ruiz, S., Sayavedra, A., Sola, T., Tran, K., Weltman, A., Wilson, D., & Zarkh, A. (2019a). Learning with and from TRU: Teacher educators and the Teaching for Robust Understanding Framework. In K. Beswick (Ed.), *International handbook of mathematics teacher education, Volume 4, the mathematics teacher educator as a developing professional* (pp. 271–304). Rotterdam, The Netherlands: Sense Publishers.

Schoenfeld, A.H., Dosalmas, A., Fink, H., Sayavedra, A., Weltman, A., Zarkh, A, Tran, K., & Zuniga-Ruiz, S. (2019b). Teaching for robust understanding with lesson study. In R. Huang, A. Takahashi, & J.P. Ponte (Eds.), *Theory and practices of lesson study in mathematics: An international perspective* (pp. 136–162). New York: Springer. ISBN 978-3-030-04031-4

Schoenfeld, A.H., Fink, H., Sayavedra, A., Weltman, R., Zarkh, A., & Zuñiga-Ruiz, S. (2023). *Mathematics teaching on target: A guide to teaching for robust understanding at all grade levels*. New York: Routledge.

Schoenfeld, A. H., Floden, R. B., & The Algebra Teaching Study and Mathematics Assessment Projects. (2018). On classroom observations. *Journal of STEM Education Research*. https://doi.org/10.1007/s41979-018-0001-7

References

Schoenfeld, A.H., & The Teaching for Robust Understanding Project. (2016). *The Teaching for Robust Understanding (TRU) observation guide for mathematics: A tool for teachers, coaches, administrators, and professional learning communities.* Berkeley, CA: Graduate School of Education, University of California, Berkeley. Retrieved from: http://TRU.framework.org

SERP. (2016). *Word Generation home.* http://wordgen.serpmedia.org/academic_vocabulary-and-apt.html

Shah, N. (2017). Race, ideology, and academic ability: A relational analysis of racial narratives in mathematics. *Teachers College Record, 119*(7), 1–42.

Shepard, L.A. (2000). *The role of classroom assessment in teaching and learning.* (CSE Technical Report 517). Los Angeles: University of California, National Center for Research on Evaluation, Standards, and Student Testing (CRESST).

Sherin, M.G. (2000). Viewing teaching on videotape. *Educational Leadership, 57*(8), 36–38.

Sherin, M.G. & Dyer, E.B. (2017). Teacher self-captured video: Learning to see. *Kappan, 98*(7), 49–54.

Sherin, M.G. & Linsenmeier, K. (2011). Pause, rewind, reflect: Video clubs throw open the class-room doors. *Journal of Staff Development, 32*(5), 38–41.

Sherin, M.G., Richards, J., & Altshuler, M. (2021). Recording one's classroom: A new source for teacher learning. *Kappan, 103*(2), 44–48.

Skinner, A., Louie, N., & Baldinger, E. (2019). Learning to see students' mathematical strengths. *Teaching Children Mathematics, 25*(6), 339–344.

Smith, M., & Stein, M. (2018). *5 practices for orchestrating productive mathematics discussions* (2nd edition). Thousand Oaks, CA: Corwin.

Smith, M., Bill, V., & Sherin, M.G. (2019). *The five practices in practice [Elementary]: Successfully orchestrating mathematics discussions in your elementary classroom.* Thousand Oaks, CA: Corwin.

Smith, M., & Sherin, M.G. (2019). *The five practices in practice [middle school]: Successfully orchestrating mathematics discussions in your middle school classroom.* Thousand Oaks, CA: Corwin.

Smith, M., Steele, M., & Sherin, M.G. (2020). *The five practices in practice [high school]: Successfully orchestrating mathematics discussions in your high school classroom.* Thousand Oaks, CA: Corwin.

Stein, M.K. & Smith, M.S. (1998). Mathematical tasks as a framework for reflection. *Mathematics Teaching in the Middle School, 3,* 268–275.

Swan, M. (2006). *Collaborative learning in mathematics: A challenge to our beliefs and practices.* London: National Institute for Advanced and Continuing Education (NIACE) for the National Research and Development Centre for Adult Literacy and Numeracy (NRDC).

Swan, M., & Burkhardt, H. (2014). Lesson design for formative assessment. *Educational Designer, 2*(7), Downloaded from http://www.educationaldesigner.org/ed/volume2/issue7/article24/index.htm.

Swan, M. (2017). Toward a task-based curriculum: Frameworks for task design and pedagogy. In T. McDougal (Ed.), *Essential mathematics for the next generation: What and how students should learn* (pp. 29–60). Tokyo, Japan: Tokyo Gagukei University.

Takahashi, A. (2015). Lesson study: An essential process for improving mathematics teaching and learning. In M. Inprasitha (Ed.), *Lesson study: Challenges in mathematics education* (pp. 51–58). Singapore: World Scientific Publishing. https://doi.org/10.1142/9789812835420_0004.

Takahashi, A., McDougal, T., Friedkin, S., & Watanabe, T. (Eds.). (2022). *Educators' learning from lesson study: Mathematics for ages 5–13.* New York: Routledge.

Takahashi, A., & Yoshida, M. (2004). How Can We Start Lesson Study?: Ideas for establishing lesson study communities. *Teaching Children Mathematics, 10*(9), 436–443.

TIMSS (Third International Math and Science Study). (2022). *Japanese mathematics lessons.* http://www.timssvideo.com/japan-mathematics-lessons

Turner, E., Dominguez, H., Maldonado, L., & Empson, S. (2013). English learners' participation in mathematical discussion. Shifting positioning and dynamic identities. *JRME, 44*(1), 199–234.

van Es, E. A., & Sherin, M. G. (2021). Expanding on prior conceptualizations of teacher noticing. *ZDM Mathematics Education, 53,* 17–27.

Vygotsky, L.S. (1978). *Mind in society: The development of higher mental processes.* Cambridge, MA: Harvard University Press.

Webb, N. (1997). *Research Monograph Number 6: Criteria for alignment of expectations and assessments on mathematics and science education.* Washington, DC: CCSSO.

Webb, N. (2002). *Depth-of-knowledge levels for four content areas.* Retrieved April 1, 2015 from http://schools.nyc.gov/NR/rdonlyres/2711181C-2108-40C4-A7F8-76F243C9B910/0/DOKFourContentAreas.pdf

Weber, K., & Dawkins, P.C. (2020). The role of mathematicians' practice in mathematics education research. *ZDM, 52*(6).

Wenger, E. (1998). *Communities of practice. Learning, meaning and identity.* Cambridge: Cambridge University Press.

Zuñiga-Ruiz, S. (2022). Towards a critical-mathematical consciousness: Understanding the construction of a counterspace for prospective maestras mexicanas. Ph.D. dissertation, University of California, Berkeley.

Index

Note: **Bold** page numbers refer to tables and *italic* page numbers refer to figures.

ability groups 11
abstraction 94
academic exercise 27
accountable talk 13
agency-building opportunities 96
agency, ownership and identity (AOI): Caleb's growth 80; classroom environment 135; gateway access 200; generating/incubating ideas 51; human interaction 199; implications 51, 80, 162, 199; individual and collective growth 137, 164; individual's participation and classroom's reaction 189; issues 20, 79, 136, 203; linguistic resources 133; mathematical content 172; mathematical learning environments 95; not-fully-correct statements 191–193; occasional frustrations 134; opportunities 12, 26, 231, 236, 240; over time 83; participants 214; potential 27, 84; precondition 202; productive mechanisms 163; productive struggle 134; questions *214*; sheltered mathematics 21; shifting authority 165; small-group interactions 190; small-group settings 12; students' abilities 133; student work 185; task construction 96; teacher learning communities 213; TRU observation guide 205, *206*; whole-class interactions 190–191; work in process 135; *see also* equitable access
Aguirre, J. 13
algebra tiles 83–85, *85*, 87, 94, 95, 99, 101, 107, 115, 119, 122, 125, 127, 128, 131, 135
Altshuler, M. 221
Analyzing Instruction in Mathematics using Teaching for Robust Understanding Framework (AIM-TRU) 221–222
Arcavi, A. 223
Arellano, A.R. 220
arithmetic modeling 36
asymptotes 149, 166, 169–170, 172, 175, 176, 183, 187, 189, 192, 194n6
authentic 80, 227, 228

Baldinger, E. 221, 242n1
belief systems 7
Bernstein, K.A. 97
Boyle's law *52*
Burkhardt, H. 15, *138*
Burton, L. 7

car value problem: artificial modeling 26; classroom episode 25; classroom visiting 38–40; cognitive demand 202; computing $f(n)/f(n-1)$ **33**; context 26; exemplary instruction 25; expanded table **68**; function $f(n)$ 33; links two numbers 67; math thinking 27–38; reflecting 73–81; reflection issues 35, 40–73; sense-making 25, 26, 35; small table on whiteboard **67**; student perspective *73*; values for seven years **33**; worksheet **61**; year-to-year ratios **34**
classroom: activity structures 131; CI 82, 84, 97, 104, 220; classic strategy *52*; contexts 200; contract 45, 98; dimensions of mathematics *224*, *234*; environment 12, 135; experience 4, 6, 12; focal group 50; formative assessment *138*, *186*; instruction 6; intellectual community 10; layout 154, *154*; learning community 163; management 162; non-performers 12; norms for visiting 19–20; observations 195, 226–227, 235; practices 5, 9, 71; problem-solving methods 51; sheltered algebra 21, 97; student thinking 14, 191; TRU observation form 210, *211*
Cobb, P. 13
cognitive demand: adjustments 66; AOI 27, 96; balancing act 78; car value problem 198, 202; CI 220; classroom discourse 138; counterproductive 51; dominos task 152; equitable access 213; focal group 76; formative assessment 26, 36, 43, 44, 76–77, 102; growth factor 65; instructional context 199; issues 20, 21; learning activity 185; level of difficulty 8; mathematical activities and work 8–9; meaningful sense-making 15; opportunities 229,

236; problem setup 75; productive struggle 13, 15, 25, 37, 72, 75, 129, 172, 188, 238; scaffolding 47, 57; sense-making 48, 236; student explanations 78; task experience 95
Cohen, E.G. 220
common ratio/multiplicative factor 70
complex instruction (CI) 5, 82, 84, 104, 210, 213, 217, 220–221
concave: downwards 155–157, 163, 166, 172, 179, 187, 191; upwards 155, 157, 163, 167, 172, 173, 187, 191
connected dots 70
context-based modeling 44
contextual problem 25, 29, 35, 44, 57, 72, 74

Dawkins, P.C. 7
debrief conversation 226–227
deep integration 215
degrees of freedom 146
disciplinary thinkers 7
discourse identities 13
dominos 147, *151*, 151–152, 176, 181, 183
Driskill, M. 217
Dyer, E.B. 221

economic segregation 12
empty symbol manipulation 127
Engle, R.A. 12
equitable access: argument *119*; authentic 80; bantering 112; CI 82; classroom activity structures 131; classroom norms 115; classroom visiting 97–98; content 230; counting *113*; demonstrate and practice instruction 79; equity sticks 79; explanation *120*, *122*; group explainer 116; group's collective responsibilities 131; inside tiles *106*; languages 111; learning environments 82–83; mathematical content and practices 131; mathematical sense-making 112; mathematical task 83–96; mathematics 239; mechanism 150; opportunity and support 131–132, 152; outside tiles *106*; ownership 79–81; participation 80; perimeter of object *85*, *126*; points of overlap *117*; reasonableness 79; reflecting 127–136; reflection 98–127; sampling participation 79; sense of agency 79; student perspective 80; students work 133; student-to-student exchanges 82; tiles pointing *104*, *105*; TRU dimensions 83; unit lengths 115; verbal participation and engagement 199; *xy* and *y* tiles *125*; *see also* agency, ownership and identity (AOI)
equitable instruction 11
equity-based practices 13
equity-focused instructional approach 220
equity sticks 79
Es, E.A. van 221
ethnic/gender/racial distribution 12
exponential equation 38, 51, 52, 56, 70, 71, 74, 75, 76

exponential growth/decay 20, 26, 36, 37, 46, 52, 64, 65, 71, 74–76
Eylon, B. 223

factor analysis 202
Fink, H. 12
Floden, R.B. 16n2
focal group 39–42, 48–53, 59, 63, 65, 66, 70, 72, 76–78, 79–80, 84, 198, 199
Forgasz, H. 223
formal academic language 128
formative assessment: asymptotes 176; case study 199; classroom discourse 185; cognitive demand 75–79, 220; complexities 47; demands 72; description 138; design tactics 15; distance-time graphs 15; implementation 14; instruction progresses 38; issues 21, 26, 36; learning process 14; mathematical muscles 8; mathematics classrooms *138*, *186*; quadratic functions 200, 202; students' thinking 135, 150, 193, 232, 241; student struggle 20; tasks 183; teacher judging 137; teacher's sense 63; TRU 136, 213
Formative Assessment Lessons (FALs) 14, 15, 21, 78, 137, 139–153, 174, 175, 181, 202, 203, 217–219, 221, 222
Foster, D. 217
Freeman, D. 97
Freeman, Y. 97
Friedkin, S. 220

Garner, B. 221
Gawande, A. 16n3
generalization 50, 139
Gonzales, G. 97
graphing quadratic functions: algebraic properties 142–143; case study 146–151, 153–154; classroom visiting 153–154; class structures 139; conversation guide *186*, *191*; coordinates assignment *176*; dynamic thoughts 144–146; equation of parabola 145–146; equations and graphs properties 141–142; FAL 137; final task, lesson segment *179*; formative assessment 138, *138*; fully annotated coordinates task *179*; goals 137; graph of parabola *144*; grappling 139; issues 140–146; mathematics 139–152; not-fully-correct statements 191–193; parabola *168*; parabolas, shapes and properties 140–141, *141*; parent functions *170*, *171*; reflecting 185–193; reflection 154–185; sketches *156–159*, *165–167*; two graphs *184*; vertex of parabola *143*, *147*; very narrow parabola *168*
Graven, M. 223
gravitational attraction 52
groupworthy 74, 84, 97, 220
Gutierrez, R. 13

Haumersen, R. 217
Hodge, L.L. 13

Hofstein, A. 223
horizontal parabolas 193n2
Horn, I.S. 221
Hurd, J. 220

identities in practice 13
independent research analyses 14
index cards 54, 55, 79, 147, 153, 155, 156, 158, 160, 161, 164, 166, 168
inspirational statements 153
instructional history 71
intellectual community 10
iterative (year-by-year) solution 70

K-8 mathematics 13
Kaiser, G. 223
Karsenty, R. 223
Katznelson, N. 97
knowledge base thinking 7
knowledge levels 8
Krashen, S. 97
Kuzniak, A. 223

learning environments 3, 79, 82, 83, 95, 124, 136, 200, 203
Lewis, C. 220
linear equations 145, 226
linguistic resources 133
Linsenmeier, K. 221
Lotan, R.A. 220
Louie, N. 217, 221, 242n1

Martin, D.B. 13
Mason, J. 7
mathematical proficiency 220
mathematical task: algebra tiles 83–85, 85, 94, 95; CI 84; commentary 93–96; follow-up question 84, 88–89, 93–95; group-worthy 84; initial 85–86; inside parts of rectangles 86, 87; long, medium and short sides 87; L-shaped piece 90; outline after unit length, repositioned 91, 92; outline of figure 91; overlaps 87–88, 88; perimeter of object 85, 85–86; remains after repositionings 92; solutions 89–90; supportive classroom norms 84; unit length, repositioned 91, 92
mathematics: algebra tiles 127; AOI 133; assertions 187; asymptotes 176; authority 165; car values problem 198; case study 146–151; classrooms 19, 234; coherent and compelling explanations 198; conjectures and conversations 153; content and practices 7, 9, 11, 43, 82, 108–109, 114, 131, 133, 136, 200, 209, 237; discipline 6–8; discourse 9, 11; equitable access 239; exponential functions 38, 198; expository standards 136; formative assessment 138, 186; gender inequities 10; graphing quadratic functions 198; ideas 12,

13, 26, 27, 43–45, 69, 70, 73–75, 83, 84, 95–97, 101, 102, 108, 111, 131, 139, 148, 152, 160, 164, 180, 181, 183, 200, 203, 205, 228, 229, 231, 233; identities 7, 11, 13–15, 27, 34, 36, 72, 79, 81–83, 95, 135, 152, 181, 184, 185, 192, 199, 202, 230, 236; knowledgeable 110; learning environments 95; modeling 7, 28, 94; ownership 111; quadratic functions formative assessment lesson 139–152; reasoning 175–176; sense-making 29, 44, 70, 95, 108, 111, 112, 219; standards 7, 128–131, 133; student perspective 235; teachers' reflection 222; teaching on target 205–207, 207; thinking 7, 88, 175, 232; thoughts 140; TRU 203
Mathematics Assessment Project (MAP) 219
Mathematics Network of Improvement Communities (Math NIC) 219
Mayfield-Ingram, K. 13
McDougal, T. 7, 220
meaningless symbol manipulation 94
Mertens, A. 217
metacognition 7
Michaels, S. 12
Miller, G. 201
model instruction 21, 200
Moses, R.P. 10
Movshovitz-Hadar, N. 223
Murray, E. 217

Narasimhan, L. 217

observation sheet 235–236, 236
O'Connell, M.P. 220
O'Connor, C. 12

participatory identities 13
partner teachers 214, 218
pedagogical content knowledge 17n5
performance gaps 10
Perry, R. 220
positional identities 13
problem-solving strategies 7, 31, 153
productive struggle 8, 9, 13, 25, 27, 37, 47, 49, 51, 53, 63, 65–66, 70, 72, 75, 76, 129, 130, 134, 136, 174, 188, 198, 216
professional development (PD): coherent focus 235; collaborating teachers 222; community 153; high-quality instructional resources 221; networks 217; pre-service teachers 210; school districts 5; synergy 210; teacher collective 212–216; teacher education courses 222; teachers' experiences and concerns 208; TRU classroom observation form 210, 211; video jigsaw 210; volume 15–16; weekly reflective journals 212
professional growth 16, 209
professional learning 195, 197, 207–210, 222, 224, 234, 235

professional vision/noticing 221
Program X 6
proportional relationships 233
Pythagorean theorem 28

racial inequities 10
reasonableness 46, 79
Reinholz, D.L. 10
Resnick, L. 12
resources: AIM-TRU 221–222;
    classroom environment 135; complex instruction
    220–221; disposal 188; formative assessment
    79; lesson study 219–220; MAP 219; math NIC
    219; national and international groups 218;
    productive collaboration 134; small groups
    152; student's progress 136; SVMI 222; TRU
    website 218–219; video clubs 221; VIDEO-LM
    222–223
Richards, J. 221

sampling participation 79
Scarloss, B.A. 220
Schoenfeld, A.H. 7, 15, 16n1, 16n2, *138*, 217
school-based sense-making 52
school math 83, 229
self-perception 116
sense-making 7–9, 15, 25–27, 34, 35, 38, 40, 45,
    46, 49–50, 52, 53, 58, 59, 70, 76, 79, 127, 130,
    136, 149, 174, 189, 193, 199, 208, 216, 220; *see
    also* mathematics, sense-making
sense of agency 12, 48, 59, 72, 77, 79, 82, 111,
    134, 185
sentence stems/starters 13
Shah, N. 10
sheltered algebra 21, 97
Sherin, M.G. 221
Silicon Valley Mathematics Initiative (SVMI) 222
Simmt, E. 223
Skinner, A. 221
small-group discussions 39, 78, 189
small-group interactions 190
Smith, M.S. 8, 9
social contract 129
Sola, T. 217
Stacey, K. 7
Stein, M.K. 8, 9
stereotypes 11, 226
stuckness 189
student: ability groups/clustered 11; achievement
    10; African-American 12; algebra skills 94;
    AOI 21, 51, 80, *95*, 133, *165*, 199; car value
    problem 72; classic classroom strategy 52;
    classroom experience 4; cognitive demand 81,
    188; contextual/modeling problem 20, 25, 35;
    engagement 15, 16, 26, 43, 72, 84, 93, 108,
    200; equitable instruction 11; ethnic/gender/
    racial distribution 12; exponential growth 37;

focal group 76; formative assessment 136;
    graphing quadratic functions 198; grappling 39;
    group dynamics 46; learning 3, 7, 14, 37, 38,
    94, 128–129, 161, 200; linear equations 145;
    mental and physical health 81; misconceptions/
    alternative conceptions 14; occasional
    discomfort 124; over time 31; participation
    13, 79, 80–81, 190; perspective 5, *73*, 80, *201*,
    234, *235*; presentation 166–167, 171–173;
    productive struggle 9, 76; sense of agency 82;
    thinking 8, 14, 15, 47, 62, 78, 84, 88, 136, 138,
    139, 147–150, 152, 153, 160, 171–175, 181,
    184, 185, 188, 191–193, 200, 201, 208, 219,
    226; work, coordinates task *177, 179*; work on
    coordinates task *177, 179*
Swan, M. 15
symbol pushing 127
symbol sense 151

Takahashi, A. 220
teacher collective: deep dives 213–215; models
    of sequencing 213; overview 212; transferring
    ownership 215; TRU 212–213; TRU-lesson
    study 215–217, **216**; TRU-with-something else
    215; video clip *214*
teacher educators 218
teacher joins focal group 42–43
teacher learning community 197
teaching for robust understanding (TRU)
    framework 4; big ideas 201–203; classroom
    experience 4; coherent professional development
    5; content issues 174; conversation and
    observation guides 203–205; differences,
    similarities and unifying themes 197–200;
    dimensions (*see* TRU dimensions); language
    4; learning environment 3; observation guide
    206; planning *204*; principles of high-quality
    instruction 5; professional development 15–16,
    210–216; research-based response 3; student
    perspective 5; teacher collective 212–213;
    website 218–219
3-point formula 146
TRU conversation guide: classroom observation,
    pre-brief 226–227; coherence 225; coherent and
    ongoing discussions 225; disconnected trainings
    225; ground discussion 226; instructional
    techniques 224; link planning 227; long-term
    learning agenda 225–226; mathematical ideas
    and practices 228–233; mathematically powerful
    classrooms *224*; menu, discussion questions 226;
    teachers' strengths 227; teaching and learning
    224; types 225; visual filter 224
TRU dimensions: agency, ownership and identity
    11–14, 79–81, 83, 133–135, 138, 189–193, 199;
    CI 220; classroom mathematical activity 235;
    cognitive demand 8–9, 75–79, 95, 129–131,
    138, 188–189, 198–199; decomposition 202;

disciplinary content and practices 10–11; equitable access 9–11, 79–81, 83, 131–133, 138, 189–193, 199; formative assessment 14–15, 75–79, 83, 135–136, 199–200; mathematics 6–9, 74–75, 83, 127–129, 138, 186–188, 198; participants 213; planning *204*; properties 3–6; weekly reflective journals 212
TRU observation guide 234–241
TRU video jigsaw 212
Turner, E. 13
tweaking 39, 40, 69, 72, 76
2-point formula 34, 146

unpacking solution 32

video clubs 16, 217, 218, 221
Viewing, Investigating and Discussing Environments of Learning Mathematics (VIDEO-LM) 222–223

Watanabe, T. 220
Webb, N. 8
Weber, K. 7
Wenger, E. 13
whole-class conversations 39–41, 78–81, 123, 139, 152, 188, 198–199
whole-class interactions 190–191
wide curve 166, 172–173, 192
Wilson, D. 217
window dressing 29, 53
wisdom of practice 208
women's participation 10

Xu, B. 223

Yarden, A. 223
year-by-year approach 32, 35, 53, 63, 64, 66, 70, 72, 74
Yoshida, M. 220

Taylor & Francis eBooks

www.taylorfrancis.com

A single destination for eBooks from Taylor & Francis with increased functionality and an improved user experience to meet the needs of our customers.

90,000+ eBooks of award-winning academic content in Humanities, Social Science, Science, Technology, Engineering, and Medical written by a global network of editors and authors.

TAYLOR & FRANCIS EBOOKS OFFERS:

A streamlined experience for our library customers

A single point of discovery for all of our eBook content

Improved search and discovery of content at both book and chapter level

REQUEST A FREE TRIAL
support@taylorfrancis.com

 Routledge
Taylor & Francis Group

 CRC Press
Taylor & Francis Group